Driving Sustainability: Innovations in Producing Eco-Friendly Substrates for Soilless Culture Systems

Driving Sustainability: Innovations in Producing Eco-Friendly Substrates for Soilless Culture Systems

Guest Editors

Wanlai Zhou
Wei Lin
Hong Wang

MDPI

Basel • Beijing • Wuhan • Barcelona • Belgrade • Novi Sad • Cluj • Manchester

Guest Editors

Wanlai Zhou
Institute of Urban Agriculture
Chinese Academy of
Agricultural Sciences
Chengdu
China

Wei Lin
Institute of Urban Agriculture
Chinese Academy of
Agricultural Sciences
Chengdu
China

Hong Wang
Institute of Urban Agriculture
Chinese Academy of
Agricultural Sciences
Chengdu
China

Editorial Office
MDPI AG
Grosspeteranlage 5
4052 Basel, Switzerland

This is a reprint of the Special Issue, published open access by the journal *Plants* (ISSN 2223-7747), freely accessible at: https://www.mdpi.com/journal/plants/special_issues/0UJN8J3OH6.

For citation purposes, cite each article independently as indicated on the article page online and as indicated below:

Lastname, A.A.; Lastname, B.B. Article Title. *Journal Name* **Year**, *Volume Number*, Page Range.

ISBN 978-3-7258-8007-2 (Hbk)
ISBN 978-3-7258-8008-9 (PDF)
https://doi.org/10.3390/books978-3-7258-8008-9

Contents

plants

Editorial

From Waste to Resource: Driving Sustainability Through Eco-Friendly Substrates for Soilless Culture

Wanlai Zhou *, Wei Lin and Hong Wang

Institute of Urban Agriculture, Chinese Academy of Agricultural Sciences, Chengdu 610213, China; linwei01@caas.cn (W.L.); wanghong07@caas.cn (H.W.)
* Correspondence: zhouwanlai@caas.cn

From high-tech greenhouses to urban vertical farms, the rapid global expansion of soilless cultivation has brought about unprecedented growing efficiency. However, this high productivity is heavily dependent on specific growing media—particularly peat and rockwool. Peat extraction destroys vital wetland carbon sinks, rockwool production is extremely energy-intensive, and its disposal also poses serious challenges. At the same time, agricultural and forestry activities generate large quantities of organic waste that are usually landfilled or incinerated, causing both resource waste and environmental burden. Converting waste into growing media could potentially address both issues. This Special Issue, entitled "Driving Sustainability: Innovations in Producing Eco-Friendly Substrates for Soilless Culture Systems", aims to explore a circular alternative: transforming waste into high-performance growing media. After a rigorous review process, we have collected 12 original research and review articles that together map the potential, limitations, and future directions of this approach.

The papers collected here cover a wide diversity of feedstocks, processing methods, and target crops. Studies show that agricultural and forestry residues—pine bark, wood fiber, coffee silverskin, brewer's spent grain, kenaf core, and green waste—as well as recycled construction waste, can serve as partial or complete substitutes for peat and perlite. Processing strategies include direct physical mixing, aging, composting, and an innovative ammonium incubation technique that rapidly eliminates phytotoxicity. Test crops include leafy vegetables (lettuce, lemon basil), a fruit vegetable (cucumber), ornamentals (sage), and woody shrubs (*Alyogyne huegelii* and *Goodenia ovata*). Thus, a unifying narrative emerges: organic and mineral waste-derived substrates can indeed drive sustainability, but their success is highly context-dependent, requiring species-specific optimization and careful management of two major bottlenecks—phytotoxicity and nitrogen immobilization.

1. Renewable Materials as Alternatives to Conventional Substrates: Feasibility Validation

Several studies evaluated locally available plant-derived materials as alternatives to conventional substrate components. In lemon basil, a mixture of 70% coconut fiber and 30% perlite significantly improved seed germination, seedling leaf number, and total leaf area [1]. In another trial, pine bark (8% v/v) was used to replace perlite in a coir-based substrate for lettuce. No negative effects on yield, chlorophyll content, or total phenolic content were observed; however, leaf iron and boron concentrations were elevated [2]. When wood fiber, coffee silverskin, and brewer's spent grain were used to replace 10–40% of peat for two Salvia genotypes, the results showed that the optimal substrate formulation depends not only on the type and rate of the amendment but also, to a large extent, on the plant genotype [3]. Overall, the suitability of plant-derived materials as growing media depends on crop species, material processing, and management [3,4]. Future research should optimize pretreatment methods for different materials and investigate long-term stability in diverse cropping systems.

Sustainability in soilless systems is not limited to organic alternatives. One study explored the possibility of using pine bark-amended recycled construction waste (recycled sand, crushed concrete, crushed rock) as a capping substrate for urban plantings [5]. The pure mineral wastes were strongly alkaline (pH 9.2–12.3) and had poor water retention; however, after adding 50% pine bark, the pH reached acceptable ranges and supported the growth of both alkaline-tolerant and alkaline-sensitive species. The best-performing mix (recycled sand +10% pine bark) produced plant biomass comparable to that of scoria, a conventional quarried material. This opens the door to large-scale reuse of demolition waste in green infrastructure.

2. Tackling Phytotoxicity: From Composting to Ammonium Incubation

A primary obstacle for fresh green waste is its inherent phytotoxicity, mainly derived from phenolics, organic acids, and other allelopathic compounds. Several papers in this Special Issue mention phytotoxicity issues in plant-derived materials such as coffee silverskin and green waste [6,7]. One research article optimized an ammonium incubation process for green waste; using 1.5% ammonium carbonate at 30 °C for five days with daily aeration reduced total phenolic content by 74% and increased the germination index (GI) from 6.4% to 127%, effectively eliminating phytotoxicity [6]. When this treated green waste replaced 75% of peat in lettuce cultivation, biomass increased nearly five-fold compared to untreated waste, though it remained below the peat control due to residual physical limitations. This low-energy, low-cost method represents a breakthrough for rapid detoxification of urban green waste.

3. Nitrogen Immobilization: A Double-Edged Sword

High-carbon substrates, especially wood fiber and fresh woody residues, often induce strong nitrogen immobilization, leading to transient nitrogen deficiency. A study on cucumber grown in wood fiber–peat mixtures showed that when the wood fiber content was 50%, and an additional 13 g N per plant was supplied, the fruit number and fruit biomass per plant were significantly higher [8]. Another study used chlorophyll fluorescence kinetics to investigate the physiological impact of wood fiber-induced nitrogen immobilization on lettuce [9]. The results indicated that wood fiber substrates significantly reduced available nitrogen, decreased chlorophyll content by 14–36%, and impaired electron transfer on the donor side of photosystem II (OEC) and the acceptor side of photosystem I. Notably, this damage was partially reversible over time, suggesting that dynamic, stage-adjusted nitrogen management can mitigate negative effects. A comprehensive review framed nitrogen immobilization as a "double-edged sword": short-term immobilization may cause crop stress, but the sequestered nitrogen, stored in microbial biomass and lignin-associated organic pools, forms a slow-release reservoir that reduces nitrate leaching [10].

4. Main Conclusions and Future Perspectives

Together, the 12 articles in this Special Issue provide compelling evidence that waste-derived materials can successfully replace peat, perlite, and rockwool in a wide range of soilless applications, provided that material-specific pretreatments (composting, ammonium incubation, aging, or blending) are applied and that nitrogen and water management are optimized according to crop growth stage. The two recurring challenges—phytotoxicity and nitrogen immobilization—are now better understood and can be addressed through the proposed technical and management frameworks.

https://doi.org/10.3390/plants15101459

Looking forward, we highlight two promising research directions that emerged from this collection but remain underexplored.

The first is the positive role of humic-like substances (e.g., fulvic acid) as biostimulants in substrate-based systems. A floating hydroponic trial with iceberg lettuce systematically evaluated the effects of fulvic acid (FA), amino acids (AA), and vermicompost (VC), alone and in combination. The treatment FA 40 ppm + VC 2 mL/L gave the highest total yield (17.15 kg/m^2), a 17.94% increase over the control. Further analysis showed that the combined application of FA, AA, and VC not only significantly increased vitamin C, total phenolics, and flavonoid contents but also effectively reduced nitrate accumulation in leaves [11]. These findings highlight the important value of biostimulants in improving yield and nutritional quality. Future research should pay more attention to humic-like substances released during the degradation of organic waste-based substrates, systematically evaluating their potential as endogenous biostimulants and their long-term effects on crop growth.

The second future research direction is the great potential of urban green waste (pruned branches, fallen leaves) as a local renewable substrate resource. Green waste is widely distributed, produced in large quantities, and has the advantage of local supply. Being mainly composed of plant fibers, it can be converted—after proper treatment—into a substrate with good water-holding capacity and aeration, while also supplying essential nutrients. Nevertheless, green waste generally suffers from severe phytotoxicity and nitrogen immobilization. The ammonium incubation technique presented in this Special Issue offers a rapid, low-cost, and scalable treatment method. However, long-term studies under commercial greenhouse conditions are needed to validate its performance over multiple crop cycles and to integrate it with closed-loop nutrient management systems.

We call upon the horticultural research community, industry stakeholders, and policymakers to actively embrace these innovations. The transition from a linear to a circular soilless cultivation model is not only an environmental necessity but also an economic opportunity—reducing dependence on peat and rockwool and creating value from local by-products. The articles in this Special Issue provide a solid scientific foundation for this transition.

Funding: This research received no external funding.

Data Availability Statement: No new data was generated for this article.

Acknowledgments: We extend our sincere gratitude to all authors for their high-quality contributions, to the peer reviewers for their rigorous and constructive feedback, and to the editorial team of *Plants* for their continuous support. We also thank the many researchers and practitioners who are dedicated to advancing sustainable soilless cultivation systems worldwide.

Conflicts of Interest: The authors declare no conflicts of interest.

References

1. Signorini, L.; Modarelli, G.C.; Di Pierro, P.; Langellotti, A.L.; Cirillo, C.; De Pascale, S.; Masi, P. Effects of Seedling Substrate and Hydroponic Versus Aquaponic Nutrient Solution on Growth, Nutrient Uptake, and Eco-Physiological Response of Lemon Basil (*Ocimum × citriodorum*). *Plants* **2025**, *14*, 1929. [CrossRef] [PubMed]
2. Dias, G.C.; Machado, R.M.A.; Alves-Pereira, I.; Ferreira, R.A.; Gruda, N.S. Potential of Pine Bark to Replace Perlite in Coir-Based Substrates: Effects on Nutrient Uptake, Growth, and Phytochemicals in Lettuce Under Two Salinity Levels. *Plants* **2025**, *14*, 2577. [CrossRef] [PubMed]
3. Sdao, A.E.; Cacini, S.; Loconsole, D.; Conversa, G.; Cristiano, G.; Elia, A.; De Lucia, B. Partial Replacement of Peat: Effects on Substrate Physico-Hydrological Properties and Sage Growth. *Plants* **2025**, *14*, 2801. [CrossRef] [PubMed]
4. Austin, C.C.; Parrish, S.B.; Clark, D.G.; Wilkie, A.C. Kenaf Core as an Alternative Soilless Growing Medium: A Review. *Plants* **2026**, *15*, 666. [CrossRef] [PubMed]

https://doi.org/10.3390/plants15101459

5. Kenefick, C.; Livesley, S.J.; Rayner, J.P.; Farrell, C. Using Recycled Construction Waste Amended with Pine Bark as a Substrate for Urban Plantings. *Plants* **2026**, *15*, 403. [CrossRef] [PubMed]
6. Cui, W.; Liu, J.; Bai, Q.; Wu, L.; Qi, Z.; Zhou, W. Rapid Reduction of Phytotoxicity in Green Waste for Use as Peat Substitute: Optimization of Ammonium Incubation Process. *Plants* **2024**, *13*, 2360. [CrossRef] [PubMed]
7. Sdao, A.E.; Gruda, N.S.; De Lucia, B. Beyond Peat: Wood Fiber and Two Novel Organic Byproducts as Growing Media—A Systematic Review. *Plants* **2025**, *14*, 1945. [CrossRef] [PubMed]
8. Čepulienė, R.; Butkevičienė, L.M.; Steponavičienė, V. Nutrient Use Efficiency and Cucumber Productivity as a Function of the Nitrogen Fertilization Rate and the Wood Fiber Content in Growing Media. *Plants* **2024**, *13*, 2911. [CrossRef]
9. Wu, L.; Li, R.; Liu, J.; Cui, W.; Qi, Z.; Zhou, W. Nitrogen Immobilization by Wood Fiber Substrates Strongly Affects the Photosynthetic Performance of Lettuce. *Plants* **2025**, *14*, 1518. [CrossRef]
10. Li, R.; Cui, W.; Zhang, M.; Qi, Z.; Zhou, W. Nitrogen Immobilization in Organic Media: A Double-Edged Sword Affecting the Utilization of Green Waste as Growing Media. *Plants* **2026**, *15*, 1298. [CrossRef]
11. Keskin, B.; Akhoundnejad, Y.; Dasgan, H.Y.; Gruda, N.S. Fulvic Acid, Amino Acids, and Vermicompost Enhanced Yield and Improved Nutrient Profile of Soilless Iceberg Lettuce. *Plants* **2025**, *14*, 609. [CrossRef] [PubMed]

https://doi.org/10.3390/plants15101459

plants

MDPI

Article

Partial Replacement of Peat: Effects on Substrate Physico-Hydrological Properties and Sage Growth

Anna Elisa Sdao [1], Sonia Cacini [2,*], Danilo Loconsole [1], Giulia Conversa [3], Giuseppe Cristiano [1], Antonio Elia [3] and Barbara De Lucia [1,*]

1 Department of Soil, Plant and Food Sciences, University of Bari Aldo Moro, Via Amendola 165/A, 70126 Bari, Italy; anna.sdao@uniba.it (A.E.S.); danilo.loconsole@uniba.it (D.L.); giuseppe.cristiano@uniba.it (G.C.)

2 CREA Research Centre for Vegetables and Ornamental Crops, Council for Agricultural Research and Economics, Via dei Fiori 8, 51017 Pescia, Italy

3 Department of Agriculture, Food, Natural Resources and Engineering (DAFNE), University of Foggia, via Napoli 25, 71100 Foggia, Italy; giulia.conversa@unifg.it (G.C.); antonio.elia@unifg.it (A.E.)

* Correspondence: sonia.cacini@crea.gov.it (S.C.); barbara.delucia@uniba.it (B.D.L.)

Abstract

The transformation of organic by-products derived from waste into value-added resources represents a promising strategy to advance circular economy principles and bolster environmental and agricultural sustainability, especially in soilless cultivation. This study evaluates the viability of three organic by-products—wood fiber (WF), coffee silverskin (CS), and brewer's spent grains (BSGs)—as partial peat replacements in horticultural substrates. Ten growing media formulations were assessed, incorporating increased doses (0–40% v/v as peat replacement-PR) of each alternative by-product. The effects on physical and hydraulic substrate properties, along with plant growth traits, were examined using two ornamental *Salvia* genotypes, 'Victoria' and 'Amistad'. To synthesize the multivariate growth data into a single, biologically meaningful metric, based on the first principal component, a Growth Index (GI), a PC1-derived index, was calculated, providing a powerful, unified metric to rank substrate efficacy. WF-based substrates exhibited increased porosity and diminished water retention, whereas media enriched with CS and BSG enhanced moisture availability, particularly at 20–40 PR. The bulk density was highest at PR40 for both WF and BSG treatments, and at PR20 in CS-based substrates. Electrical conductivity increased in CS and BSG treatments with rising PR levels. The results on the vegetative growth of ornamental sages have highlighted that differential PR rates are required depending on the specific organic by-product and plant genotype. In 'Victoria', GI indicates that a 20% replacement of peat with BSG provided the optimal conditions for holistic plant development; the lowest GI for WF substrates across nearly all peat replacement levels indicated that it was the most detrimental alternative for this cultivar. In 'Amistad', the analysis of the GI scores revealed that the CS20 and BSG20 of peat replacement yielded the highest overall growth, with GI scores significantly greater than those of the peat control. CS10 and BSG40 also showed high GI scores in 'Amistad'. WF10 had GI scores similar to those of the peat control. In general, the GI-based approach confirms that moderate inclusion of brewer's spent grain (BSG20) is a highly effective peat replacement for both genotypes. At the same time, coffee silverskin (CS) is particularly effective for the 'Amistad' genotype. This analysis underscores that optimal substrate formulation is not only dependent on the amendment type and rate but also critically on the plant genotype.

Keywords: brewer's spent grain; by-product valorization; circular alternative; coffee silverskin; potted ornamentals; soilless cultivation; sustainable growing medium; wood fiber

1. Introduction

The ornamental industry produces potted plants, grown in soilless cultivation, for their esthetic value from a wide range of species [1], which are distinguished by broad agronomic practices with specific requirements in terms of substrate. This is also referred to as growing medium, which consists of solid and porous materials of natural or synthetic origin, mineral or organic. They form the structural and functional basis for root development [2]. Schindler et al. [3] defined the growing medium as a man-made environment designed to optimize water and nutrient availability, provide mechanical support, and promote microbial interactions essential for plant growth and productivity. To fulfill these roles effectively, substrates must satisfy several key requirements. Physically, it should have a porous and stable structure to ensure adequate aeration and water-holding capacity [4–6]. Chemically, it must offer consistency, maintaining stable chemical properties over time and be free of toxic compounds [7]. Biologically, it should support beneficial microorganisms and suppress pathogens [8]. In addition, practical attributes such as lightweight, affordability, ease of handling and local availability are equally critical for sustainable production [9,10]. Historically, peat has served as the primary component of commercial horticultural substrates due to its superior moisture retention, structural stability, and compatibility with a wide range of crops [11,12]. Global peat extraction is estimated to be approximately 90 million m^3 annually, with nearly 40 million m^3 allocated to horticulture. In Europe alone, approximately 20 million tons are harvested annually [13,14]. However, growing environmental concerns, such as greenhouse gas emissions and the destruction of wetland ecosystems, have prompted regulatory action [15,16]. Countries such as the United Kingdom have announced bans on peat use in amateur gardening by 2026 and in the professional horticulture sector by 2030 [17,18]. Despite these policy efforts, the transition away from peat remains technically challenging. Commercial growers require alternative substrates that can match the performance of peat across multiple agronomic parameters without compromising crop yield or quality. In this context, one promising strategy involves the valorization of organic by-products generated by the forestry and agri-food sectors [19]. These materials, often produced in large volumes and locally available, align well with circular economy principles and offer significant potential as sustainable substrate components [20,21].

1.1. Wood Fiber

Global demand for wood fiber (WF) is projected to increase by approximately 1000% by 2050, accompanied by an estimated requirement of 65 Mm^3 $year^{-1}$ of novel materials to support the expanding adoption of soilless cultivation systems worldwide [13]. WF, a by-product of wood processing, is a growing material in the substrate industry, although it is usually not more than 30% (v/v) with peat [22]. As a renewable and biodegradable resource derived from products of wood manufacture, it exhibits several favorable properties for use in soilless cultivation. It has a low bulk density, good drainability and wettability, and can optimize the physical properties and reduce the weight of the growing media [7,23]. WF is a stable, pathogen-free material that could help reduce dependence on non-renewable substrates [24]. However, the increasing need for biomass for energy and heat production presents significant competition for the substrate sector in securing access to this raw material [25].

1.2. Coffee Silverskin

Coffee silverskin (CS), which represents 4.2% of coffee green beans' total weight, is a by-product of coffee roasting and is currently discarded as waste, but exhibits several favorable properties for substrate utilization. It is lightweight, biodegradable, pathogen-free, and renewable, and may contribute functional compounds, such as antioxidants, to the growing medium [21,26].

1.3. Brewer's Spent Grain

Similarly, brewer's spent grain (BSG), which accounts for up to 85% of brewery by-product, is another abundant organic residue [27]; it consists of layers of peel, pericarp, and seeds with residual amounts of endosperm and aleurone from barley. With global production estimated at 37.2 million tons in 2021—approximately 21–22 kg generated per 100 L of beer—BSG is microbiologically sterile post-production due to thermal processing. However, its high moisture and nutrient content make it highly perishable [28,29]. When stabilized and incorporated into substrates in small proportions (<5%), BSG may act as a biofertilizer and enhance the organic content of the medium [30].

1.4. Hydrophysical Recommended Ranges

The EU is promoting research on sustainable practices to recover high-quality resources from organic residues and by-products, aligning with the Green Deal objectives [31], the Farm to Fork (F2F) Strategy [32] and the new Circular Economy Action Plan [33]. The effectiveness of any growing medium, including those incorporating organic by-products, is closely linked to its physical and hydraulic characteristics, particularly the balance between water availability and air-filled porosity. These properties are primarily influenced by the material composition, particle size distribution, and container geometry [34–36].

The recommended ranges for the commercial production of soilless crops are as follows: dry bulk density (BD) < 0.4 g cm^{-3}, total porosity space (TPS) > 85%, air content (AC) 20–30% vol., total water-holding capacity (WHC) 550–700 mL L^{-1} and a water volume content at −1 kPa in the range 55–70% vol. [37]. Maintaining substrates within the recommended physical property ranges [38] is expected to reduce the sensitivity of plant production management; however, outcomes may still vary depending on the specific cultivation practices of the growers. Assessing the similarities and differences in hydrophysical properties between forestry and beverage by-products, as well as conventional components such as peat, constitutes a crucial step in enhancing their potential as components of potting media.

1.5. Ornamental Sages

Salvia is the largest genus in the Lamiaceae family, with 1036 accepted species to date [39], widely cultivated for its ornamental, medicinal, and aromatic properties [40,41], and constitutes almost one-quarter of this family [42]. It is widely distributed in temperate, subtropical, and tropical regions worldwide [43]. *Salvia* spp. are also important ornamental plants due to their beautiful leaves, inflorescence variations, and hardiness in dry and hot [44,45] or wet [46] and cold [47] environments. Ornamental sages are propagated asexually both in vivo [48,49] and in vitro [50]. The floriculture industry continuously strives to introduce new cultivars into the market. In this context, cultivars and interspecific hybrids among *Salvia* species offer the potential to exploit the extensive variability in leaf morphology, fragrance, flower color, and inflorescence architecture, as well as for the agronomic needs of cultivation and planting in green spaces.

Despite the great importance of partial peat replacement in growing media, in recent years, few studies have investigated the species and cultivars of *Salvia*. Greco et al. [51]

in potted *S. officinalis* L., experimented with the use of vermicompost, compost, and solid digestate as valid peat alternatives. In *S. rosmarinus* Spenn., the application of 10% smoked biochar displayed a subtle stimulating effect compared to plants grown in substrates composed solely of 100% peat [52].

1.6. Aim of Research

Many studies have focused on the role of physico-hydraulic properties of substrates with reduced peat content in influencing the agronomic performance of ornamental plants [53–57]. Additionally, other studies have examined physical characteristics [58–60] to assess the feasibility of successfully replacing peat in ornamental plant production.

Building on this background, the present study evaluated the feasibility of incorporating forestry and drink industry by-products—WF, CS, and BSG—as partial peat replacement (PR) at increasing doses, monitoring the growth of potted ornamental sages. The focus was on assessing their physico-hydraulic behavior and crop growth in the context of a circular, sustainable and high-performance greenhouse pot-based growing system.

2. Results

2.1. Physical and Hydrological Characterization

Table 1 presents the ANOVA results for the effects of peat replacement (PR) with wood fiber (WF), coffee silverskin (CS), and brewer's spent grain (BSG) on physical properties of the substrates assessed before cultivation.

Table 1. ANOVA results for the effects of peat replacement (PR) with wood fiber (WF), coffee silverskin (CS), or brewer's spent grain (BSG) on the physical properties of the substrates assessed before cultivation. Orthogonal polynomial contrasts were employed to evaluate linear (*lin*) and quadratic (*qua*) trends. WV = Water Volume at −1 kPa; AC = Air Content at −1 kPa; SM = Substrate Moisture at −1 kPa; EAW = Easily Available Water; BC = Water Buffering Capacity; TAW = Total Available Water; TPS = Total Porosity Space; BD = Dry Bulk Density.

	WV	AC	SM	EAW	BC	TAW	TPS	BD
				Wood fiber				
Peat replacement (PR)	* [a]	ns	*	*	ns	*	***	***
PR*lin*	**	ns	*	*	ns	*	***	***
PR*qua*	*	ns	ns	ns	ns	*	ns	ns
				Coffee silverskin				
PR	*	*	***	**	ns	**	***	**
PR*lin*	ns	ns	***	***	ns	***	ns	ns
PR*qua*	ns	ns	**	ns	ns	*	**	**
				Brewer's spent grain				
PR	*	*	*	**	*	***	***	***
PR*lin*	ns	ns	**	***	*	***	***	***
PR*qua*	ns	ns	ns	***	ns	***	***	***

[a] ns, *, **, and ***, not significant or significant at $p \leq 0.05$, $p \leq 0.01$, or $p \leq 0.001$, respectively.

Water volume at −1 kPa (WV) (Figure 1) was affected differently by peat replacement (PR) matrices: both wood fiber (WF) and brewer's spent grain (BSG) substrates exhibited an asymptotic approach; coffee silverskin (CS) did not significantly alter WV across the tested range. When WF replaces peat at 20%, as expected, the WV reaches a lower value than that of the 0PR. With BSG, at 0 and 40% PR, the WV values appear comparable, therefore difficult to interpret

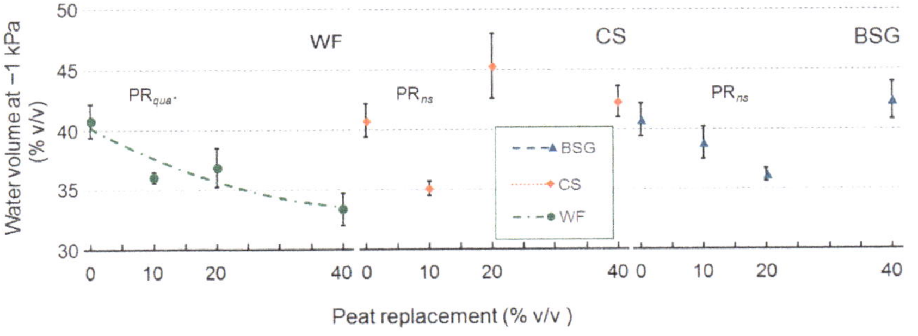

Figure 1. Effect of peat replacement level (PR, 0–40% v/v) with three by-product materials—wood fiber (WF), coffee silverskin (CS), or brewer's spent grain (BSG)—on water volume at −1 kPa. Data points represent means ± standard error (n = 3). Statistical analysis performed through ANOVA: ns and *, not significant or significant at $p \leq 0.05$, respectively. Orthogonal polynomial contrasts were employed to evaluate quadratic (*qua*) trend.

Substrate moisture (SM) at −1 kPa, which is a parameter close to field capacity (Figure 2) when BSG was used as the substrate for PR, was not significant. In contrast, with WF, values decreased as replacement levels increased. With CS as PR, SM increased quadratically, reaching its most significant peak at doses between 20 and 40% PR.

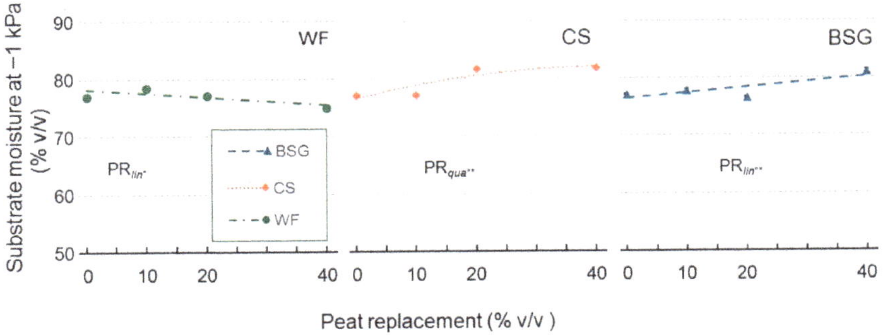

Figure 2. Effect of peat replacement level (PR, 0–40% v/v) with three by-product materials—wood fiber (WF), coffee silverskin (CS), or brewer's spent grain (BSG)—on substrate moisture at −1 kPa. Data points represent means ± standard error (n = 3). Statistical analysis performed through ANOVA: * and **, significant at $p \leq 0.05$ or $p \leq 0.01$, respectively. Orthogonal polynomial contrasts were employed to evaluate linear (*lin*) and quadratic (*qua*) trends.

PR generally reduced the amount of easily available water (EAW) (Figure 3), although the effect varied depending on the substrate type. With WF and CS as PR, easily available water decreased linearly as the percentage of peat replacement increased. In contrast, the use of BSG resulted in a distinct non-linear, quadratic response. The relationship for BSG was convex, with EAW declining to a minimum at approximately 20% PR. Crucially, beyond this point, the EAW values appeared to stabilize, suggesting that a plateau was reached rather than a continued decline. This indicates that BSG's impact on water availability is limited beyond a certain application rate.

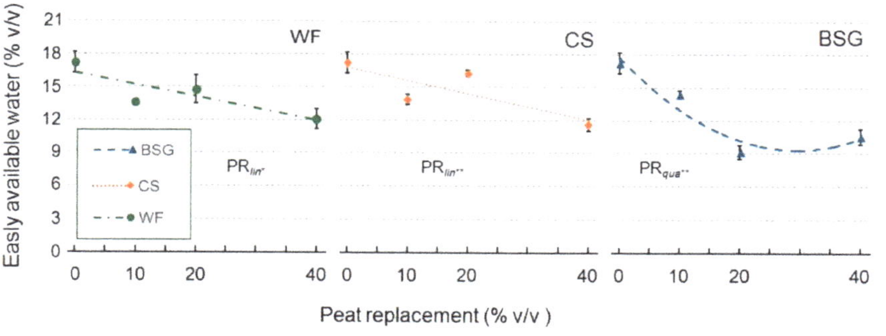

Figure 3. Effect of peat replacement level (PR, 0–40% v/v) with three by-product materials—wood fiber (WF), coffee silverskin (CS), or brewer's spent grain (BSG)—on easily available water. Data points represent means ± standard error (n = 3). Statistical analysis performed through ANOVA: * and **, significant at $p \leq 0.05$ or $p \leq 0.01$, respectively. Orthogonal polynomial contrasts were employed to evaluate linear (*lin*) and quadratic (*qua*) trends.

The available water results (Figure 4) followed the same trend as those relating to EAW.

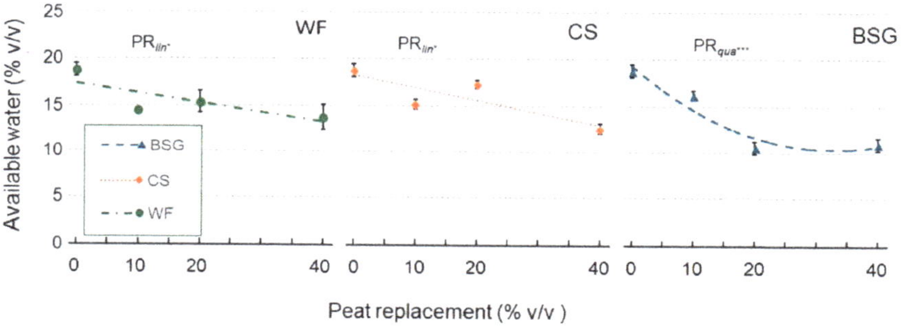

Figure 4. Effect of peat replacement level (PR, 0–40% v/v) with three by-product materials—wood fiber (WF), coffee silverskin (CS), or brewer's spent grain (BSG)—on available water. Data points represent means ± standard error (n = 3). Statistical analysis performed through ANOVA: * and ***, significant at $p \leq 0.05$ or $p \leq 0.001$, respectively. Orthogonal polynomial contrasts were employed to evaluate linear (*lin*) and quadratic (*qua*) trends.

The water buffering capacity (BC) (Figure 5) was not significantly affected when WF and CS replaced the peat. However, when using BSG as a substrate, the BC exhibited a negative response as the PR level increased. Trend lines are shown for consistency and to aid in the visual interpretation of potential linear or quadratic trends, even where ANOVA results indicate non-significance (WF and CS).

When WF was used as the substrate, the dry bulk density (BD) (Figure 6) showed a linear increase, reaching the highest value at the 40% PR level. In contrast, with CS and BSG as substrates, a quadratic response with a concave course in the CS and a convex course in the BSG was observed. The reverse trend is apparent in the air spaces, so this is an effect of compaction driven by the 20% replacement pots.

Wood fiber as a substrate reduced the total porosity space (TPS) linearly as the PR percentage increased (Figure 7). With CS and BSG substrates, this property exhibited an asymptotic trend, convex in the CS and concave in the BSG.

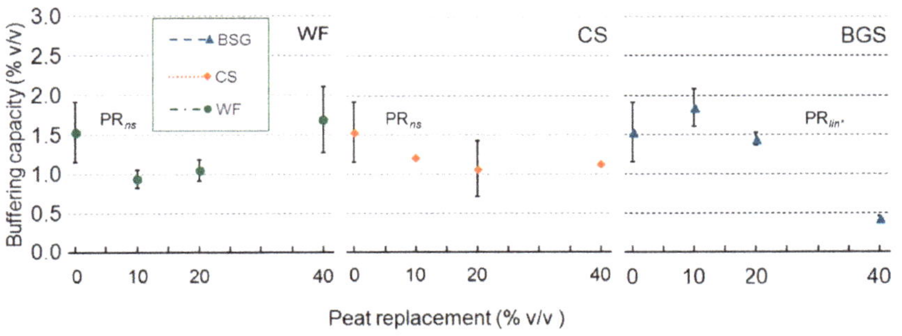

Figure 5. Effect of peat replacement level (PR, 0–40% v/v) with three by-product materials—wood fiber (WF), coffee silverskin (CS), or brewer's spent grain (BSG)—on buffering capacity. Data points represent means ± standard error (n = 3). Statistical analysis performed through ANOVA: ns and *, not significant or significant at $p \leq 0.05$, respectively. Orthogonal polynomial contrasts were employed to evaluate linear (*lin*) trend.

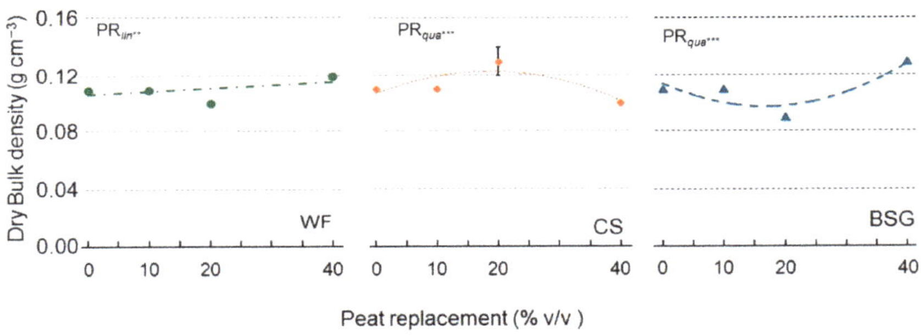

Figure 6. Effect of peat replacement level (PR, 0–40% v/v) with three by-product materials—wood fiber (WF), coffee silverskin (CS), or brewer's spent grain (BSG)—on bulk density. Data points represent means ± standard error (n = 3). Statistical analysis performed through ANOVA: ** and ***, significant at $p \leq 0.01$, or $p \leq 0.001$, respectively. Orthogonal polynomial contrasts were employed to evaluate linear (*lin*) and quadratic (*qua*) trends.

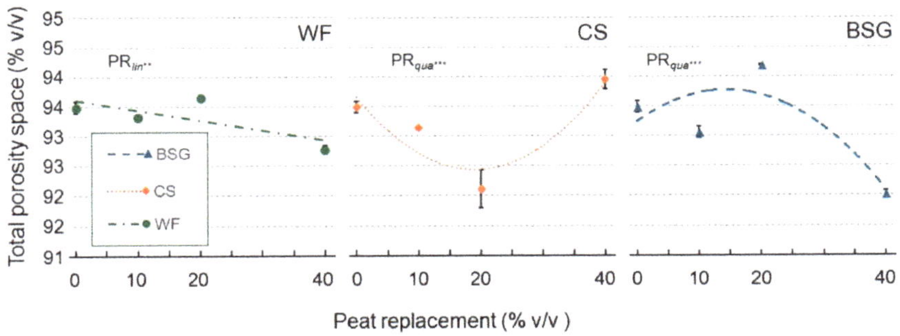

Figure 7. Effect of peat replacement level (PR, 0–40% v/v) with three by-product materials—wood fiber (WF), coffee silverskin (CS), or brewer's spent grain (BSG)—on total porosity. Data points represent means ± standard error (n = 3). Statistical analysis performed through ANOVA: ** and ***, significant at $p \leq 0.01$, or $p \leq 0.001$, respectively. Orthogonal polynomial contrasts were employed to evaluate linear (*lin*) and quadratic (*qua*) trends.

2.2. Volumetric Water Content

Volumetric water content (VWC) shows the behavior of the peat-based substrate at increasing percentages of WF (Figure 8A), CS (Figure 8B) and BSG (Figure 8C).

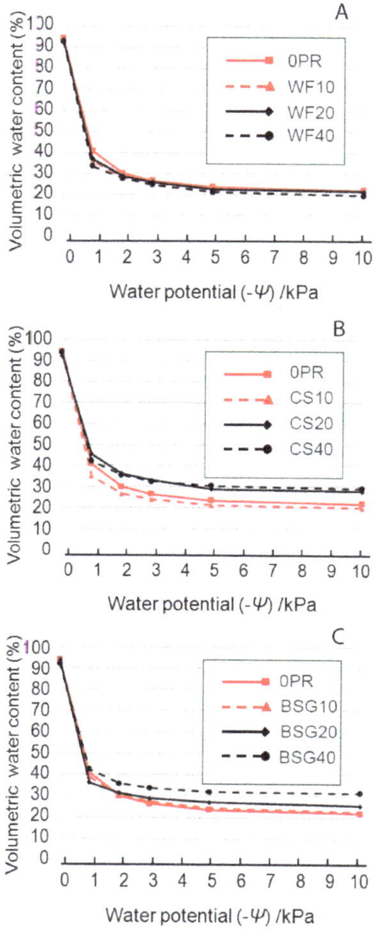

Figure 8. Effect of partial peat replacement (PR at 0, 10, 20 and 40% levels v/v) with wood fiber (WF, (**A**)), coffee silverskin (CS, (**B**)), or brewer's spent grain (BSG, (**C**)) on the water retention curve. Each water retention curve was determined on the basis of 3 replicates.

Figure 8A presents the VWC across varying matric suctions (0 to −10 kPa) for peat control (0PR) and wood fiber-based substrates (10, 20 and PR40). The water retention curves shows clear and gradual shifts in hydraulic behavior as the proportion of WF increases, with significant implications for substrate water availability. At 0 kPa, all substrates displayed high VWC values, with a gradual decline from 0PR (94.3%) to WF40 (92.8%). Substrates did not differ significantly in Air Content (AC) at −1 kPa, which ranged from 44.9% to 59.5%, indicating a similar amount of gravitational water regardless of WF content. Conversely, the Total Available Water (TAW) decreased significantly, with the highest value (18.8%) recorded for 0PR and the lowest (13.8%) for WF40.

Figure 8B shows the water retention curves for the 0PR and CS substrates with PR of 10, 20, and 40% (v/v). The water retention curves show distinct changes in substrate hydrophysical behavior as CS content increases, suggesting a potential impact on water availability and substrate performance. Under saturated conditions, all substrates exhibited high VWC, with the 0PR showing the highest value (94.3%), followed by CS40 (93.9%), CS10 (93.1%), and CS20 (92.1%).

Differences became more evident as matric suction increased with a substantial reduction in VWC at −1 kPa observed across all treatments. The 0PR retained 40.8% VWC, while CS40 showed a more pronounced decline to 33.1%, highlighting reduced water retention with a higher CS level. Intermediate treatments (CS10 and CS20) exhibited moderate VWC losses (38.4 and 36.1%, respectively), consistent with their CS proportions. As tension further increased to −2 and −3 kPa, the VWC continued to decrease, with CS-based substrates consistently retaining less water than the peat control. By −5 kPa tension, the gap among treatments narrowed slightly but remained relevant. At −10 kPa, all substrates had reached the hygroscopic water range, yet differences persisted: 0PR retained the highest residual water content (22.1%), while CS40 showed the lowest (18.9%).

Figure 8C shows the water retention curves for 0PR and brewer's spent grain-based substrates (BSG 10, 20 and 40% PR). At 0 kPa tension, all substrates demonstrated high VWC under saturated conditions, with the highest value recorded in 0PR at 94.3%, followed by BSG40 (93.9%), BSG10 (93.1%), and BSG20 (92.1%), indicating a slight initial decrease in VWC with increasing BSG percentage. At tension of −1 kPa, a sharp drop in VWC occurred across all substrates: 0PR maintained the highest value (40.8%), whereas BSG10 showed the lowest (35.1%), followed by BSG40 (42.1%) and BSG20 (45.3%). At −2 and −3 kPa tension, the decrease in VWC continued, with values consistently distributed in proportion to the BSG percentage. At −2 kPa, the values were 0PR (30.1%), BSG10 (26.6%), BSG20 (36.2%), and BSG40 (35.5%). At −3 kPa, the values were 0PR (26.4%), BSG10 (23.9%), BSG20 (33.1%), and BSG40 (32.4%). At a tension of −5 kPa, the rate of water loss slowed down, with values of 0PR (23.5%), BSG10 (21.1%), BSG20 (29.0%), and BSG40 (30.4%). Finally, at the highest tension of −10 kPa, the differences between the substrates further narrowed, with 0PR (22.1%) maintaining the highest value and BSG10 (19.9%) the lowest. BSG20 (27.9%) and BSG40 (29.2%) exhibited significantly higher water retention.

2.3. Chemical Characteristics of Substrates

Table 2 presents the ANOVA results for the effects of peat replacement (PR) with wood fiber (WF), coffee silverskin (CS), and brewer's spent grain (BSG) on chemical properties, pH and EC (electrical conductivity) of the substrates assessed before cultivation.

Table 2. ANOVA results for the effects of peat replacement (PR) with wood fiber (WF), coffee silverskin (CS), or brewer's spent grain (BSG) on chemical properties, pH and EC (electrical conductivity) of the substrates assessed before cultivation. Orthogonal polynomial contrasts were employed to evaluate linear (*lin*) and quadratic (*qua*) trends.

Treatment	pH	EC
	Wood fiber	
PR	ns	***
PR*lin*	ns	**
PR*qua*	ns	**
	Coffee silverskin	
PR	ns	**
PR*lin*	ns	**
PR*qua*	ns	ns

Table 2. *Cont.*

Treatment	pH	EC
	Brewer's spent grain	
PR	ns	***
PR*lin*	ns	***
PR*qua*	ns	ns

ns, **, and ***, not significant or significant at $p \leq 0.01$, or $p \leq 0.001$, respectively.

Moving on to examine Table 2, no significant effect on pH was observed following the PR with any of the substrate types. The pH values remained relatively stable across all PR levels for WF, CS, and BSG, with statistical analysis confirming no significant trends (PR_{ns}) in all cases (Figure 9).

When WF was used as a substrate (Figure 10), the trend was asymptotic; its interpretation remains uncertain, probably due to the limited dataset available. For CS, EC remains relatively stable at lower PRs; the 20% PR appears to be an outlier. EC begins to increase more noticeably beyond 30%, following a general linear pattern. In contrast, BSG shows a strong linear increase in EC with increasing peat replacement, reaching the highest EC value at PR40 (Figure 10).

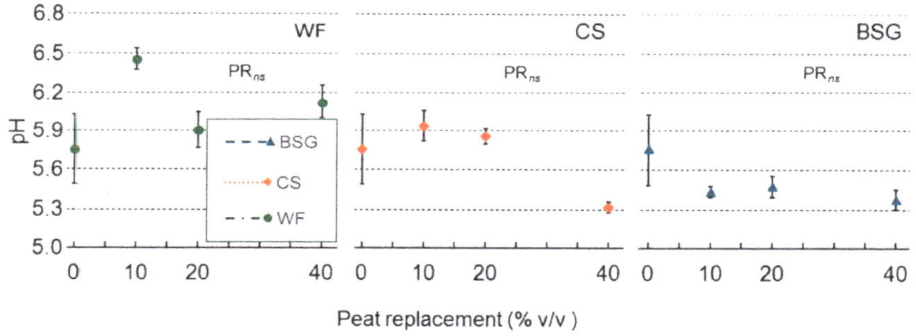

Figure 9. Effect of peat replacement level (PR, 0–40% *v/v*) with three by-product materials—wood fiber (WF), coffee silverskin (CS), or brewer's spent grain (BSG)—on pH of the final substrate mixture. Data points represent means ± standard error (n = 3). Statistical analysis performed through ANOVA: ns, not significant.

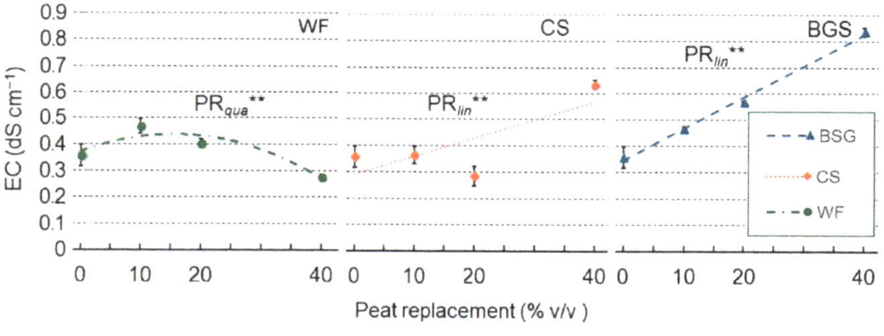

Figure 10. Effect of peat replacement level (PR, 0–40% *v/v*) with three by-product materials—wood fiber (WF), coffee silverskin (CS), or brewer's spent grain (BSG)—on the electrical conductivity (EC) of the final substrate mixture. Data points represent means ± standard error (n = 3). Statistical analysis performed through ANOVA: **, significant at $p \leq 0.01$. Orthogonal polynomial contrasts were employed to evaluate linear (*lin*) and quadratic (*qua*) trends.

2.4. Sage Growth

Table 3 (cv. *Victoria*) and Table 4 (cv. *Amistad*) report the ANOVA results for the effect of Substrate Type (ST) and Peat Replacement (PR) on sage growth parameters at the end of the vegetative phase (14 January 2024).

Table 3. ANOVA results showing the effect of Substrate Type (ST) and Peat Replacement (PR) on the main growth parameters in 'Victoria' genotype at the end of the vegetative phase (January 14, 2024). FW: Fresh Weight, DW: Dry Weight, DM: Dry Matter, no.: number; ChlC: Chlorophyll Concentration; LA: leaf area. Orthogonal polynomial contrasts were employed to evaluate linear (*lin*) and quadratic (*qua*) trends.

Treatment	Plant Height (cm)	Leaves (no./Plant)	Leaf Area (cm^2/Plan)	ChlC (µmol m^{-2} LA)	Plant FW (g/Plant)	Shoot DW (g/Plant)	Root DW (g/Plant)	Plant DM (g 100 g^{-1} FW)
Substrate Type (ST)	*** [a]	***	***	***	***	***	ns	***
Peat replacement (PR)	***	***	***	***	***	***	***	**
PR*lin*	ns	***	***	***	***	***	***	ns
PR*qua*	ns	***	***	***	ns	*	***	ns
ST × PR	***	***	***	***	***	***	ns	ns
ST × PR*lin*	ns	***	***	***	*	***	ns	ns
ST × PR*qua*	***	ns	***	***	***	***	ns	ns

[a] ns, *, **, and ***, not significant or significant at $p \leq 0.05$, $p \leq 0.01$, or $p \leq 0.001$, respectively.

Table 4. ANOVA results showing the effect of Substrate Type (ST) and Peat Replacement (PR) on plant growth parameters in 'Amistad' genotype at the end of the vegetative phase (14 January 2024). FW: Fresh Weight, DW: Dry Weight, DM: Dry Matter, no.: number; ChlC: Chlorophyll Concentration, LA: leaf area. Orthogonal polynomial contrasts were employed to evaluate linear (*lin*) and quadratic (*qua*) trends.

Treatment	Plant Height (cm)	Leaves (no./Plant)	Leaf Area (cm^2/Plant)	ChlC (µmol m^{-2} LA)	Plant FW (g/Plant)	Shoot DW (g/Plant)	Root DW (g/Plant)	Plant DM (g 100 g^{-1} FW)
Substrate type (ST)	ns [a]	*	***	***	***	***	***	*
Peat Replacement (PR)	ns	ns	***	***	***	***	*	ns
PR*lin*	ns	ns	***	*	***	***	**	ns
PR*qua*	ns	*	***	***	***	***	ns	ns
ST × PR	*	***	***	***	***	***	***	***
ST × PR*lin*	ns	***	***	***	**	ns	***	***
ST × PR*qua*	ns	*	***	**	***	**	*	ns

[a] ns, *, **, and ***, not significant or significant at $p \leq 0.05$, $p \leq 0.01$, or $p \leq 0.001$, respectively.

In 'Victoria' (Figure 11A), plant height remained relatively stable across all PR levels when WF was used as the substrate. With the BSG, the plant height showed a quadratic response with a slightly concave pattern. The CS substrate demonstrated similar behavior, but the asymptotic and convex trend, reaching its highest height value at 20–25% PR. In 'Amistad', the ST × PR*qua* interaction was not significant (Figure 11B).

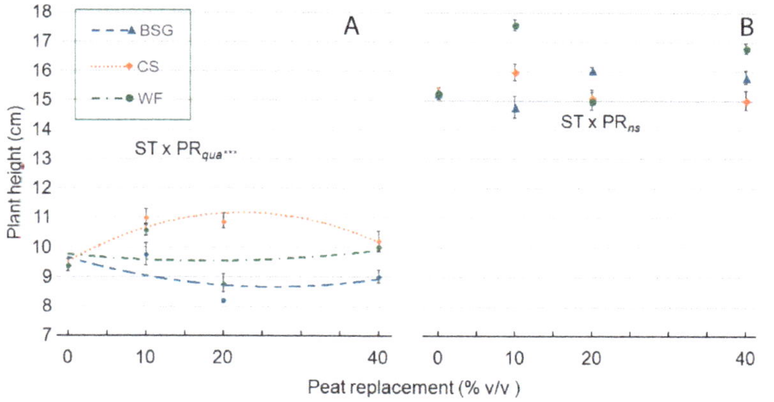

Figure 11. Effect of peat replacement level (PR, 0–40% v/v) with three organic materials (ST)—brewer's spent grain (BSG), coffee silverskin (CS), or wood fiber (WF)—on plant height (cm) in Victoria (**A**) and Amistad (**B**) genotypes at the end of the vegetative phase (14 January 2024). Data points represent means ± standard error (n = 3). Statistical analysis performed through ANOVA: ns, and ***, not significant or significant at $p \leq 0.001$, respectively. Orthogonal polynomial contrasts were employed to evaluate quadratic (*qua*) trend.

In 'Victoria', PR generally reduced leaf number per plant, although the effect varied by substrate type. With WF, leaf number decreased linearly as the PR percentage increased. In contrast, the CS substrate initially showed a slight increase in leaf number at PR10, followed by a clear linear decline at higher PR levels. The BSG substrate had the least impact, with leaf numbers remaining relatively stable across all PR levels (Figure 12A). Similarly to 'Victoria', increasing PR in 'Amistad' generally reduced leaf number per plant, with substrate-specific asymptotic trends. (Figure 12B). In 'Victoria' (Figure 12A), the number of leaves per plant was negatively affected by wood fiber (WF) at all substitution levels, from −11% at PR10 to −14% at PR40, whereas coffee silverskin (CS) strongly increased leaf number at PR10 and PR20 (+54 and +34%), but declined at PR40 (−17%). Brewer's spent grain (BSG) enhanced leaf formation at PR10 and PR20 (+52% and +38%), with only a minor increase at PR40 (+11%). In 'Amistad' (Figure 12B), the number of leaves per plant, when WF was PR, slightly increased under PR10 (+9%), and decreased at PR40 (−8%). CS reduced leaf number both at PR10 and PR40 (−13%). BSG decreased leaf numbers at PR10 and PR20 (−18 and −15%) but markedly increased at PR40 (+14%).

Regarding leaf area per plant, as the primary metric of leaf development, when WF was used as the substrate, it consistently increased with higher PR levels. However, this effect was less pronounced than that of the other treatments. The response followed a near-linear to slightly quadratic positive trend in both genotypes, indicating that WF moderately benefited leaf area development. With a BSG substrate, leaf area exhibited a quadratic response, peaking at approximately 20–30% PR in 'Victoria' and up to PR40 in 'Amistad.' The trend was convex for 'Victoria' and slightly concave for 'Amistad.' CS exhibited a positive trend similar to that of BSG but required lower PR levels to achieve comparable effects. In 'Victoria', leaf area reached its maximum at approximately 20% before declining, whereas in 'Amistad', it peaked at 30%. (Figure 13A,B). In 'Victoria' (Figure 13A), WF reduced plant leaf area to PR10 (−13%) but markedly increased it at 20 and 40 PR levels (+50% and +61%, respectively). CS, on the one hand, strongly promoted it at PR10 (+115%), while on the other hand, it was moderately promoted at PR20 (+34%) and PR40 (+69%). BSG caused substantial increases at PR10 (+88%) and PR20 (+134%), followed by a less pronounced improvement at PR40 (+57%). Conversely, 'Amistad' (Figure 13B) responded positively to all substrate types. WF showed strong increases at both PR20 (+93%) and PR40 (+81%), while CS induced the most significant gains at both PR10 (+170%) and PR20 (+153%), and a moderate increase at PR40 (+69%). BSG promoted progressive improvements up to PR40 (+173%).

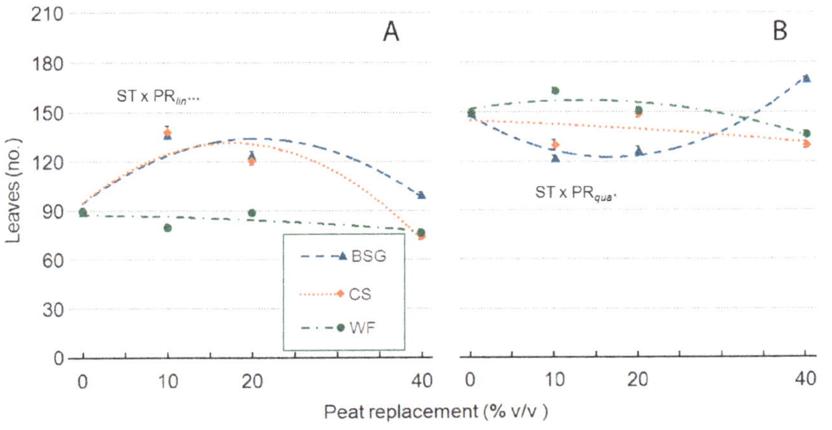

Figure 12. Effect of peat replacement level (PR, 0–40% v/v) with three alternative organic matrices (ST)—brewer's spent grain (BSG), coffee silverskin (CS), or wood fiber (WF)—on leaves per plant (no.: number) in 'Victoria' (**A**) and 'Amistad' (**B**) genotypes at the end of the vegetative phase (14 January 2024). Data points represent means ± standard error (n = 3). Statistical analysis performed through ANOVA: * and ***, significant at $p \leq 0.05$, or $p \leq 0.001$, respectively. Orthogonal polynomial contrasts were employed to evaluate linear (*lin*) and quadratic (*qua*) trends.

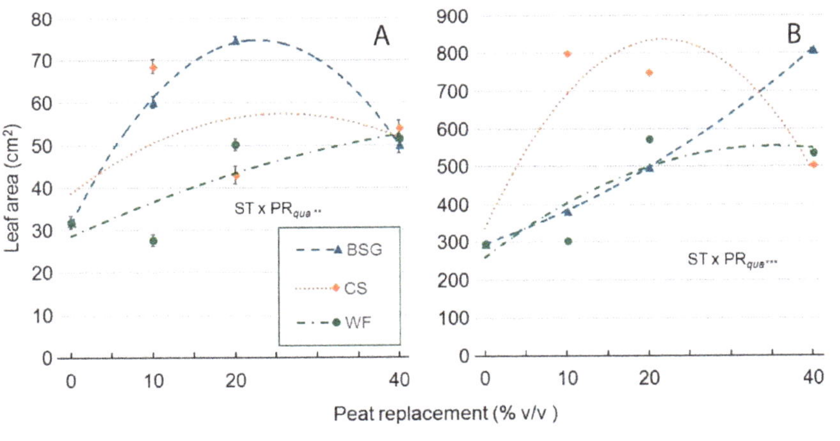

Figure 13. Effect of peat replacement level (PR, 0–40% v/v) with three alternative organic matrices (ST)—brewer's spent grain (BSG), coffee silverskin (CS), or wood fiber (WF)—on leaf area (cm$^{2/\text{plant}}$) in 'Victoria' (**A**) and 'Amistad' (**B**) genotypes at the end of the vegetative phase (14 January 2024). Data points represent means ± standard error (n = 3). Statistical analysis performed through ANOVA: **, and ***, significant at $p \leq 0.01$, or $p \leq 0.001$, respectively. Orthogonal polynomial contrasts were employed to evaluate quadratic (*qua*) trend.

With WF as substrate, 'Victoria' Chlorophyll Concentration (ChlC), exhibited a quadratic response, with values peaking at 20% PR before declining at higher replacement levels (Figure 14A). In 'Amistad', ChlC showed a slight near-linear increase as the

WF proportion increased, with no observed peak (Figure 14B). With CS as substrate, both genotypes reached maximum ChlC values at PR20. At PR40, the ChlC of 'Victoria' remained slightly above the control (0PR), while in 'Amistad', it dropped below control levels. With BSG, both genotypes displayed an asymptotic response, with ChlC values peaking at PR20, followed by a sharp decline at higher doses (Figure 14A,B). In 'Victoria' (Figure 14A), WF produced consistent increases across all levels: +10% (PR10), +20% (PR20), and +9% (PR40). CS stimulated growth moderately, from +4% (PR10) to +16% (PR40). BSG, however, showed a slight effect at PR10 (+2%), a more substantial increase at PR20 (+16%), and a truly impressive increase at PR40 (+25%). 'Amistad' (Figure 14B), instead, displayed more variable responses. WF reduced ChlC at PR10 and PR20 (−2% and −7%, respectively) but slightly promoted it at PR40 (+1%). CS markedly enhanced ChlC at PR10 (+13%), while producing small reductions at PR20 and PR40 (−2% and −6%, respectively). BSG decreased ChlC at PR10 (6%) but strongly stimulated it at higher levels (+29% at PR20 and +40% at PR40).

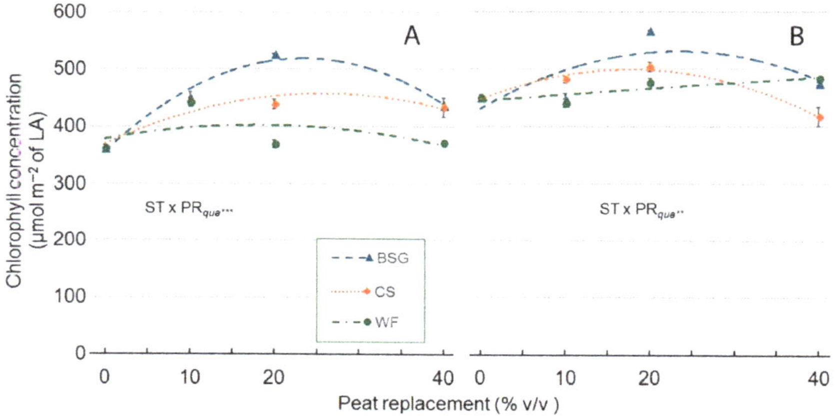

Figure 14. Effect of peat replacement level (PR, 0–40% v/v) with three alternative organic matrices (ST) —brewer's spent grain (BSG), coffee silverskin (CS), or wood fiber (WF)—on Chlorophyll Concentration (ChlC) in 'Victoria' (**A**) and 'Amistad' (**B**) genotypes at the end of the vegetative phase (14 January 2024). LA: leaf area. Data points represent means ± standard error (n = 3). Statistical analysis performed through ANOVA: ** and ***, significant at $p \leq 0.01$, or $p \leq 0.001$, respectively. Orthogonal polynomial contrasts were employed to evaluate quadratic (*qua*) trend.

Plant fresh weight (Plant FW) is a key trait in ornamentals as it integrates growth and resource-use efficiency. Higher values reflect greater biomass, visual quality, and vigor, directly enhancing marketability and consumer appeal. In both genotypes, plant fresh weight was significantly influenced by substrate type and PR level, with notable differences between 'Victoria' and 'Amistad' (Figure 15A,B). In the 'Victoria' genotype (Figure 15A), increasing the PR level with the three organic by-products generally decreased plant fresh weight; the response was non-linear for all substrates, with trends suggesting an asymptotic approach. In the 'Amistad' genotype (Figure 15B), the trend was the opposite. With WF as substrate, fresh weight peaked at 40% PR. CS and BSG, on the other hand, showed a different trend from WF as described by the curves: the fresh weight value increased steadily reaching a maximum at 25% PR and then decreased slightly. In 'Victoria' (Figure 15A), the replacement of peat with wood fiber (WF) caused a progressive plant fresh weight decline already at PR10 (−23%), with further reduction at PR20 (−31%) and the most severe decrease at PR40 (−42%). A similar trend was observed when coffee silverskin (CS) replaced peat, resulting in a reduction from 6% (PR10) to 35% (PR40). Brewer's spent grain

(BGS) showed a more variable effect, with a reduction at PR10 (−19%), a slight increase at PR20 (+6%), and a strong decrease at PR40 (−29%). Conversely, 'Amistad' (Figure 15B) demonstrated a more favorable and consistent response to substrate replacement: WF increased both to PR20 (+9%) and PR40 (+14%); CS promoted significant stimulation at PR10 (+25%), maximum at PR20 (+38%), and still positive effects at PR40 (+30%). When BGS replaced peat, it also enhanced plant fresh weight, with a modest increase at PR10 (+7%), a maximum at PR20 (+38%), and sustained improvement at PR40 (+22%).

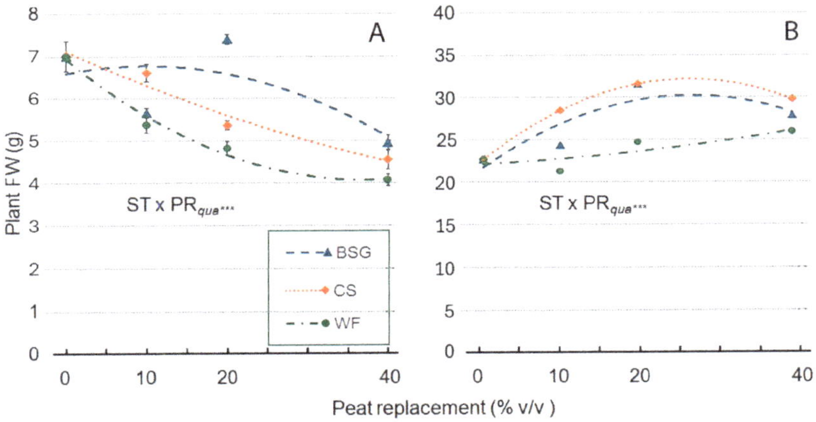

Figure 15. Effect of peat replacement level (PR, 0–40% *v*/*v*) with three alternative organic matrices (ST) —brewer's spent grain (BSG), coffee silverskin (CS), and or wood fiber (WF)—on Plant Fresh Weight (FW) in 'Victoria' (**A**) and 'Amistad' (**B**) genotypes at the end of the vegetative phase (14 January 2024). Data points represent means ± standard error (n = 3). Statistical analysis performed through ANOVA: ***, significant at $p \leq 0.001$. Orthogonal polynomial contrasts were employed to evaluate the quadratic (*qua*) trend.

With WF substrate, shoot dry weight (shoot DW) exhibited a quadratic response: in 'Victoria', it consistently decreased with the highest PR level (40%), whereas in 'Amistad', the trend was the opposite (Figure 16A,B). The CS substrate, initially, in 'Victoria', showed a slight increase in shoot DW at 10% PR, followed by a clear quadratic decline at higher PR levels (Figure 16A). In 'Amistad', shoot DW peaked at approximately PR20 (Figure 15B). When BSG was used as the substrate, shoot DW exhibited a different quadratic response: in both genotypes, 20PR showed the highest value; over PR20, in 'Victoria', it decreased to a value below the control (0PR); in 'Amistad', shoot DW decreased slightly with PR40 (Figure 16A,B). In the 'Victoria' genotype (Figure 16A), WF resulted in a slight increase in shoot dry weight at PR10 (+3%), but it decreased dry weight at both PR20 (−16%) and PR40 (−25%). CS moderately stimulated dry weight at PR10 (+28%) but caused reductions at PR20 (−6%) and PR40 (−30%). BSG had a mild positive effect on above-ground dry weight at PR10 (+5%), a substantial increase at PR20 (+37%), and a slight reduction at PR40 (−9%). In contrast, the 'Amistad' genotype (Figure 16B) showed overall stimulation across the different substrates. WF increased biomass at PR20 and PR40 (+28% each), although it led to a slight decrease at PR10 (−9%). CS significantly promoted growth at PR10 (+55%) and PR20 (+79%), with a moderate increase at PR40 (+43%). BSG demonstrated gradual improvements in biomass, with increases of +17% at PR10, +55% at PR20, and +42% at PR40.

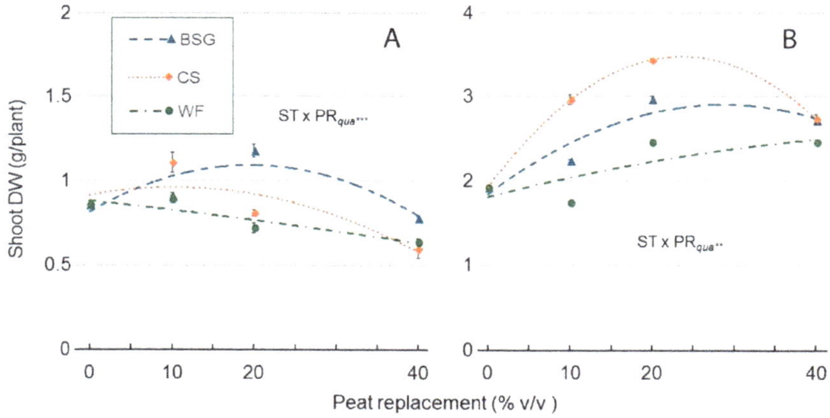

Figure 16. Effect of peat replacement level (PR, 0–40% v/v) with three alternative organic matrices (ST) —brewer's spent grain (BSG), coffee silverskin (CS), or wood fiber (WF)—on shoot dry weight (g) in 'Victoria' (**A**) and 'Amistad' (**B**) genotypes at the end of the vegetative phase (14 January 2024). Data points represent means ± standard error (n = 3). Statistical analysis performed through ANOVA: ** and ***, significant at $p \leq 0.01$, or $p \leq 0.001$, respectively. Orthogonal polynomial contrasts were employed to evaluate quadratic (*qua*) trend.

In both genotypes, root dry weight (Root DW) was significantly influenced by PR and substrate type, with notable differences between 'Victoria' and 'Amistad' (Figure 17A,B). In 'Victoria', root DW showed a concave trend in all substrates as PR levels increased (Figure 17A). In 'Amistad', the response was substrate-dependent and varied significantly. With BSG, root DW showed a consistently positive trend, increasing with higher PR percentages and peaking at PR40. In contrast, root dry weight remained relatively stable across PR levels for WF and CS, with only minor fluctuations in the latter.

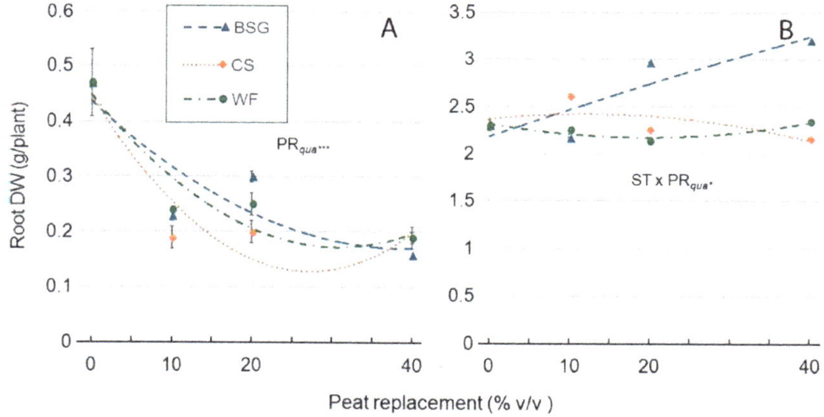

Figure 17. Effect of peat replacement level (PR, 0–40% v/v) with three alternative organic matrices (ST)—brewer's spent grain (BSG), coffee silverskin (CS), or wood fiber (WF)—on root dry weight (g) in 'Victoria' (**A**) and 'Amistad' (**B**) genotypes at the end of the vegetative phase (14 January 2024). Data points represent means ± standard error (n = 3). Statistical analysis performed through ANOVA: * and ***,significant at $p \leq 0.05$, or $p \leq 0.001$, respectively. Orthogonal polynomial contrasts were employed to evaluate quadratic (*qua*) trend.

In 'Victoria' (Figure 17A), root dry weight was consistently reduced under all treatments, with WF causing declines of −50% (PR10), −48% (PR20), and −61% (PR40). Similarly, CS induced strong reductions (−60%, −58%, −59%), while BSG led to decreases of −51% (PR10), −36% (PR20), and −67% (PR40), highlighting overall limitations in root development regardless of substrate. On the other hand, 'Amistad' (Figure 17B) showed a more resilient response: WF slightly decreased at PR10 (−2%) and PR20 (−7%) but slightly increased at PR40 (+1%). CS promoted a moderate increase at PR10 (+13%), while PR20 (−2%) and PR40 (−6%) remained nearly unchanged or experienced slight reductions. BSG, however, clearly enhanced root weight, with improvements at PR20 (+29%) and PR40 (+39%), despite a minor reduction at PR10 (−6%).

Regarding plant dry matter (Plant DM) content, the ST × PR interaction was not significant in 'Victoria' (Figure 18A,B). In 'Amistad', the WF substrate had a relatively stable plant DM across all PR levels (Figure 18A). In contrast, the CS substrate initially showed a clear linear decline at a 10% PR level. With BSG substrate as PR, plant DM exhibited a linear response as PR levels increased (Figure 18B).

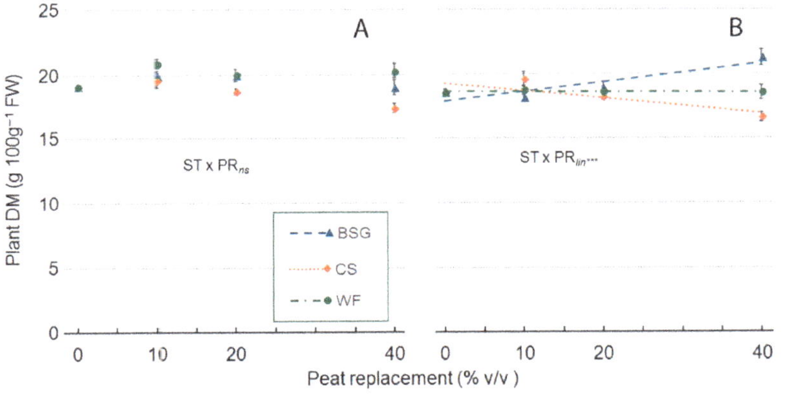

Figure 18. Effect of peat replacement level (PR, 0–40% v/v) with three alternative organic matrices (ST)—brewer's spent grain (BSG), coffee silverskin (CS), or wood fiber (WF)—on plant dry matter content (g 100 g^{-1} FW) in 'Victoria' (**A**) and 'Amistad' (**B**) genotypes at the end of the vegetative phase (14 January 2024). Data points represent means ± standard error (n = 3). Statistical analysis performed through ANOVA: ns, and ***, not significant or significant at $p \leq 0.001$, respectively. Orthogonal polynomial contrasts were employed to evaluate linear (*lin*) trend.

2.5. Leachate pH and EC

Tables 5 and 6 show the ANOVA results on the effect of Substrate type (ST) and PR on leachate pH and EC values on 15 December 2023, in 'Victoria' and 'Amistad' genotypes.

In the 'Victoria' genotype, highly significant effects were observed for both main factors and their interactions, with significance levels ($p \leq 0.001$) for the pH. Regarding the EC parameter, only ST × PR*qua* was significant at $p \leq 0.05$, while all the others were significant at $p \leq 0.001$ (Table 5).

Regarding the 'Amistad' genotype, Table 6 shows a highly significant main effect of PR on both pH and EC, while ST significantly influenced pH but not EC. Polynomial contrasts were employed, indicating strong linear and quadratic trends for both response variables. Interaction effects between ST and PR, including their respective linear and quadratic components, were not statistically significant, suggesting that the influence of peat replacement was consistent across substrate types.

Table 5. ANOVA results showing the effect of Substrate type (ST) and peat replacement (PR) on leachate pH and EC (electrical conductivity) values in 'Victoria' genotype. (Data collected on 15 December 2023). Orthogonal polynomial contrasts were employed to evaluate linear (*lin*) and quadratic (*qua*) trends.

'Victoria'	pH	EC (dS m^{-1})
Substrate type (ST)	*** [a]	***
Peat replacement (PR)	***	***
PR*lin*	***	***
PR*qua*	***	***
ST × PR	***	***
ST × PR*lin*	***	***
ST × PR*qua*	***	*

[a] * and ***, significant at $p \leq 0.05$, or $p \leq 0.001$, respectively.

Table 6. ANOVA results showing the effect of Substrate type (ST) and peat replacement (PR) on leachate pH and EC (electrical conductivity) values in 'Amistad' genotype. (Data collected on 15 December 2023). Orthogonal polynomial contrasts were employed to evaluate linear (*lin*) and quadratic (*qua*) trends.

'Amistad'	pH	EC (dS m^{-1})
Substrate type (ST)	*** [a]	ns
Peat replacement (PR)	***	**
PR*lin*	***	**
PR*qua*	***	**
ST × PR	ns	ns
ST × PR*lin*	ns	ns
ST × PR*qua*	ns	ns

[a] ns, **, and ***, not significant or significant at $p \leq 0.01$ or $p \leq 0.001$, respectively.

In the 'Victoria' genotype (Figure 19A), increasing the PR level with the three organic by-products generally increased the leachate pH. The response was non-linear for all substrates, with trends suggesting an approach to an asymptotic maximum. The highest pH value was observed with BSG at PR40. For WF, the pH increased sharply to a maximum at PR25-30 and then appeared to plateau. For CS, the pH increased steadily, reaching its highest value at PR35-40. Similarly, in the 'Amistad' genotype (Figure 19B), increasing PR doses raised the leachate pH, with distinct patterns for each substrate. When WF was used, the pH increased in a strong linear pattern with increasing PR percentage. In contrast, the responses for CS and BSG were non-linear and best described by curves. The pH in the CS treatment rose steadily to a maximum at PR40, while the pH in the BSG treatment increased rapidly, peaking at PR30-35 before showing signs of plateauing.

In the 'Victoria' genotype, the leachate EC value was influenced by ST, and the response was much more varied (Figure 20A). When BSG partially replaced peat, the EC of the leachate showed a asymptotic maximum trend. In contrast, the EC values showed a quadratic response when WF and CS were used as substrate. The quadratic trend was convex in CS, peaking at around 20%, and slightly concave in WF. In the 'Amistad' genotype, in general, PR increased the EC values of the leachate: in CS as PR, it increased steadily with higher PR percentages until reaching its maximum at CS40. When WF and BSG replaced peat, the response was quadratic with a convex trend for both, peaking at PR25 in both WF and BSG treatments (Figure 20B).

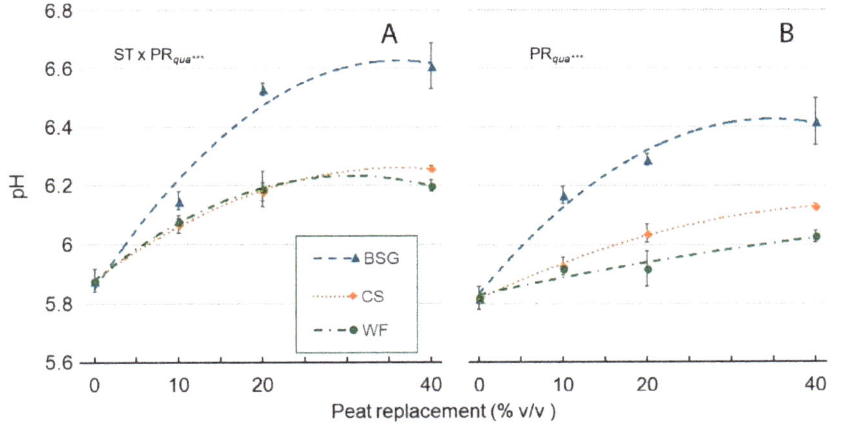

Figure 19. Effect of peat replacement level (PR, 0–40% v/v) with three alternative organic matrices (ST)—brewer's spent grain (BSG), coffee silverskin (CS), or wood fiber (WF)—on leachate pH values in 'Victoria' (**A**) and 'Amistad' (**B**) genotypes. (Data collected on 15 December 2023). Data points represent means ± standard error (n = 3). Statistical analysis performed through ANOVA: ***, significant at $p \leq 0.001$. Orthogonal polynomial contrasts were employed to evaluate quadratic (*qua*) trend.

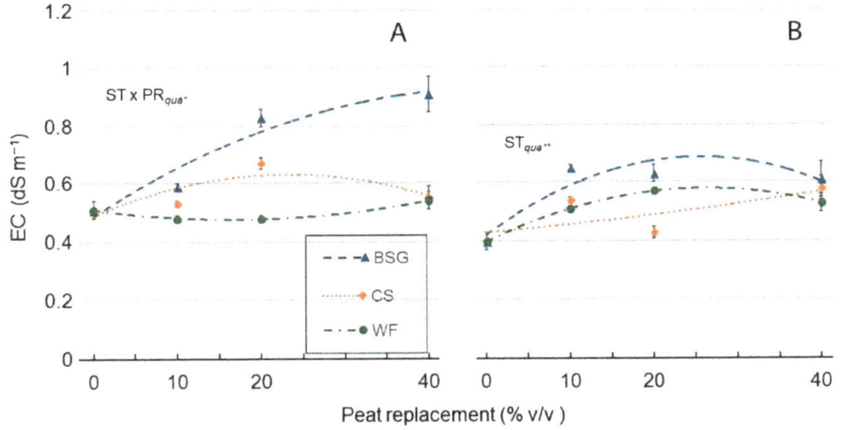

Figure 20. Effect of peat replacement level (PR, 0–40% v/v) with three alternative organic matrices (ST)—brewer's spent grain (BSG), coffee silverskin (CS), or wood fiber (WF)—on leachate EC values in 'Victoria' (**A**) and 'Amistad' (**B**) genotypes. (Data collected on 15 December 2023). Data points represent means ± standard error (n = 3). Statistical analysis performed through ANOVA: * and **, significant at $p \leq 0.05$ or $p \leq 0.01$, respectively. Orthogonal polynomial contrasts were employed to evaluate quadratic (*qua*) trend.

2.6. Principal Component Analysis (PCA) in Growth Parameters

Principal component analysis (PCA) provided an overview of the bio-morphological traits of 'Victoria' (Figure 21A,B) and 'Amistad' (Figure 22A,B) sage genotypes.

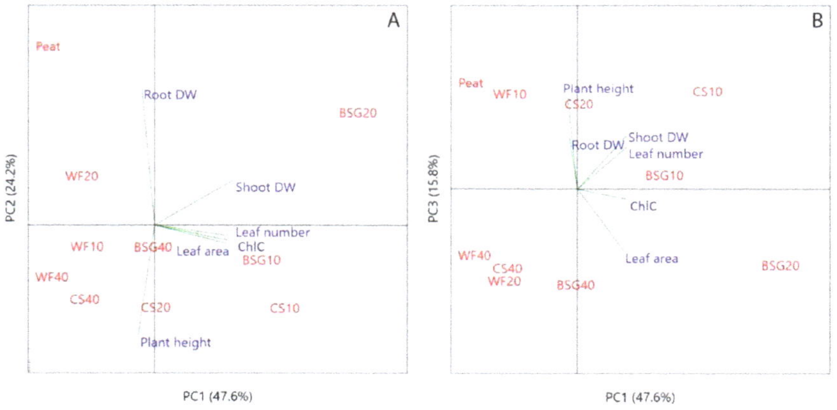

Figure 21. Principal component analysis (PCA) biplot of the PC1 vs. PC2 (**A**) and PC1 vs. PC3 (**B**) principal components, illustrating the relationships between growth parameters and peat replacement treatments in the 'Victoria' genotype. The analysis includes shoot dry weight (Shoot DW), root dry weight (Root DW), leaf area per plant, leaf number per plant, plant height, and chlorophyll concentration (ChlC). Substrate treatments are denoted as follows: Peat (0% replacement, control), wood fiber (WF10, WF20, WF40 for 10, 20, and 40% peat replacement, respectively), coffee silverskin (CS10, CS20, CS40 peat replacement), and brewer's spent grain (BSG10, BSG20, BSG40 peat replacement).

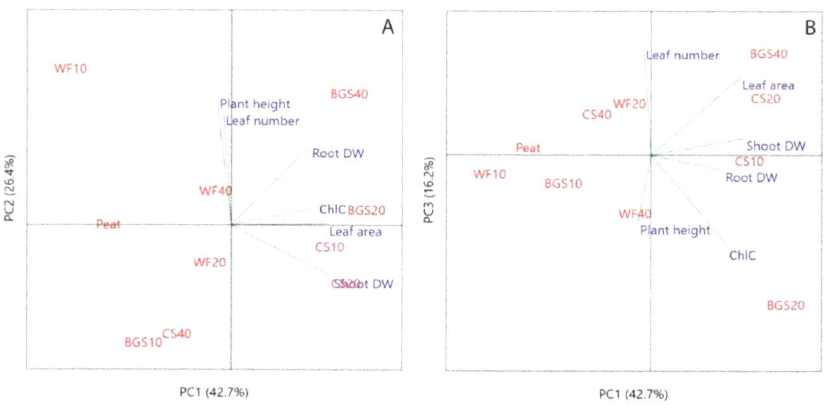

Figure 22. Principal component analysis (PCA) biplot of the PC1 vs. PC2 (**A**) and PC1 vs. PC3 (**B**) principal components, illustrating the relationships between growth parameters and peat replacement treatments in the 'Amistad' genotype. The analysis includes shoot dry weight (Shoot DW), root dry weight (Root DW), leaf area per plant, leaf number per plant, plant height, and chlorophyll concentration (ChlC). Substrate treatments are denoted as follows: Peat (0% replacement, control), wood fiber (WF10, WF20, WF40 for 10, 20, and 40% replacement, respectively), coffee silverskin (CS10, CS20, CS40), and brewer's spent grain (BSG10, BSG20, BSG40).

The PCA performed on the normalized 'Victoria' data revealed three principal components (PCs) with eigenvalues >1 that explained approximately 97.6% of the total variance in the dataset, attributing 81.8% to PC1 and PC2. In the depiction of components 1 and 2 (Figure 21A), most variables were positively correlated with PC1, specifically leaf number (r = 0.51), leaf area (r = 0.47), chlorophyll concentration (r = 0.50), and shoot dry weight (r = 0.50). Root dry weight (Root DW) and plant height were negatively correlated with PC1. PC2 positively correlated with shoot dry weight (Shoot DW) and particularly root DW (0.75). A negative correlation with PC2 was detected mainly for plant height (−0.60).

The PC3 (Figure 21B) showed a positive correlation mainly with plant height (0.66) and a negative correlation with leaf area per plant (−0.44).

The PCA performed on the normalized 'Amistad' data revealed three PCs with eigenvalues >1 that explained approximately 75.3% of the total variance in the dataset, attributing 69.1% to PC1 and PC2. In the depiction of components 1 and 2 (Figure 22A), leaf area per plant (0.52), ChlC (0.47), shoot DW (0.56) and root DW (0.43) were positively correlated with PC1. PC2 showed positive correlations with plant height (r = 0.64) and leaf number per plant (r = 0.59). The PC3 (Figure 22B) showed a positive correlation mainly with leaf number (0.61) and leaf area (0.43), while exhibiting a negative correlation with leaf ChlC (−0.53) and plant height (−0.39).

The analysis of the PC1-derived Growth Index (GI) scores revealed for 'Victoria', significant differences between treatments. The BSG20 treatment yielded the highest mean GI score, significantly outperforming the peat control (0PR) and all other amendments. Treatments such as CS10 and BSG10 showed moderate GI scores; in contrast, high replacement rates (WF40, CS40, BSG40), 0PR and WF resulted in the lowest GI scores (Figure 23A). For 'Amistad', the analysis of the GI scores revealed that the CS20 and BSG20 treatments yielded the highest overall growth, with GI scores significantly greater than those of the 0PR. CS10 and BSG40 also showed strong performance, with high GI scores. WF10 had GI scores similar to PR0, whereas the GI of WF20 and WF40 were slightly higher (Figure 23B).

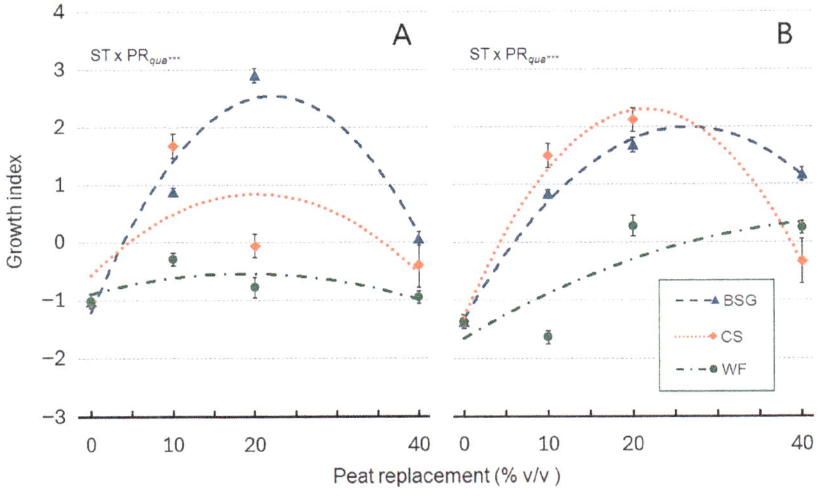

Figure 23. Effect of peat replacement level (PR, 0–40% v/v) with three alternative organic matrices (ST)—brewer's spent grain (BSG), coffee silverskin (CS), or wood fiber (WF)—on growth index in 'Victoria' (**A**) and 'Amistad' (**B**) genotypes. (Data collected on 15 December 2023). Data points represent means ± standard error (n = 3). Statistical analysis performed through ANOVA: ***, significant at $p \leq 0.001$, respectively. Orthogonal polynomial contrasts were employed to evaluate quadratic (*qua*) trends.

3. Discussion

3.1. Physical and Hydrological Characterization

The hydraulic properties of growing media provide precise information about their ability to provide good plant growth conditions [61,62].

3.1.1. Wood Fiber as Peat Replacement

Wood's by-products, in the form of processed wood fractions, have been tested and applied, for some time now, as promising growing media ingredients and their market has been expanded worldwide [63] due to renewability [7], reduced carbon footprint compared to peat or other materials [64]. Furthermore, it also has a low production cost [65], especially for conifer by-products, which show a lower phytotoxic molecule content compared to hardwood species [66]. In our results, water volume (WV) at –1 kPa, reflecting the moisture retained against minimal suction and hence indicative of easily plant-available water, exhibited a quadratic response to increasing wood fiber (WF) rates (Figure 1). This pattern is consistent with quicker drainage and higher air content (AC) [67–70] compared to peat. This behavior would appear to result from the coarser and more heterogeneous particle size distribution of WF, which accelerates drainage and reduces water retention at low tension levels [71,72]. Complementary research [54] regarding a 60:40 peat: WF blend revealed that WF exhibits hydrophilic behavior across a range of moisture contents, maintaining contact angles below 90°, unlike peat, improving rewetting dynamics under dry conditions. Substrate moisture (SM), measured at –1 kPa, declined linearly with increasing WF content (Figure 2). This contrasts with peat, which possesses superior water-holding capacity due to its high surface area and microporous structure [13]. The reduced moisture in WF-based media reflects its lower capacity for capillary water retention and is consistent with findings by Gruda [7], who noted a faster drying rate in WF-amended substrates. Both easily available water (Figure 3) and available water (Figure 4) demonstrated a linear decline with increasing levels of wood fiber. While peat typically holds large volumes of water in the range of −1 to −10 kPa—considered optimal for plant uptake—WF was less effective in maintaining this fraction, resulting in a lower easily available water value. Fields et al. [73] found that blends containing pine wood chips or shredded pine wood exhibited steeper moisture retention curves, indicating a sharper decline in volumetric water content (WVC) as suction increased. Pure wood substrates displayed higher drainage rates and reduced retention in the −1 to −30 kPa range. These findings reinforce that wood components have lower water-holding capacity (WHC) and faster drying tendencies than peat blends. Durand et al. [74] and Atzori et al. [23] explain this behavior: WF substitution often leads to reduced water availability due to diminished micropore volume and lower capillary retention. According to Michiels et al. [75], an ideal substrate has less easily available water of <40 vol.%. The values of all WF-based substrates for nursery production were within the range of an ideal substrate, suggesting that they can be effectively integrated without severely compromising substrate functionality. Unlike other peat substitutes, such as coffee silverskin (CS) or brewer's spent grain (BSG), WF did not significantly affect buffering capacity (BC) (Figure 5). The inherent lignocellulosic structure and minimal nutrient content of WF may contribute to its low interaction with substrate pH and nutrient fluxes, corroborating Zaccheo et al. [24]. A linear increase in dry bulk density (BD, Figure 6), accompanied by a corresponding linear decrease in total porosity, a key determinant of air–water balance in container media (Figure 7), was observed with increasing levels of WF replacement. While peat possesses a naturally spongy and lightweight structure [76], WF—although also lightweight—contains more rigid particles that may compact under load, thereby reducing pore volume [5]. Despite these changes, dry BD values remained within the acceptable horticultural range (<0.4 g·cm^{-3}), indicating no significant risk of substrate compaction. The water retention profiles (Figure 8A) further illustrated that increasing WF content led to a steeper decline in volumetric water content (WVC) at increasing tensions. While the control 0PR maintained higher WVC across all tensions, the WF-based substrate (especially at 40% of peat replacement) exhibited more rapid water loss between −1 and −5 kPa tension. This underscores the limited capillary continuity and weaker water-holding ability of WF, consistent with findings from Gruda and Schnitzler [68], Wallach [2] and Muhammed et al. [36]. Eveleens et al. [77], too, when they added 40 and 50% v/v WF to peat, showed that these substrates reduce WVC by 16% v/v and increase AC by the same amount in the range of -6 to -100 cm pressure head. Our results regarding the water retention curves (Figure 8A) reflect the lower capillary water retention capacity in substrates with a higher WF replacement level. WF's coarse texture and lower micropore volume limit its ability to retain water against even modest tension, consistent with prior findings [54,73,77]. The smaller differential at high tension suggests that WF substitution primarily affects water availability within the plant-easily available range (–1 to –5 kPa), while the residual water at high tension remains similar across media types. Our findings are in line with previous studies showing that high proportions of WF (>30%) significantly reduce easily available water (EAW) and total available water (TAW) due to increased macroporosity and limited fine-pore storage [7,70]. Notably, Michel et al. [54] reported that while WF improved rewetting dynamics, it did so at the cost of moisture retention in the root-active zone. Regarding chemical characteristics, pH and electrical conductivity (EC) are two important factors affecting the growing medium due to the direct effect on nutrient availability and status in the growing medium [78]. In our research, the pH value in the WF-based substrate was n.s. (Figure 12); this agrees with Woznicki et al. [79], who report that substrates with over 50% (v/v) WF showed a rapid pH increase.

3.1.2. Coffee Silverskin as Peat Replacement

In our study, the proportional peat replacement (PR) with CS underscores a potential decline in capillary WHC due to the inclusion of CS. References reported that CS has a high content of recalcitrant lignocellulosic compounds that contribute to bulk and physical structure in substrates [80,81]. Other authors [82,83] have linked the WHC of CS alone to the soluble fiber content. The physical behavior of substrates incorporating CS is strongly influenced by its unique particle morphology. As reported by Miler et al. [26], the flat and delicate flake-like structure of CS, in contrast to more fibrous peat or WF, contributes to increased compaction within the substrate matrix. This morphological trait hinders the optimization of the water–air balance, a key determinant of substrate performance. Although CS exhibits low BD, its limited structural resilience may reduce pore continuity and restrict gaseous exchange in the root zone. This is consistent with previous findings that link excessive substrate compaction and insufficient macroporosity to reduced oxygen availability and impaired root respiration [76,84]. Similar trends have been observed with other finely textured or plate-like organic materials, which often exhibit poor structural stability under cultivation conditions [85,86]. Consequently, substrates with high CS content may limit root zone aeration, potentially compromising plant growth and quality.

3.1.3. Brewer's Spent Grain as Peat Replacement

To date, no studies have comprehensively characterized the physical-hydrological properties of growing media formulated with BSG, highlighting a significant gap in the current literature. In contrast, existing research on raw material itself provides some foundational insights. For instance, Naibaho et al. [87] reported bulk density values for BSG ranging between 0.129 and 0.159 g cm^{-3}, while Castro and Colpini [88] documented a consistently high moisture content, typically between 78% and 84%. This substantial water content is primarily attributed to the structural and compositional characteristics of BSG. The fibrous matrix is notably rich in cellulose, hemicellulose, and lignin—components that enhance water retention through both hydrogen bonding and capillary entrapment mechanisms. Among these, hemicellulose is particularly hydrophilic, enabling substantial water adsorption within the porous network of the BSG [89]. Furthermore, the mashing process in brewing, which partially disrupts plant cell walls, increases the porosity and specific surface area of the biomass, thereby improving its WHC [90]. In addition to structural polysaccharides, residual proteins and gelatinized starch fractions present in BSG may contribute to moisture stabilization by forming hydrated matrices that interact with water molecules [91].

3.2. Chemical Characteristics of Growing Media

The chemical stability and ionic activity of growing media are pivotal factors influencing plant performance in containerized horticultural systems. Among the key physicochemical parameters, pH and EC serve as critical indicators of substrate suitability, directly affecting nutrient solubility, microbial dynamics, and root development. Across all tested PR levels (10–40%), no statistically significant variation in pH was observed for WF, CS, or BSG treatments (Figure 9). This pH stability underscores the dominant BC of peat, which remains effective even when partially substituted with organic residues [92]. The lignocellulosic nature of WF and CS contributes to their low acid dissociation potential, thereby exerting minimal influence on substrate pH. Although BSG is rich in proteins and amino acids, its limited mineralization before cultivation is likely to delay any acidifying effects, further supporting pH consistency across treatments. In contrast, EC exhibited substrate-specific and dose-dependent variability (Figure 10), reflecting differences in soluble salt concentrations and rates of organic matter decomposition. Notably, BSG treatments demonstrated a strong linear increase in EC with rising replacement levels, reaching the highest values at 40%. This trend is attributed to the elevated concentrations of organic nitrogen, phosphorus, and potassium in BSG, which undergo rapid mineralization and contribute to increased ionic activity in the substrate [93].

3.3. Principal Component Analysis (PCA) in Growth Parameters

The principal component analysis (PCA) for 'Victoria' sage highlighted that the first two PCA components (Figure 21A) enable the distinction between mixtures with PR10 of both BSG and CS, and those with PR20 of BSG. The latter, located in the first quadrant (upper right), clustered according to shoot DW. While for BSG10 and CS10 mixtures (lower-right quadrant) clustering was driven by leaves leaf number and area, and chlorophyll concentration (ChlC). The PC3 also supported the best performance of BSG20, primarily in terms of leaf area (Figure 21B). These findings could be related to the nutrient availability of BSG, which is rich in organic matter, proteins, nitrogen, phosphorus, and micronutrients. At PR20, BSG may provide an optimal nutrient balance that enhances shoot development without causing substrate compaction or nutrient imbalance. Furthermore, moderate inclusion of BSG can improve aeration (Figure 7), enhance water retention, and increase water availability (Figure 8), thereby supporting shoot expansion and leaf development compared to 0PR. The 'Victoria' genotype may be particularly responsive to the nutrient profile or physical properties of BSG at this concentration, suggesting their successful use in the cultivation. Located on the left upper side of the PC1 component, WF20 and mainly peat displayed a high correlation with root DW, while other mixtures (left lower side) mainly improved the plant height. Furthermore, PC3 separated peat (0PR), CS20 and WF10 clustering for root DW and plant height (Figure 21B). WF20 could create a more oxygen-rich rhizosphere, encouraging root elongation and biomass accumulation; on the other hand, BSG may release nutrients more rapidly, favouring shoot elongation and vertical growth rather than root expansion. Peat-based substrates consistently yield higher or comparable root DWs and overall plant growth, especially when compared to high-percentage WF or alternative organic amendments [94–96]. In general, our results show that in 'Victoria', 0PR clustered distinctly, reflecting optimal growth conditions. When included at moderate levels (10–20 PR), WF maintains structural stability and high macroporosity, supporting strong root aeration and growth [68], alignment with WF10-20 closeness to control in PCA. At higher proportions (WF40 as PR), excessive porosity may reduce water retention and nutrient availability, echoing the PCA displacement toward less favourable axes [85,86]. The flake-like, delicate morphology of CS leads to high compaction and a poor air-filled pore space, despite its low Bulk Density—consistent with PCA clusters for CS20 and CS40, which are located far from the growth vectors [26,80]. High recalcitrant lignocellulose content may immobilize nitrogen and slow decomposition, further contributing to poor plant performance at higher inclusion rates [97]. With moderate hydrophilic fiber content, BSG10-20 PR occupies intermediate positions in PCA space, offering improved moisture retention without severe structural issues [89,90]. However, high BSG PR (BSG40) may lead to compaction, rapid microbial decay, and salinity buildup—explaining its divergence in PCA from positive growth traits [98].

The Growth Index (GI) (Figure 23), which represents a PC1-derived index, provided a powerful, unified metric to rank substrate efficacy. It indicates that a 20% replacement of peat with BSG provided the optimal conditions for holistic plant development in 'Victoria' (Figure 23A). The biplot (Figure 21A) illustrates how this treatment is positioned farthest along the positive direction of the PC1 axis.

Moderate GI scores of CS10 and BSG10 were due to clustering in the biplot based on strong performance in specific traits like leaf area and chlorophyll, but not all traits simultaneously. The peat control (0PR), the WF, and high replacement rates, with the lowest GI scores, clustered on the negative side of PC1, aligning with their observed poor performance and displacement from positive growth vectors in the biplot. In general, in 'Victoria', the lowest GI for WF substrate across nearly all peat replacement levels indicated that it was the most detrimental alternative for this cultivar.

The first two PCA components (Figure 22A) for 'Amistad' genotype allow for the distinction of mixtures PR at 10–20 CS, and 20–40 BSG. The BSG20 and BSG40, located in the first quadrant (upper right), clustered according to root DW and ChlC. The findings related to BSG substrates could be related to nitrogen availability, thus richness in organic nitrogen and amino acids, which can stimulate root development, chlorophyll biosynthesis [99] and photosynthetic capacity [93]. In contrast, for CS10 and CS20 PR (lower-right quadrant), clustering was driven by leaf area and shoot DW. This finding could be related to a rapid nutrient release: Melo et al. [100] confirmed that a high caffeine and polyphenolic content can act as biostimulants, promoting shoot elongation and leaf expansion. Furthermore, at low PR levels, CS improves substrate structure without compromising water retention, supporting vigorous shoot development. Located on the left upper side of Component 1, WF10 showed a high correlation with plant height (0.64) and plant leaf number. Furthermore, PC3 separated WF20 for leaf number and WF40 for plant height. WF40 and CS40, located distantly from the 0PR, showed reduced association with both shoot and root biomass, indicating potential limitations in substrate structure or nutrient availability at higher PR. In 'Amistad', the best substrates appear to be CS10 and CS20 because they increase shoot weight and plant leaf area, although BSG20 and BSG40 appear to improve root growth and leaf color. Wood fiber at all three concentrations appears to improve some aspects of the above-ground plant, such as height and plant leaf number. In any case, the mixtures seem to perform better than 0PR.

Analysis of GI scores showed a markedly different response from 'Victoria'. Multiple amendment treatments outperformed the peat control. The CS20 and BSG20 treatments produced the highest overall growth (Figure 23B). CS10 and BSG40 also showed strong performance, with high GI scores, demonstrating 'Amistad's' greater responsiveness and adaptability to peat replacement. This is visually supported by their position in the positive

PC1 space of the biplot (Figure 22A). The 10PR treatment with wood fiber generated GI scores comparable to those of the peat control. However, GI for WF20 and WF40 was slightly better than PR0, consistent with their central positioning in the biplot.

In general, the GI-based approach confirms that moderate inclusion of brewer's spent grain (BSG20) is a highly effective peat replacement for both genotypes, while coffee silverskin (CS) is particularly effective for the 'Amistad' genotype. This analysis underscores that optimal substrate formulation is not only dependent on the amendment type and rate but also critically on the plant genotype.

4. Materials and Methods

A greenhouse pot experiment was conducted to evaluate the physical characteristics and agronomic performance of alternative substrates, utilizing organic by-products as a partial replacement for peat. Two sage cultivars, characterized by different growth habits, were used as plant test subjects.

4.1. Substrate Treatments

Ten substrates were used as treatment. Each substrate included 15% of the total volume of an inorganic fraction and an organic fraction (85% of the total volume). The ten substrates were created by partially replacing peat in the organic fraction with alternative organic matrices. These matrices, selected from by-products of the agricultural industry, included wood fiber (WF), coffee silverskin (CS), and brewer's spent grain (BSG). Each matrix was used at four levels of peat replacement (0, 10, 20, and 40% v/v of the organic fraction), resulting in the treatments summarized in Table 7.

Table 7. Substrate treatments used in the trial. The mixtures were formulated using peat as the primary organic component, with increasing levels of peat replacement (ranging from 0% to 40%) by alternative organic matrices: wood fiber (WF), coffee silverskin (CS), and brewer's spent grain (BSG).

Substrate Treatment Code	Peat	Peat Replacement	Matrices Used as Peat Replacement
	(% v/v of the Organic Fraction of the Substrate)		
0PR	100	0	None
WF10	90	10	Wood fiber
WF20	80	20	
WF40	60	40	
CS10	90	10	Coffee silverskin
CS20	80	20	
CS40	60	40	
BSG10	90	10	Brewer's spent grain
BSG20	80	20	
BSG40	60	40	

The sources of the organic matrices were: (i) peat (P, Plantaflor®, Vechta, Germany); (ii) WF (Sunfiber®, Nuova Flesan S.r.l., Milano, Italy), (iii) CS (Brunocaffè roasting company, Bari, Italy) and (iv) BSG (Peroni brewery, Bari, Italy). WF derived mainly from conifers, such as pine and fir, by mechanical defibrillation of virgin wood chips; CS from a blend of *Coffea arabica* L. (20%) and *C. robusta* Pierre ex A. Froehner (syn. *C. canephora* 80%), BSG was obtained from the production of a lager beer brewed with barley malt (70%) and maize (30%) and does not contain spent yeast.

Regarding some physical characteristics, such as particle size or fraction, in WF, there was a blend of 3 (10–16 mm), 25 (4–10 mm), 25 (2–4 mm), 9 (1–2 mm), and 38% (<1 mm).

Regarding the average particle size, as diameter, in CS it was 1.5–2.0 mm and in BSG it was 3.5 mm. Sdao et al. [98] analyzed the four organic matrices and provided values for EC, TNK and P content in experiments involving bedding plant species in a potted greenhouse cultivation.

4.2. Experimental Setup

The on-field cultivation experiment was carried out from 16 September 2023 to 12 February 2024 (150 days) in an unheated and naturally lit greenhouse, covered with ethyl vinyl acetate (EVA) plastic film, located in a commercial ornamental nursery (Monopoli, Bari, Italy, 40°54′19.1″ N, 17°18′21.4″ E; 66 m a.s.l.). The site has a typical Mediterranean climate (climate type "Csa" of the Köppen Classification) and it is characterized by hot, dry summers and cool, wet winters. During the experimental period, the Monopoli climatic station reported average monthly temperatures ranging from 25 to 26 °C (max) and 18–20 °C (min) in September, declining to 12–13 °C (max) and 7–8 °C (min) by February.

Two sage cultivars ('Victoria' and 'Amistad') were grown in 1.2 L recycled plastic pots filled with the substrates described above. Both cultivars represent high-market value ornamental crops and were also selected for their widespread ornamental application and proven adaptability to Mediterranean climatic conditions. Those two genotypes were chosen for their contrasting growth habits, with 'Victoria' exhibiting more compact shoot and root development and 'Amistad' showing a more vigorous and expansive growth, long stems with numerous leaves. Table S1 shows the primary botanical and ornamental characteristics, as well as the agronomic needs for the utilization of green spaces, of the two genotypes.

Forty-day-old young transplants (approx. 4.0 cm height), grown in small plastic pots (4 cm × 4 cm × 4 cm) filled with *Sphagnum* peat and obtained from rooted stem soft cutting, were used as plant starting material.

One plant per pot was transplanted on 16 September 2023, using 480 sage plants per genotype. Sixteen pots were used per experimental unit. Plants were spaced 30 cm × 30 cm apart and were arranged on growing benches in a randomized block design with three replications per treatment. Irrigation was supplied through one pressure-compensated dripper per pot, delivering 2 L h^{-1}. Raw water from the experimental farm (electrical conductivity = 0.45–0.65 dS m^{-1}; pH= 6.3–6.5) was used. Irrigation frequency and duration were controlled by a timer and adjusted throughout the growing period according to prevailing weather conditions, to replenish moisture losses in the substrates, bringing the pots up to a certain weight. Peat was pre-enriched with 4.5 kg m^{-3} calcium carbonate. For mineral nutrition, each pot received a controlled-release fertilizer (N:P:K = 16:8:12 + 2MgO + micronutrients, 6-month release) at a rate of 3 g L^{-1}, following the manufacturer's recommendations (Compo Expert, Italy).

4.3. Physical and Hydrological Substrates Characterization

Before transplanting, the substrates were characterized for their main physical properties. These included WV = Water Volume at −1 kPa; AC = Air Content at −1 kPa; SM = Substrate Moisture at −1 kPa; EAW = Easily Available Water; BC = Water Buffering Capacity; TAW = Total Available Water; TPS = Total Porosity Space; BD = Dry Bulk Density. These parameters were determined by constructing water retention curves using a tension table according to the method described by De Boodt et al. [84], and further applied by Abad et al. [86] and Noguera et al. [37] for the physical evaluation of organic soilless substrates. All analyses were conducted in triplicate under controlled laboratory conditions using stainless cylinders and porous plates.

4.4. Chemical Substrates Characterization

In this study, pH and electrical conductivity (EC) determinations of the peat control (0PR) substrate and each mixture obtained by WF, CS and BSG as partial peat replacement (PR) were determined before substrate fertilization. Three replications were used to determine the EC [101] and pH [102] of the growing media, which were analyzed in a water extract (substrate sample: distilled water, 1:5 v/v). Values of pH and EC were measured using a portable XS PC 7 VIO Professional multi-parameter meter (XS Instruments, Carpi, Italy).

4.5. Chemical Leachate Substrate Fraction Characterization

The pH and EC of the leached substrate fraction were monitored. The leachate was collected from the pots using the pour-through method [103] one hour after irrigation. A volume of 100 mL of distilled water was poured onto the surface of each pot by hand, and the extract was allowed to drain for 10 min in saucers placed under each container until 50 mL of leachate was obtained. The pH and EC values were measured using a portable multiparameter meter XS PC 7 VIO Professional (XS Instruments, Carpi, Italy) with three replications.

4.6. Plant Growth and Biomass Measurements

At the end of the experiment, once the vegetative phase was over, five pots were randomly selected and collected for each replication and each treatment to determine plant growth parameters via non-destructive and destructive sampling. Plant height (cm) and total leaf number (no.) were determined at first. The height of the plant was determined by direct dimensional measurement using a ruler graduated in cm, from the surface substrate level to the apex of the sages. Leaf area per plant (cm^2) was measured using a Li-Cor area meter (LI-3000, Licor, Lincoln, Nebraska, USA). Chlorophyll concentration (ChlC) was measured using a Chlorophyll Content Meter (Apogee, North Logan, UT, USA) on five selected plants, with three measurements taken per leaf, on six fully intermediate and expanded, mature leaves. Roots were meticulously washed under running tap water, using a sieve to recover any detached fragments. Plant fresh weight (roots and shoots) was recorded using a three-decimal analytical balance and subsequently oven-dried at 80 °C for 48 h in a ventilated oven until constant weight was achieved. Root and shoot dry weights (g) were recorded too. Finally, plant dry matter content (Plant DM) (g 100 g^{-1} fresh weight), was calculated by measuring the fresh weight of individual plant parts (shoots and roots) at the onset of destructive sampling, alongside their corresponding dry weights, using the following formula:

$$\text{Dry Matter Content (\%)} = [\text{Dry Weight (g)}/\text{Fresh Weight (g)}] \times 100$$

4.7. Statistical Analysis

All data were analyzed using ANOVA through the GLM Procedure in SAS software (Version 9.0) [104]. Mean values are presented in graphs together with their corresponding standard errors (SE). Orthogonal polynomial contrasts were employed to evaluate linear, quadratic, and cubic trends in response to peat replacement and its interaction with substrate type (ST × PR). Due to the unequal spacing of PR levels (0, 10, 20, and 40%), the trends were analyzed using the following contrast coefficients: linear (−7, −3, 1, 9), quadratic (7, −4, −8, 5), and cubic (3, −8, 6, −1), calculated using the NumPy library in Python (version 3.12.0). Based on theoretical relevance, only biologically meaningful contrasts (e.g., linear or quadratic trends) were reported, even when higher-order trends (e.g., cubic) were statistically significant.

The physicochemical properties of the final substrate mixtures, measured before sage's transplant, were analyzed separately for each organic matrix type used (WF, CS, or BSG) as PR. A one-way ANOVA was performed with PR level (0, 10, 20, and 40%) as the treatment (three replicates per treatment). The control (0PR) was consistent across all matrices.

Plant growth parameters were evaluated separately for the two genotypes ('Victoria' and 'Amistad') using a two-way randomized block design, with substrate type (ST) and PR level as factors (three replications per treatment). The control treatment (0PR) was shared across all substrate types.

Plant growth responses were further subjected to Principal Component Analysis (PCA) using PAST 4.03 software (http://folk.uio.no/ohammer/past accessed on 9 July 2025) [105]. Before PCA, all variables were standardized using the formula [(x − mean)/standard deviation] to ensure comparability across measurements. To synthesize the multivariate growth data into a single, biologically meaningful metric, a Growth Index was calculated for each experimental replicate. This index was constructed based on the first principal component (PC1). To ensure the index represented a conserved axis of growth across both cultivars, we included only traits that demonstrated positive loadings on PC1 in both: namely, leaf area, chlorophyll concentration, and shoot dry weight. The final index was computed as the weighted sum of the standardized values (z-scores) for these three traits, with each trait's contribution weighted by the magnitude of its PC1 loading. The Growth Index data were further subjected to ANOVA.

5. Conclusions

The results on the vegetative growth of ornamental sages have highlighted that differential PR rates are required depending on the specific organic by-product and plant genotype. In 'Victoria', GI indicates that a 20% replacement of peat with BSG provided the optimal conditions for holistic plant development; the lowest GI for WF substrate across nearly all peat replacement levels indicated that it was the most detrimental alternative for this cultivar. In 'Amistad', the analysis of the GI scores revealed that the CS20 and BSG20 of peat replacement yielded the highest overall growth, with GI scores significantly greater than those of the peat control. CS10 and BSG40 also showed high GI scores. WF10 had GI scores similar to those of the peat control.

In general, the GI-based approach confirms that moderate inclusion of brewer's spent grain (BSG20) is a highly effective peat replacement for both genotypes. At the same time, CS and BSG were particularly effective for the 'Amistad' genotype, which was more flexible in adapting to cultivation using the new proposed by-products than the 'Victoria' genotype. This analysis underscores that optimal substrate formulation is not only dependent on the amendment type and rate but also critically on the plant genotype. Future research should prioritize analyzing the influence of the chemical characteristics of these by-products on the mineral content of sage genotypes and evaluate sustainability also from the point of view of eco-efficiency and profitability.

Supplementary Materials: The following supporting information can be downloaded at: https://www.mdpi.com/article/10.3390/plants14172801/s1, Table S1: Botanical and ornamental characteristics in cv. Victoria and Amistad®.

Author Contributions: Formal analysis, data curation, writing—original draft, writing—review and editing, A.E.S.; conceptualization, methodology, validation, data curation, investigation, writing—original draft, S.C., G.C. (Giulia Conversa), and G.C. (Giuseppe Cristiano); A.E.; investigation, data curation, writing—review and editing, D.L.; conceptualization, methodology, writing—original draft and editing, supervision, project administration, funding acquisition, B.D.L. All authors have read and agreed to the published version of the manuscript.

Funding: This research was funded by the project: "Innovazione di prodotto e di processo nel floro-vivaismo biologico per il verde pubblico, healing e kitchen garden (BIOGARDEN) (MASAF, Italy)".

Data Availability Statement: Data are contained within the article and Supplementary Materials.

Acknowledgments: The authors are grateful to Beniamino Leoni (Department of Soil, Plants and Food Sciences, University of Bari, Italy) for his valuable support and for providing research facilities during the experiment, and to Giovanni Fanizzi (Vivai Capitanio, Monopoli, Italy) for offering the necessary infrastructure. They thank Nuova Flesan (Milano, Italy), Brunocaffè s.r.l. (Modugno, Bari, Italy), and Birra Peroni (Bari, Italy) for the supply of by-products.

Conflicts of Interest: The authors declare no conflicts of interest.

References

1. Trivellini, A.; Toscano, S.; Romano, D.; Ferrante, A. LED lighting to produce high-quality ornamental plants. *Plants* **2023**, *12*, 1667. [CrossRef] [PubMed]
2. Wallach, R. Physical characteristics of soilless media. In *Soilless Culture*, 2nd ed.; Raviv, M., Lieth, J.H., Bar-Tal, A., Eds.; Elsevier: Amsterdam, The Netherlands, 2019; pp. 33–112. [CrossRef]
3. Schindler, U.; Müller, L.; Eulenstein, F. Measurement and evaluation of the hydraulic properties of horticultural substrates. *Arch. Agron. Soil Sci.* **2016**, *62*, 806–818. [CrossRef]
4. Al Naddaf, O.; Livieratos, I.; Stamatakis, A.; Tsirogiannis, I.; Gizas, G.; Savvas, D. Hydraulic characteristics of composted pig manure, perlite, and mixtures of them, and their impact on cucumber grown on bags. *Sci. Hortic.* **2011**, *129*, 135–141. [CrossRef]
5. Banitalebi, G.; Mosaddeghi, M.R.; Shariatmadari, H. Feasibility of agricultural residues and their biochars for plant growing media: Physical and hydraulic properties. *Waste Manag.* **2019**, *87*, 577–589. [CrossRef]
6. Paradelo, R.; Basanta, R.; Barral, M.T. Water-holding capacity and plant growth in compost-based substrates modified with polyacrylamide, guar gum or bentonite. *Sci. Hortic.* **2019**, *243*, 344–349. [CrossRef]
7. Gruda, N.S. Increasing sustainability of growing media constituents and stand-alone substrates in soilless culture systems. *Agronomy* **2019**, *9*, 298. [CrossRef]
8. Van Gerrewey, T.; Vandecruys, M.; Ameloot, N.; Perneel, M.; Van Labeke, M.-C.; Boon, N.; Geelen, D. Microbe-plant growing media interactions modulate the effectiveness of bacterial amendments on lettuce performance inside a plant factory with artificial lighting. *Agronomy* **2020**, *10*, 1456. [CrossRef]
9. Erdal, İ.; Aktaş, H. Comparison of the Perlite, Leonardite, Vermicompost and Peat Moss and Their Combinations with Cocopeat as Tomato Growing Media. *J. Soil Sci. Plant Nutr.* **2025**, *25*, 2726–2741. [CrossRef]
10. Balawejder, M.; Matłok, N.; Szostek, M.; Kuboń, M. The Influence of a Biopolymer Coating on Selected Surface Properties of Soilless Substrates Made from Coconut Fibre. *Appl. Sci.* **2025**, *15*, 7039. [CrossRef]
11. Barrett, G.E.; Alexander, P.D.; Robinson, J.S.; Bragg, N.C. Achieving environmentally sustainable growing media for soilless plant cultivation systems—A review. *Sci. Hortic.* **2016**, *212*, 220–234. [CrossRef]
12. Fesendouz, S.O.; Ebrahimzadeh, A.; Rasouli, F. Using agricultural waste as an alter-native growing medium for cultivating Cucumis sativus L. greenhouse transplants. *Sci. Rep.* **2025**, *15*, 14899. [CrossRef]
13. Blok, C.; Eveleens, B.; van Winkel, A. Growing media for food and quality of life in the period 2020-2050. *Acta Hortic.* **2021**, *1305*, 341–356. [CrossRef]
14. Hirschler, O.; Ostenburg, B. Peat extraction, trade and use in Europe: A material flow analysis. *Mires Peat* **2022**, *28*, 27. [CrossRef]
15. Salimi, S.; Almuktar, S.A.; Scholz, M. Impact of climate change on wetland ecosystems: A critical review of experimental wetlands. *J. Environ. Manag.* **2021**, *286*, 112160. [CrossRef] [PubMed]
16. Leifeld, J.; Menichetti, L. The underappreciated potential of peatlands in global climate change mitigation strategies. *Nat. Commun.* **2018**, *9*, 1071. [CrossRef] [PubMed]
17. DEFRA. Ending the Retail Sale of Peat in Horticulture in England and Wales. Available online: https://consult.defra.gov.uk/soils-and-peatlands/endingtheretailsaleofpeatinhorticulture/ (accessed on 12 May 2025).
18. Koseoglu, M.N.; Roberts, M. Supply Chain Dynamics of Moving from Peat-Based to Peat-Free Horticulture. *Sustainability* **2025**, *17*, 6159. [CrossRef]
19. Sdao, A.E.; Gruda, N.S.; De Lucia, B. Beyond Peat: Wood Fiber and Two Novel Organic Byproducts as Growing Media—A Systematic Review. *Plants* **2025**, *14*, 1945. [CrossRef] [PubMed]
20. Taparia, T.; Hendrix, E.; Nijhuis, E.; de Boer, W.; van der Wolf, J. Circular alternatives to peat in growing media: A microbial perspective. *J. Clean. Prod.* **2021**, *327*, 129375. [CrossRef]
21. Poláková, K.; Bobková, A.; Demianová, A.; Bobko, M.; Lidiková, J.; Jurčaga, L.; Belej, Ľ.; Mesárošová, A.; Korčok, M.; Tóth, T. Quality attributes and sensory acceptance of different botanical coffee co-products. *Foods* **2023**, *12*, 2675. [CrossRef]

22. Reineke, T.; Muhammed, H.H.A.; Anlauf, R.; Daum, D. Impact of thermo-hydrolytically treated wood fibers as a substrate component on the growth of petunias. *Acta Hortic.* **2024**, *1389*, 105–112. [CrossRef]
23. Atzori, G.; Pane, C.; Zaccardelli, M.; Cacini, S.; Massa, D. The role of peat-free organic substrates in the sustainable management of soilless cultivations. *Agronomy* **2021**, *11*, 1236. [CrossRef]
24. Zaccheo, P.; Cattivello, C.; Longo, C.; Crippa, L.; Notaristefano, P.; Orfeo, D. A comparative study on some locally available organic materials for their potential utilization in sustainable growing media blends. *Acta Hortic.* **2024**, *1389*, 123–130. [CrossRef]
25. Gruda, N.S.; Hirschler, O.; Stuart, J. Peat reduction in horticulture: An overview of Europe. *Acta Hortic.* **2024**, *1391*, 545–560. [CrossRef]
26. Miler, N.; Wojewódzki, P.; Woźny, A.; Rymarz, D.; Gołębiewska, A. Exploring Coffee Silverskin as a Sustainable Peat Additive in the Plant Nursery Production. *Agronomy* **2025**, *15*, 1633. [CrossRef]
27. Hejna, A.; Aniśko-Michalak, J.; Skórczewska, K.; Barczewski, M.; Sulima, P.; Przyborowski, J.A.; Cieśliński, H.; Marć, M. Brewers' Spent Grain from Different Types of Malt: A Comprehensive Evaluation of Appearance, Structure, Chemical Composition, Antimicrobial Activity, and Volatile Emissions. *Molecules* **2025**, *30*, 2809. [CrossRef]
28. Paciulli, M.; Sogari, G.; Rodolfi, M.; Parenti, O.; Andreani, G.; Chiavaro, E. Fostering circular economy: Brewing by-products as innovative ingredients for cereal bar formulation. *Foods* **2024**, *13*, 2355. [CrossRef] [PubMed]
29. Paksin, P.; Tangjaidee, P.; Klangpetch, W.; Unban, K.; Khumsap, T.; Klunklin, W.; Yawootti, A.; Jantanasakulwong, K.; Rachtanapun, P.; Phongthai, S. Quality Attributes, Structural Characteristics, and Functional Properties of Brewer's Spent Grain Protein Concentrates as Affected by Alkaline and Pulsed Electric Field-Assisted Extraction. *Foods* **2025**, *14*, 1515. [CrossRef]
30. Tumbure, A.; Pulver, C.; Black, L.; Walsh, L.; Prasad, M.; Leahy, J.J.; Corbett, E.; Gaffney, M.T. Bio-resource availability in Ireland: A practical review of potential replacement materials for use in horticultural growth media. *Horticulturae* **2025**, *11*, 378. [CrossRef]
31. European Commission. Communication on the European Green Deal. 2019. Available online: https://eur-lex.europa.eu/legal-content/EN/TXT/?qid=1588580774040&uri=CELEX%3A52019DC0640 (accessed on 11 June 2025).
32. European Commission. A Farm to Fork Strategy for a Fair, Healthy and Environmentally-Friendly Food System. 2020. Available online: https://eur-lex.europa.eu/legal-content/EN/TXT/?uri=CELEX:52020DC0381 (accessed on 11 June 2025).
33. European Commission. A New Circular Economy Action Plan for a Cleaner and More Competitive Europe. 2020. Available online: https://eur-lex.europa.eu/legal-content/EN/TXT/?uri=COM:2020:98:FIN (accessed on 25 June 2025).
34. Caron, J.; Pepin, S.; Periard, Y. Physics of growing media in a green future. *Acta Hortic.* **2014**, *1034*, 309–317. [CrossRef]
35. Schindler, U.; Lischeid, G.; Müller, L. Hydraulic performance of horticultural substrates—3. Impact of substrate composition and ingredients. *Horticulturae* **2017**, *3*, 7. [CrossRef]
36. Muhammed, H.H.; Anlauf, R.; Daum, D. Estimation of Hydraulic Properties of Growing Media from Numerical Inversion of Mini Disk Infiltrometer Data. *Hydrology* **2025**, *12*, 100. [CrossRef]
37. Noguera, P.; Abad, M.; Puchades, R.; Maquieira, A.; Noguera, V. Influence of particle size on physical and chemical properties of coconut coir dust as container medium. *Commun. Soil Sci. Plant Anal.* **2011**, *34*, 593–605. [CrossRef]
38. Bilderback, T.E.; Warren, S.L.; Owen, J.S., Jr.; Albano, J.P. Healthy substrates need physicals too! *HortTechnology* **2005**, *15*, 747. [CrossRef]
39. Plants of the World Online (POWO). Available online: https://powo.science.kew.org/taxon/urn:lsid:ipni.org:names:30000096-2 (accessed on 28 July 2025).
40. Afonso, A.F.; Pereira, O.R.; Fernandes, Â.S.F.; Calhelha, R.C.; Silva, A.M.S.; Ferreira, I.C.F.R.; Cardoso, S.M. The health-benefits and phytochemical profile of *Salvia apiana* and *Salvia farinacea* var. Victoria Blue decoctions. *Antioxidants* **2019**, *8*, 241. [CrossRef] [PubMed]
41. De Mastro, G.; El Mahdi, J.; Ruta, C. Bioherbicidal Potential of the Essential Oils from Mediterranean Lamiaceae for Weed Control in Organic Farming. *Plants* **2021**, *10*, 818. [CrossRef] [PubMed]
42. Kamatou, G.P.; Makunga, N.P.; Ramogola, W.P.; Viljoen, A.M. South African Salvia species: A review of biological activities and phytochemistry. *J. Ethnopharmacol.* **2008**, *119*, 664–672. [CrossRef]
43. Sharifi-Rad, M.; Ozcelik, B.; Altın, G.; Daşkaya-Dikmen, C.; Martorell, M.; Ramirez-Alarcon, K.; Sharifi-Rad, J. *Salvia* spp. plants-from farm to food applications and phytopharmacotherapy. *Trends Food Sci. Technol.* **2018**, *80*, 242–263. [CrossRef]
44. Lin, K.-H.; Lin, T.-Y.; Wu, C.-W.; Chang, Y.-S. Protective Effects of Salicylic Acid and Calcium Chloride on Sage Plants (*Salvia officinalis* L. and *Salvia elegans* Vahl) under High-Temperature Stress. *Plants* **2021**, *10*, 2110. [CrossRef]
45. Papafotiou, M.; Martini, A.N.; Papanikolaou, E.; Stylias, E.G.; Kalantzis, A. Hybrids development between Greek *Salvia* species and their drought resistance evaluation along with *Salvia fruticosa*, under attapulgite-amended substrate. *Agronomy* **2021**, *11*, 2401. [CrossRef]
46. Moughan, J.; McGinn, K.J.; Jones, L.; Rich, T.C.; Waters, E.; de Vere, N. Biological flora of the British Isles: *Salvia pratensis*. *J. Ecol.* **2021**, *109*, 4171–4190. [CrossRef]
47. Mossi, A.; Cansian, R.; Paroul, N.; Toniazzo, G.; Oliveira, J.; Pierozan, M.; Serafini, L. Morphological characterization and agronomical parameters of different species of *Salvia* sp. (Lamiaceae). *Braz. J. Biol.* **2011**, *71*, 121–129. [CrossRef]

48. Nanos, C.; Tsoulpha, P.; Kostas, S.; Hatzilazarou, S.; Michail, I.; Anastasiadi, V.; Pipinis, E.; Gklavakis, E.; Kanellis, A.K.; Nianiou-Obeidat, I. Asexual propagation of Greek *Salvia officinalis* L. populations selected for ornamental use. *Horticulturae* **2023**, *9*, 847. [CrossRef]
49. Scaltrito, E.; Cristiano, G.; Sdao, A.E.; Gruda, N.S.; Loconsole, D.; De Lucia, B. Influence of water spraying intervals and indole-3-butyric acid concentrations on *Salvia* rooted cuttings quality in a closed aeroponics system. *Sci. Hortic.* **2024**, *337*, 113452. [CrossRef]
50. Gokdogan, E.Y; Burun, B. The studies on seed germination and in vitro cultures of *Salvia* L. species from Turkish flora. *Nat. Prod. Biotechnol.* **2022**, *2*, 60–73. [CrossRef]
51. Greco, C.; Comparetti, A.; Fascella, G.; Febo, P.; La Placa, G.; Saiano, F.; Mammano, M.M.; Orlando, S.; Laudicina, V.A. Effects of vermicompost, compost and digestate as commercial alternative peat-based substrates on qualitative parameters of *Salvia officinalis*. *Agronomy* **2021**, *11*, 98. [CrossRef]
52. Iacomino, G.; Cozzolino, A.; Idbella, M.; Amoroso, G.; Bertoli, T.; Bonanomi, G.; Motti, R. Potential of biochar as a peat substitute in growth media for *Lavandula angustifolia, Salvia rosmarinus* and *Fragaria × ananassa*. *Plants* **2023**, *12*, 3689. [CrossRef]
53. Salvador, E.D.; Balas, J. Physical characteristics of soilless growing media: Basis for the development of methods to formulate substrates for ornamental plants in Brazil. In *Floriculture, Ornamental and Plant Biotechnology: Advances and Topical Issues*, 1st ed.; Teixeira da Silva, J.A., Ed.; Global Science Books: London, UK, 2006; pp. 161–171.
54. Michel, J.-C.; Kerloch, E. Evolution of hydraulic properties and wettability of organic growing media during cultivation according to irrigation strategies. *Sci. Hortic.* **2017**, *217*, 28–35. [CrossRef]
55. Chrysargyris, A.; Antoniou, O.; Tzionis, A.; Prasad, M.; Tzortzakis, N. Alternative soilless media using olive-mill and paper waste for growing ornamental plants. *Environ. Sci. Pollut. Res.* **2018**, *25*, 35915–35927. [CrossRef]
56. Grosbellet, C.; Beaudet, E.; Duclaud, E. Performances of new materials for peat substitution in horticulture. *Acta Hortic.* **2019**, *1266*, 315–322. [CrossRef]
57. Thakur, T.; Garg, A. Growth and flowering of ornamental pot plants influenced by growing media—A review. *J. Ornam. Hortic.* **2025**, *27*, 95–106. Available online: https://epubs.icar.org.in/index.php/JOH/article/view/164891 (accessed on 29 July 2025).
58. Awang, Y.; Shaharom, A.S.; Mohamad, R.B.; Selamat, A. Chemical and Physical Characteristics of Cocopeat-Based Media Mixtures and Their Effects on the Growth and Development of *Celosia cristata*. *Am. J. Agric. Biol. Sci.* **2009**, *4*, 63–71. [CrossRef]
59. Di Lonardo, S.; Massa, D.; Orsenigo, S.; Zubani, L.; Rossi, G.; Fascella, G.; Cacini, S. Substitution of peat in the cultivation of two shrub species used for ecological restoration and recovery of degraded green areas. *Acta Hortic.* **2021**, *1305*, 97–102. [CrossRef]
60. Di Lonardo, S.; Cacini, S.; Becucci, L.; Lenzi, A.; Orsenigo, S.; Zubani, L.; Rossi, G.; Zaccheo, P.; Massa, D. Testing new peat-free substrate mixtures for the cultivation of perennial herbaceous species: A case study on *Leucanthemum vulgare* Lam. *Sci. Hortic.* **2021**, *288*, 110472. [CrossRef]
61. Caron, J.; Nkongolo, N.V. Assessing Gas Diffusion Coefficients in Growing Media from in situ Water Flow and Storage Measurements. *Vadose Zone J.* **2004**, *3*, 300–306. [CrossRef]
62. Cannavo, P.; Michel, J.-C. Peat particle size effects on spatial root distribution, and changes on hydraulic and aeration properties. *Sci. Hortic.* **2013**, *151*, 148–154. [CrossRef]
63. Mariotti, B.; Oliet, J.A.; Andivia, E.; Tsakaldimi, M.; Villar-Salvador, P.; Ivetić, V.; Cocozza, C. A global review on innovative, sustainable, and effective materials composing growing media for forest seedling production. *Curr. For. Rep.* **2023**, *9*, 413–428. [CrossRef]
64. Carlile, B.; Coules, A. Towards sustainability in growing media. *Acta Hortic.* **2013**, *1013*, 341–349. [CrossRef]
65. Schmilewski, G. Growing media constituents used in the EU in 2013. *Acta Hortic.* **2017**, *1168*, 85–92. [CrossRef]
66. Jackson, B.E. The Role of Wood Substrates in Controlled Environment Vegetable and Fruit Production Systems. Available online: https://www.producegrower.com/article/the-role-of-wood-substrates-in-controlled-environment-vegetable-and-fruit-production-systems/ (accessed on 1 July 2025).
67. Lemaire, F.; Dartigues, A.; Riviere, L.M.; Charpentier, S. *Cultures en pots et Conteneurs: Principes Agronomiques et Applications*, 1st ed.; INRA: Paris, France, 1989; ISBN 2-7380-0157-2.
68. Gruda, N.; Schnitzler, W.H. Suitability of wood fiber substrate for production of vegetable transplants: I. Physical properties of wood fiber substrates. *Sci. Hortic.* **2004**, *100*, 309–322. [CrossRef]
69. Domeño, I.; Irigoyen, I.; Muro, J. New Wood Fibre Substrates Characterization and Evaluation in Hydroponic Tomato Culture. *Eur. J. Hortic. Sci.* **2010**, *75*, 89. [CrossRef]
70. Anlauf, R.; Muhammed, H.H.A.; Reineke, T.; Daum, D. Water retention properties of wood fiber based growing media and their impact on irrigation strategy. *Acta Hortic.* **2024**, *1389*, 227–236. [CrossRef]
71. Durand, S.; Jackson, B.E.; Fonteno, W.C.; Michel, J.C. Advances in substrate particle characterization using dynamic image analysis compared to sieving procedure for predicting water retention properties. *Acta Hortic.* **2023**, *1377*, 537–544. [CrossRef]

72. Caron, J.; Parent, L.E.; Elrick, D.E.; Naasz, R.; Carter, M.R.; Gregorich, E.G. Physical properties of organic soils and growing media: Water and air storage and flow dynamics. In *Soil Sampling and Methods of Analysis*, 2nd ed.; Carter, M.R., Gregorich, E.G., Eds.; CRC Press: Boca Raton, FL, USA, 2008; pp. 885–912.
73. Fields, J.S.; Fonteno, W.C.; Jackson, B.E.; Heitman, J.L.; Owen, J.S. Hydrophysical properties, moisture retention, and drainage profiles of wood and traditional components for greenhouse substrates. *HortScience* **2014**, *49*, 827–832. [CrossRef]
74. Durand, S.; Jackson, B.E.; Fonteno, W.C.; Michel, J.-C. The Use of Wood Fiber for Reducing Risks of Hydrophobicity in Peat-Based Substrates. *Agronomy* **2021**, *11*, 907. [CrossRef]
75. Michiels, P.; Hartmann, R.; Coussens, C. Physical properties of peat substrates in an ebb/flood irrigation system. *Acta Hortic.* **1993**, *342*, 205–220. [CrossRef]
76. Raviv, M.; Lieth, J.H.; Bar-Tal, A. *Soilless Culture: Theory and Practice*; Elsevier: Amsterdam, The Netherlands, 2019.
77. Eveleens, B.; van Winkel, A.; Blok, C. Wood fiber in pot plant culture; peat replacement up to 50% in volume? *Acta Hortic.* **2021**, *1317*, 165–174. [CrossRef]
78. Arancon, N.Q.; Pant, A.; Radovich, T.; Hue, N.V.; Potter, J.K.; Converse, C.E. Seed germination and seedling growth of tomato and lettuce as affected by vermicompost water extracts (teas). *HortScience* **2012**, *47*, 1722–1728. [CrossRef]
79. Woznicki, T.L.; Sønsteby, A.; Aurdal, S.M.; Kusnierek, K.; Haraldsen, T.K. Optimizing Peat and Wood Fiber Blends: Impacts of Liming and Fertilization on Growth of Petunia (*Petunia x hybrida* Vilm.) and Basil (*Ocimum basilicum* L.). *Horticulturae* **2024**, *10*, 895. [CrossRef]
80. Ballesteros, L.F.; Teixeira, J.A.; Mussatto, S.I. Chemical, functional, and structural properties of spent coffee grounds and coffee silverskin. *Food Bioproc. Technol.* **2014**, *7*, 3493–3503. [CrossRef]
81. Campos-Vega, R.; Loarca-Piña, G.; Vergara-Castañeda, H.A.; Oomah, B.D. Spent coffee grounds: A review on current research and future prospects. *Trends Food Sci. Technol.* **2015**, *45*, 24–36. [CrossRef]
82. Pourfarzad, A.; Mahdavian-Mehr, H.; Sedaghat, N. Coffee silverskin as a source of dietary fiber in bread-making: Optimization of chemical treatment using response surface methodology. *LWT-Food Sci. Technol.* **2013**, *50*, 599–606. [CrossRef]
83. Martuscelli, M.; Esposito, L.; Di Mattia, C.D.; Ricci, A.; Mastrocola, D. Characterizationof coffee silver skin as potential food-safe ingredient. *Foods* **2021**, *10*, 1367. [CrossRef]
84. De Boodt, M.; Verdonck, O.; Cappaert, I. Method for measuring the water release curve of organic substrates. *Acta Hortic.* **1974**, *37*, 2054–2063. [CrossRef]
85. De Boodt, M.A.V.O.; Verdonck, O. The physical properties of the substrates in horticulture. In Proceedings of the III Symposium on Peat in Horticulture 26, Dublin, Ireland, 28 June–3 July 1971; pp. 37–44.
86. Abad, M.; Noguera, P.; Burés, S. National inventory of organic wastes for use as growing media for ornamental potted plant production: Case study in Spain. *Bioresour. Technol.* **2001**, *77*, 197–200. [CrossRef]
87. Naibaho, J.; Korzeniowska, M. The Variability of Physico-Chemical Properties of Brewery Spent Grain from 8 Different Breweries. *Heliyon* **2021**, *7*, e06329. [CrossRef]
88. Castro, L.E.N.; Colpini, L.M.S. All-Around Characterization of Brewers' Spent Grain. *Eur. Food Res. Technol.* **2021**, *247*, 3013–3021. Available online: https://link.springer.com/article/10.1007%2Fs00217-021-03860-5 (accessed on 4 June 2025). [CrossRef]
89. Mussatto, S.I.; Dragone, G.; Roberto, I.C. Brewers' spent grain: Generation, characteristics and potential applications. *J. Cereal Sci.* **2006**, *43*, 1–14. [CrossRef]
90. Xiros, C.; Christakopoulos, P. Biotechnological potential of brewers' spent grain and its recent applications. *Waste Biomass Valorization* **2012**, *3*, 213–232. [CrossRef]
91. Beldman, G.; Hennekam, J.; Voragen, A.G. Enzymatic hydrolysis of beer brewers' spent grain and the influence of pretreatments. *Biotechnol. Bioeng.* **1987**, *30*, 668–671. [CrossRef] [PubMed]
92. Hauck, D.; Lohr, D.; Meinken, E.; Schmidhalter, U. Plant availability of secondary phosphates depending on pH in a peat-based growing medium. *Acta Hortic.* **2021**, *1305*, 437–442. [CrossRef]
93. Cacace, C.; Cocozza, C.; Traversa, A.; Coda, R.; Rizzello, C.G.; Pontonio, E.; De Mastro, F.; Brunetti, G.; Verni, M. Potential of Native and Bioprocessed Brewers' Spent Grains as Organic Soil Amendments. *Front. Sustain. Food Syst.* **2022**, *6*, 1010890. [CrossRef]
94. Kyriacou, M.; El-Nakhel, C.; Pannico, A.; Graziani, G.; Soteriou, G.; Giordano, M.; Palladino, M.; Ritieni, A.; De Pascale, S.; Rouphael, Y. Phenolic Constitution, Phytochemical and Macronutrient Content in Three Species of Microgreens as Modulated by Natural Fiber and Synthetic Substrates. *Antioxidants* **2020**, *9*, 252. [CrossRef] [PubMed]
95. Zawadzińska, A.; Salachna, P.; Nowak, J.S.; Kowalczyk, W. Response of Interspecific Geraniums to Waste Wood Fiber Substrates and Additional Fertilization. *Agriculture* **2021**, *11*, 119. [CrossRef]
96. Dickson, R.W.; Helms, K.M.; Jackson, B.E.; Machesney, L.M.; Lee, J.A. Evaluation of Peat Blended with Pine Wood Components for Effects on Substrate Physical Properties, Nitrogen Immobilization, and Growth of Petunia (*Petunia* × *hybrida* Vilm.-Andr.). *HortScience* **2022**, *57*, 304–311. [CrossRef]

97. Iriondo-DeHond, A.; Uranga, J.A.; del Castillo, M.D.; Abalo, R. Effects of Coffee and Its Components on the Gastrointestinal Tract and the Brain–Gut Axis. *Nutrients* **2021**, *13*, 88. [CrossRef]
98. Sdao, A.E.; Mordelli, D.; Piscitelli, L.; Scaltrito, E.; Leoni, B.; Cristiano, G.; De Lucia, B. Morpho-physiological response of bedding plants quality to unconventional agro-industrial organic matrices to peat replacing: Preliminary outcomes. *Acta Hortic.* **2025**, *1417*, 125–132. [CrossRef]
99. Yang, H.; Yang, J.; Lv, Y.; He, J. SPAD Values and Nitrogen Nutrition Index for the Evaluation of Rice Nitrogen Status. *Plant Prod. Sci.* **2014**, *17*, 81–92. [CrossRef]
100. Cangussu, L.B.; Melo, J.C.; Franca, A.S.; Oliveira, L.S. Chemical Characterization of Coffee Husks, a By-Product of *Coffea arabica* Production. *Foods* **2021**, *10*, 3125. [CrossRef]
101. *European Standard 13038*; Determination of Electrical Conductivity, in Soil Improvers and Growing Media. European Committee for Standardization: Brussels, Belgium, 1999.
102. *European Standard 13037*; Determination of pH, in Soil Improvers and Growing Media. European Committee for Standardization: Brussels, Belgium, 1999.
103. Cavins, T.; Whipker, B.; Fonteno, W. PourThru: A Method for Monitoring Nutrition in the Greenhouse. *Acta Hortic.* **2008**, *779*, 289–298. [CrossRef]
104. SAS Institute Inc. *SAS/STAT® User's Guide, Version 8*; SAS Institute Inc.: Cary, NC, USA, 1999.
105. Hammer, O.; Harper, D.A.T.; Ryan, P.D. PAST: Paleontological Statistics Software Package for Education and Data Analysis. *Palaeontol. Electron.* **2001**, *4*, 9.

plants

MDPI

Article

Potential of Pine Bark to Replace Perlite in Coir-Based Substrates: Effects on Nutrient Uptake, Growth, and Phytochemicals in Lettuce Under Two Salinity Levels

Gonçalo C. Dias [1], Rui M. A. Machado [1,*], Isabel Alves-Pereira [2,*], Rui A. Ferreira [2] and Nazim S. Gruda [3]

[1] MED—Mediterranean Institute for Agriculture, Environment and Development & CHANGE—Global Change and Sustainability Institute, Crop Science Department, School of Sciences and Technology, University of Évora, Pólo da Mitra, Ap. 94, 7006-554 Évora, Portugal; goncasdias123@hotmail.com

[2] MED—Mediterranean Institute for Agriculture, Environment and Development & CHANGE—Global Change and Sustainability Institute, Chemistry and Biochemistry Department, School of Sciences and Technology, University of Évora, Colégio Luís António Verney, Ap. 94, 7006-554 Évora, Portugal; raf@uevora.pt

[3] Division of Horticultural Sciences, University of Bonn, Auf dem Hügel 6, 53121 Bonn, Germany; ngruda@uni-bonn.de

* Correspondence: rmam@uevora.pt (R.M.A.M.); iap@uevora.pt (I.A.-P.)

Abstract

Enhancing the sustainability of growing media is an important objective in soilless vegetable cultivation. Here, we evaluated the potential of pine bark to replace perlite in coir-based substrates for lettuce (*Lactuca sativa* L. cv. 'Godzilla') cultivation. The experiment followed a factorial design with two coir-based substrate blends—one amended with perlite and the other with pine bark—and two nutrient solution EC levels (1.5 $\pm$ 0.2 and 2.5 $\pm$ 0.2 dS m^{-1}). The plants were cultivated in Styrofoam containers containing a substrate mix of 80% coir, 12% compost, and 8% perlite or pine bark (v/v). Replacing perlite with pine bark did not affect leaf macronutrient concentrations but increased leaf Fe and B levels. Increasing the EC of the nutrient solution increased leaf N, P, and K, with a significant rise in nitrogen. The substitution of perlite with pine bark in coir-based substrates did not affect leaf dry weight, head fresh weight, or chlorophyll content, total phenols, ascorbic acid, or proline, even under different salinity levels. The findings indicate the pine bark is an alternative to perlite, supporting comparable agronomic and quality outcomes in lettuce. Further research is recommended to confirm these results in crops with longer growing cycles.

Keywords: *Lactuca sativa* L.; soilless cultivation; eco-friendly substrate; salinity; total phenols; ascorbate

1. Introduction

Sustainable production practices are increasingly crucial in soilless culture, particularly regarding the choice of substrate components [1–3]. Utilizing renewable, locally obtained resources can reduce both environmental harm and manufacturing expenses while promoting circular approaches within horticulture [4]. Consequently, carefully selecting and preparing substrates is crucial for achieving success in soilless cultivation, particularly in high-value crop sectors [5].

Perlite, a commonly used substrate component, is both energy-intensive to produce and relatively expensive [6]. Perlite is mined (mainly in countries such as the United States, Greece, and Turkey), thermally expanded, and shipped over long distances, resulting in considerable carbon emissions. The energy costs to produce and transport raise concerns about its environmental and economic sustainability. According to [7], milling alone requires approximately 6.70 kWh of electricity per metric ton, excluding the energy used in mining and transportation. Furthermore, spent perlite is difficult to recycle due to contamination with organic residues [8]. Pine bark can be an alternative, making it a more economical choice compared to the relatively expensive substrate, perlite [9]. It is a renewable byproduct of the forestry industry, widely available in countries like Portugal. A life cycle assessment made by Vinci and Mattia [10] has shown that bark-based substrates tend to have significantly lower environmental impacts than perlite. Overall, both materials are used to improve substrate aeration and drainage, but their physical and chemical properties differ significantly. Perlite is an inert, stable material with a pH typically ranging from 7.0 to 7.5. It does not provide mineral nutrients or buffering capacity, but it exhibits high capillarity [11]. Although its water-holding capacity depends on particle size, it is generally lower than that of most organic materials [12,13]. Pine bark, on the other hand, has a low cation exchange capacity [14], although it varies with particle size [15], and typically has a low pH. Pine bark has a negative charge due to the ionization of functional groups like carboxylic and phenolic [16]. Additionally, pine bark can degrade over time, leading to nitrogen immobilization [17,18]. When not composted or well-aged, it may contain phytotoxins [19]. These differences, alongside the characteristics of the nutrient solution, can affect water availability, pH, and electrical conductivity (EC), all of which are essential for nutrient uptake and plant performance [20].

Salinity in the root zone may reduce growth and enhance quality. Although generally applicable to leafy vegetables, exceptions such as lettuce and wild rocket may exhibit variations in consistency, which can potentially compromise certain quality traits [21]. In lettuce, eustress due to increased salinity in nutrient solution enhanced the accumulation of phenolic compounds and other secondary metabolites [22,23]. The objective of this study was to assess the agronomic and physiological responses of lettuce grown in coir-based substrates amended with either perlite or pine bark, under two nutrient solution salinity levels. Specifically, this study was to assess whether pine bark could serve as an alternative to perlite while maintaining nutrient absorption, growth, and quality attributes. We hypothesized that pine bark has the potential to replace perlite in coir-based substrates and would support comparable nutrient uptake, biomass production, and quality traits in lettuce. Additionally, we expected that this substitution would remain effective under increased salinity, indicating its potential as a sustainable component in soilless culture systems.

2. Results

2.1. Initial Growing Media Physicochemical Characteristics

The substitution of perlite by pine bark increased the electrical conductivity (EC), bulk density, and moisture content of the mix while decreasing its pH and total porosity (Table 1). The average EC values of the mixes with pine bark and perlite were 3.01 and 2.55 dS m^{-1}, respectively (Table 1).

The pH of the mixes with pine bark (7.6) and perlite (7.5) was higher than the range considered suitable for cultivating vegetables (5.5 to 6.8) [24]. Pine bark increased the bulk density of the mix from 0.08 to 0.11 g/cm^3 and decreased the total porosity from 91.41% to 89.66% (Table 1). Pine bark decreased the mass wetness of the mix, whereas the mix containing perlite exhibited a lower bulk density (0.08 g/cm^3) and a higher moisture content.

Table 1. Initial mix physicochemical characteristics.

Parameter	Mix		Significance
	C [1] + Comp + P	C + Comp + PB	
pH	7.6 b [2]	7.5 a	*
Electrical conductivity (dS m^{-1})	2.55 a	3.01 b	***
Bulk density (g cm^{-3})	0.08 b	0.11 a	***
Mass wetness (g water/g substrate)	5.22 a	4.60 b	***
Total porosity (%)	91.4 a	89.7 b	*
Moisture content (%, *w*/*w*)	79.56 b	81.06 a	*

[1]—C—coir, Comp—compost, P—perlite, PB—pine bark; [2]—means followed by different letters within a column are significantly different. * and *** significant at $p < 0.05$ and 0.001 levels, respectively. Mean separation was performed using Duncan's multiple-range test. Means are based on five replicates.

2.2. Leachate EC and pH

The electrical conductivity of leachate above the initial EC of the nutrient solution was not affected by the interaction between treatments ($p > 0.05$) on any of the sampling dates (Figure 1). Replacing perlite with pine bark in the substrate mix had no significant effect on leachate EC when the nutrient solution had an EC of 1.5 dS m^{-1}. In this treatment, leachate EC was slightly higher than the feed solution, ranging from 0.2 to 0.7 dS m^{-1} (Figure 1), which can be considered an acceptable level.

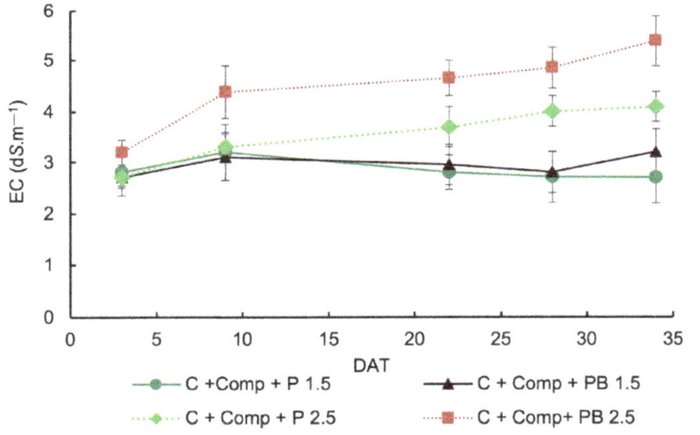

Figure 1. Effects of cultivation substrate mix and nutrient solution EC on average EC values of leachate above the initial electrical conductivity of the nutrient solution. Each symbol represents the mean of five replicates, and the error bars represent ±1 standard error. (C—coir, Comp—compost, P—perlite, PB—pine bark, DAT—days after transplantation).

When the nutrient solution EC was 2.5 dS m^{-1}, leachate EC remained consistently above the initial EC. EC values under this treatment were higher than those recorded under the 1.5 dS m^{-1} solution. In both substrate mixes, a clear upward trend in leachate EC was observed over time, indicating a loss of nutrients.

Except for the first sampling date, leachate EC was higher in the pine bark mix than in the perlite mix (Figure 1).

Leachate pH was not significantly affected by the interaction between treatments ($p > 0.05$) on any sampling date (Figure 2). Although pine bark is not inert like perlite, its inclusion in the mix did not significantly affect leachate pH. However, a significant decrease in pH ($p < 0.001$) was observed on the last two sampling dates, accompanied by an increase in EC in the nutrient solution. Across all treatments and dates, leachate pH values remained consistently higher than the pH of the incoming nutrient solution, which averaged 6.4 ± 0.5.

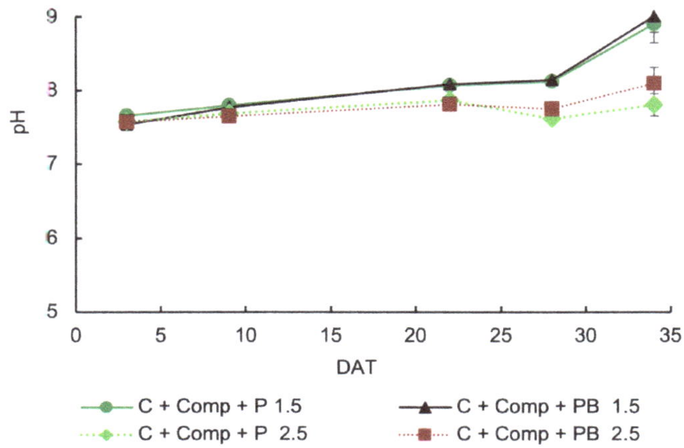

Figure 2. Effects of cultivation substrate mix and nutrient solution EC on pH of leachate. Each symbol represents the mean of five replicates, and the error bars represent ±1 standard error. (C—coir, Comp—compost, P—perlite, PB—pine bark, DAT—days after transplantation).

2.3. Growing Media pH and EC

Substrate solution pH and EC collected at 9, 17, and 29 days after transplanting (DAT) were not significantly affected by the interaction between substrate mix and EC of the nutrient solution, indicating independent effects (Figure 3).

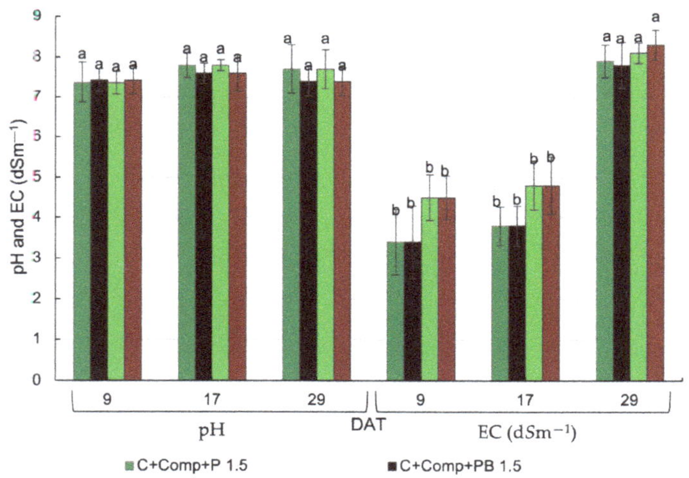

Figure 3. Effects of cultivation substrate mix and nutrient solution EC on pH and electrical conductivity of substrate solution (C—coir, Comp—compost, P—perlite, PB—pine bark, DAT—days after transplantation). Different letters for the same parameter indicate statistically significant differences ($p < 0.05$).

The substitution of perlite by pine bark in the substrate did not significantly influence the pH or EC of the substrate solution at any of the sampling dates (Figure 3). However, the nutrient solution EC had a significant effect on the EC of the substrate solution at 9 and 17 DAT. Specifically, higher EC values were observed in the treatment receiving the 2.5 dS m^{-1} nutrient solution compared to the 1.5 dS m^{-1} treatment (Table 2). On average, EC values measured across all three sampling dates were relatively high—particularly at

17 and 29 DAT—exceeding the recommended EC threshold for lettuce of 2.0 dS m^{-1} [25]. Despite these high EC average values, plants did not exhibit visible symptoms of salinity stress.

Table 2. Effect of sampling location on pH and EC of the cultivation substrate mixes at harvest time, measured near the plant stem (center) and on the lateral side of the container.

Sampling Location	pH	EC (dS m^{-1})
Center (near plant stem)	7.0 b [1]	0.84 b
Lateral (perpendicular to the stem)	7.9 a	1.80 a
Significance	**	***

[1]—Means followed by different letters within a column are significantly different. ** and *** significant at $p < 0.01$ and 0.001 levels, respectively. Mean separation was performed using Duncan's multiple-range test. Means are based on five replicates. EC = electrical conductivity.

The pH and EC of the substrate, measured using the saturation extract method (1:5 v/v) after lettuce harvest, were not significantly affected by the interaction between treatments, mix composition, or the EC of the nutrient solution (Table 2). However, both pH and EC were significantly influenced by the sampling location ($p < 0.001$) (Table 2). pH and EC were higher on the lateral side perpendicular to the plant stem than on the center, near the stem (Table 2).

2.4. Leaf Nutrient Concentration

Leaf concentrations of N, P, K, Ca, and Mg were not influenced by the interaction of substrate and nutrient solution treatments (Table 3).

Table 3. Effects of cultivation substrate mix and nutrient solution EC on leaf nutrient content.

Treatments	Leaf Macronutrients (%)					Leaf Micronutrients (μg·g^{-1})				
	N	P	K	Ca	Mg	Fe	B	Mn	Zn	Na [3]
Mix (M)										
C+ Comp + P [1]	4.0 b [2]	0.75 b	6.29 b	0.85 a	0.29 a	90.9 b	34.6 b	55.7 a	67.3 a	0.36 a
C+ Comp + PB	4.1 b	0.67 b	6.33 b	0.81 a	0.28 a	106.6 a	43.1 a	52.5 a	65.2 a	0.38 a
EC (dS m^{-1})										
1.5	3.6 c	0.65 b	6.01 b	0.85 a	0.28 a	99.8 b	35.9 b	52.5 a	63.7 a	0.39 a
2.5	4.5 a	0.78 a	6.60 a	0.81 a	0.29 a	97.8 b	41.8 a	55.8 a	68.7 a	0.35 a
Significance										
M	NS	NS	NS	NS	NS	*	*	NS	NS	NS
EC	***	**	**	NS	NS	NS	*	NS	NS	NS
M x EC	NS	NS	*	NS	NS	NS	NS	NS	NS	NS

[1]—C—coir, Comp—compost, P—perlite. PB—pine bark; [2]—means followed by different letters within a column are significantly different. *, **, and *** significant at $p < 0.05$, 0.01, and 0.001 levels, respectively, NS—not significant. Mean separation was performed using Duncan's multiple-range test. [3]—Even though sodium is not considered a micronutrient, it has been included here for convenience.

The replacement of perlite by pine bark in the substrate mix did not affect leaf macronutrient concentrations. In contrast, leaf N, K, and P content increased with the rise in the EC of the nutrient solution. The application of a nutrient solution with 2.5 dS m^{-1} increased leaf N content by 25% relative to plants grown with a nutrient solution of 1.5 dS m^{-1}. The rise in leaf N content (0.9%) was greater than the increases in K (0.59%) and P (0.13%) (Table 3). Leaf Ca and Mg were not affected by the EC of the nutrient solution ($p < 0.05$) (Table 3).

Leaf micronutrient and Na content were not affected by the interaction of treatments (Table 3). Plants grown with pine bark had higher leaf Fe and B than those grown in perlite (Table 4). The nutrient solution EC only affected leaf B content, which increased with the rise in EC.

Table 4. Effects of cultivation substrate mix and nutrient solution EC on leaf dry weight, leaf number, area, and head fresh yield.

Treatments	Leaf Dry Weight (g/Plant)	Leaves (n°/Plant)	Leaf Area (cm^2/Plant)	Head Fresh Yield ($kg{\cdot}m^{-2}$)
Mix (M)				
C [1] + Comp + P	24.5 a [2]	37.2 a	8854.9 a	9.5 a
C + Comp + PB	23.8 a	37.4 a	8486.9 a	9.8 a
EC (dS m^{-1})				
1.5	23.9 a	36.8 a	8806.4 a	9.7 a
2.5	24.0 a	37.8 a	8534.8 a	9.6 a
Significance				
M	NS	NS	NS	NS
EC	NS	NS	NS	NS
M x EC	NS	NS	NS	NS

[1]—C—coir, Comp—compost, P—perlite, PB—pine bark; [2]—means followed by different letters within a column are significantly different. NS = not significant.

2.5. Photosynthetic Pigments

Leaf Chl a, Chl b, and Cc levels were unaffected by the interaction of treatments (Figure 4). This indicates that the effect of the EC of the nutrient solution on photosynthetic pigments has not been affected by the presence of perlite or pine bark in the mix. Chl a, b, and Cc content were not affected by the mix or the EC of the nutrient solution.

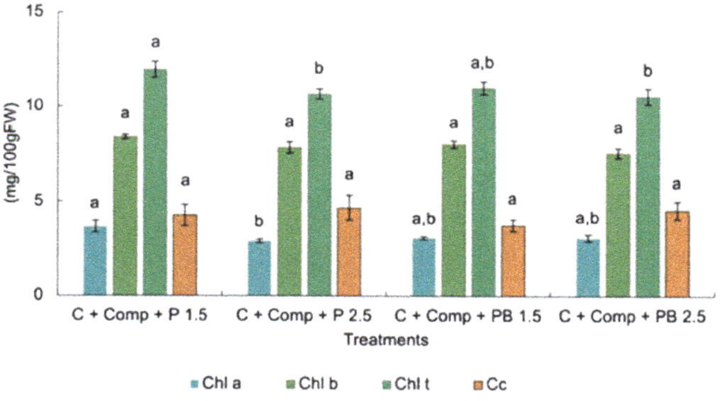

Figure 4. Effect of cultivation substrate mix and nutrient solution EC on photosynthetic pigments' content. Chl a—chlorophyll a, Chl b—chlorophyll b, Chl t—total chlorophyll, Cc—carotenoids, C—coir, Comp—compost, P—perlite. PB—pine bark. Different letters for the same pigment indicate statistically significant differences ($p < 0.05$).

2.6. Plant Growth and Yield

Leaf dry weight, leaf number, foliar area, and fresh yield (kg m^{-2}) were not affected by the interaction of treatments, nor by mix or nutrient solution EC (Table 4).

The nutrient solution EC and pine bark affected leaf nutrient content (Table 4); however, as plant growth was not affected, this may indicate that the leaf nutrient contents were within the sufficiency range for lettuce growth. The fresh yield of lettuce was high, ranging from 9.5 to 9.8 kg/m^2 (Table 4).

2.7. Phytochemical Accumulation

Leaf phytochemical accumulation, including total phenolic content (TPC), ascorbate (AsA), glutathione (GSH), proline (Pro), water-soluble protein and proline dehydrogenase (PDH) (Table 5), was not significantly affected by the interaction between treatments, nor by the substrate mix or the nutrient solution EC (Table 5). The average TPC ranged from

22.24 to 28.66 mg GAE/100 g fresh weight (FW). The average AsA content ranged between 3.10 and 3.46 mg/100 g FW.

Table 5. Effects of cultivation substrate mix and nutrient solution EC on leaf total phenols, protein, AsA—ascorbate, GSH—glutathione, Pro—proline, PDH—proline dehydrogenase activity.

Treatments	TPC (mg GAE/100 g FW)	WS-Protein [3]	AsA (mg/100 g FW)	GSH	Proline	PDH (nmol min^{-1}/mg)
Mix (M)						
C + comp + P [1]	27.69 a [2]	195.9 a	3.46 a	1.22 a	1.51 a	29.40 a
C + comp + PB	22.48 a	224.1 a	3.10 a	1.15 a	1.44 a	32.30 a
EC (dS m^{-1})						
1.5	25.43 a	214.3 a	3.35 a	1.20 a	1.44 a	28.02 a
2.5	24.66 a	205.7 a	3.20 a	1.17 a	1.51 a	27.55 a
Significance						
M	[3] NS	NS	NS	NS	NS	NS
EC	NS	NS	NS	NS	NS	NS
M x EC	NS	NS	NS	NS	NS	NS

[1]—C—coir, Comp—compost, P—perlite, PB—pine bark; [2]—means followed by different letters within a column are significantly different. [3]—WS—water-soluble; NS = not significant; Duncan's multiple range test was used to distinguish between means ($p < 0.05$).

3. Discussion

The substitution of perlite with pine bark affected the EC values of the mix, but they remained within the range considered acceptable for soilless substrates [14,26]. In contrast, the pH values of both mixes exceed the optimal range for vegetable cultivation (5.5 to 6.8) [24], which may negatively affect lettuce plant nutrition. The increase in bulk density and decrease in total porosity with pine bark are consistent with its physical nature. However, both parameters remained within the acceptable range for substrate use, as total porosity values above 85% are considered ideal [26]. The lower mass wetness observed in the pine bark mix indicates reduced water retention capacity per unit of substrate weight, possibly due to pine bark's coarser texture and lower intrinsic water-holding capacity compared to perlite [14]. Despite the pine bark affecting the mix properties, the changes were not substantial, indicating that a mix containing pine bark can be suitable for growing lettuce.

Replacing perlite with pine bark did not significantly alter leachate EC under low salinity conditions (1.5 dS m^{-1}). In both substrate mixes, the leachate EC remained close to that of the nutrient solution, indicating that plants are properly taking up the nutrients. This implies that under moderate salinity, pine bark can function comparably to perlite in supporting nutrient dynamics within the root zone. However, under an EC of nutrient solution of 2.5 dS m^{-1}, the pine bark mixes consistently exhibited higher leachate EC values compared to the mix with the perlite blend, especially in the last sampling dates. This is likely due to its higher initial EC (3.01 dS m^{-1}) than perlite (2.55 dS m^{-1}) (Table 1) and/or possibly greater retention of ions. Unlike perlite, which is inert, pine bark may contribute additional soluble salts or influence ion exchange processes within the substrate, thereby elevating EC in the leachate over time. Despite differences in leachate EC values, both mixes showed higher EC levels than the nutrient solution (2.5 dS m^{-1}), indicating a loss of nutrients.

Leachate pH was not significantly affected using pine bark compared to perlite, suggesting that the organic nature of pine bark did not affect root-zone pH in the coir-based mix. This could be due to the relatively low proportion of pine bark in the mix (8% *v*/*v*). Both substrates increased leachate pH values, which were consistently higher than those of the incoming solution. According to [27], the pH of the drained solution can be either higher or lower than that of the nutrient solution. The uptake of nitrate can contribute to the release of OH^- ions, increasing the pH [28,29]. In this study, approximately 70% of the N in the nutrient solution was supplied as NO_3^-. The pH levels of the leachate—which may represent the pH of the solution within the growing medium in the two mixes at the different sampling dates—were higher than the optimal range for lettuce growth [30], which can negatively affect plant growth. A significant decline in leachate pH was observed at the final two sampling dates under high EC of nutrient solution (2.5 dS m^{-1}). This may result from an increased release of protons from the soil exchange complex triggered by the higher concentration in the incoming nutrient solution.

Despite the average EC values of the substrate solution recorded on three sampling dates (Table 3) being too high, mainly at 17 and 29 DAT, exceeding the EC threshold for lettuce (2 dS m^{-1}) [25], the plants did not exhibit visible symptoms of salinity stress. This may be related to the location of sample collection and/or the method of EC determination. The EC could be high because the substrate solution was collected some distance away from the emitter, at the periphery of the wetting bulb, where salt tends to accumulate. However, nutrient uptake occurs primarily at the center of the bulb, where moisture levels are higher and EC is lower [31], thereby reducing the salinity impact on the plants. Furthermore, the EC of the substrate solution collected with Rhizon samplers may be higher than that obtained using the saturated substrate extract method (SME), as Rhizon samplers extract the solution without additional dilution, preserving equilibrium conditions and maintaining higher concentrations of freely dissolved ions. The EC of the growing media was affected by the determination method [32,33]. Jeong et al. (2012) [33] reported that EC values from saturated media extract solutions (SME) were approximately 43% lower than those obtained by the Rizhon method. Further research is needed to clarify how sampling location and measurement method influence EC readings and their correlation with species-specific salinity thresholds.

The observed variation in pH and EC measured using the saturation extract method within the substrate profile can be attributed to salt accumulation patterns associated with the wetting front created during drip irrigation. When drip irrigation is used, salts tend to accumulate at the edges of the wet bulb, which likely explains the higher pH and EC values observed in the lateral. The placement of emitters near the plant stem may have helped to expand the wet bulb, thus preventing salt build-up around the stem. The placement of emitters near the plant's stem prevents salt accumulation around the stem [34]. According to Ondrasek et al. [35], the location of emitters inside the pot can greatly influence the way moisture and salts are dispersed throughout the substrate. On the one hand, this study utilized two emitters, which may also contribute to the lack of difference in EC of the substrate. The use of more than one emitter contributes to the flushing of the salts [36] because the water and the salts extend more horizontally [37]. The configuration of the wet bulb formed by each emitter is determined by the physical properties of the substrate, including its texture and hydraulic conductivity, together with the emitter's rate [38]. Despite the observed spatial differences in pH and EC, the lack of significant differences in mean values across treatments may contribute to the absence of treatment effects on lettuce yield and phytochemical accumulation.

The replacement of perlite with pine bark in a coir-based substrate does not affect leaf macronutrient content, supporting the feasibility of this substitution. In contrast, the N, P, and K content increased with nutrient solution EC, likely due to the higher availability resulting from the elevated nutrient concentration in the solution and/or the lower pH of the root medium (Figure 2).

Low pH can enhance the availability of P in the substrate [39]. Plants grown with pine bark had higher leaf Fe and B than those grown in perlite (Table 3). Fe and B concentrations in blueberry leaves also increased with the addition of Douglas fir bark to the media [40]. The absence of visual symptoms of nutrient disorders, combined with unaffected yield across treatments (Table 5), indicates that leaf nutrient levels remained within the sufficiency range for lettuce growth. The average values of leaf N, P, and K content determined in this study were above the lower sufficiency thresholds reported by [41] (3.3% for N, 0.35% for P, and 2.9% for K) at the pre-harvest stage.

Photosynthetic pigment concentrations were also unaffected by the treatments, indicating that the conditions in the pine bark mix were comparable to those in the perlite-based substrate. Leaf Chl b levels were higher than Chl a, which deviates from the typical pattern observed in most plant species. However, similar findings have been reported in lettuce by [42–45], which may be attributed to environmental conditions.

The absence of significant differences in plant growth and TPC, AsA, GSH, Pro, and PHD means that pine bark can replace coir-based substrates. The lack of response to nutrient solution salinity may be related to the fact that EC levels within the substrate were similar across treatments, possibly due to greater nutrient leaching when the EC of the nutrient solution was higher. This leaching could have been influenced by the type of container used, which allowed for lateral drainage. Katsoulas and Voogt [46] also noted that the effects of salinity strongly depend on the irrigation and drainage management methods applied.

The average TPC ranged between 22.24 and 28.66 mg GAE/100 g fresh weight (FW). These values fall within the ranges reported by Petropoulos et al. (18 to 203 mg GAE/100 g FW) and [47] (18.2–65 mg GAE/100 g FW) for green leaf lettuce. In contrast, ref. [42] observed higher average leaf total phenol concentrations in lettuce of the same cultivar, ranging from 54.52 to 138.96 mg GAE/100 g FW, in Batavia-type green leaf lettuce (Godzilla') cultivated under similar cultural practices. This discrepancy could be attributed to differences in climatic conditions. Environmental factors such as temperature and radiation are known to influence the accumulation of phenolic compounds in vegetables [48]. The radiation and temperature are positively correlated with the content of phenolic acids and flavonoids in pigmented baby leaf lettuce, with levels increasing as the season progresses [49].

The novelty of this study lies in demonstrating that pine bark, a renewable and environmentally friendly by-product, has the potential to partially replace perlite in coir–compost substrates for lettuce cultivation. A key limitation, however, is that the experimental design evaluated only one substitution level (8% v/v), which restricts the assessment of dose–response effects and a full evaluation of its impact on lettuce yield and quality. Future studies should therefore test a wider range of pine bark proportions under different salinity levels, determine their effects on lettuce growth and quality, and assess whether this substitution remains suitable for crops with longer vegetative cycles.

4. Materials and Methods

4.1. Growth Conditions and Substrates

The experiment was carried out in a polycarbonate-covered greenhouse at the Herdade Experimental da Mitra (38°57′ N, 8°32′ W), University of Évora, Portugal. No supplemental lighting was used. Air temperature inside the greenhouse ranged from 8 to 27 °C, and daily solar radiation ranged from 34 to 248 W $m^{-2}{\cdot}d^{-1}$.

The experiment comprised four different treatments: two coir-based mixes + compost with pine bark or perlite and two electrical conductivities (EC) of nutrient solution (1.5 ± 0.2 and 2.5 ± 0.2 dS m^{-1}). Both mixes were composed of 80% coir pith (C) and 12% compost (Comp), expressed as volume per volume (v/v). One mix included 8% (v/v) of perlite (P), while the other included 8% (v/v) of pine bark (PB).

The following are the physical and chemical characteristics of the materials used, as provided by the manufacturer. The coir pith had a pH of 5.5 to 6.0, an EC greater than 1.5 dS m^{-1}, granulometry 0–10 mm, total porosity = 95% v/v, air = 25% v/v, and cation exchange capacity (CEC) within the range of 60–120 meq/100 g. Compost (Nutrimais, Lipor Company, Baguim do Monte, Portugal) was prepared from horticultural materials, selected food waste from restaurants and cafeterias, forest harvest residues (branches and foliage), and green waste including flowers, grasses, and pruning debris. The EC and pH (1:5 compost: distilled water, w/v) were 5.4 dS m^{-1} and 9.0, respectively. According to the manufacturer, the compost used in this study is free of pathogens. Perlite (Knauf, Dortmund, Germany) has particles from 2 to 6 mm (coarse perlite), is pH-neutral, and is chemically inert. The pine bark (Siro, Mira, Portugal), obtained as a by-product of the country's large maritime pine (*Pinus pinaster*), contained particles with a diameter of 8 to 15 mm and a pH (in $CaCl_2$ solution) of 4.5. Pine bark had (expressed as a percentage of dry weight): organic matter (99.1%), C/N ratio (278), C (55.6%), N (0.20%), P_2O_5 (0.04%), K_2O (0.11%), and Mg (0.05%).

On March 15, seedlings of lettuce (*Lactuca sativa* L. cv. Godzilla), type Batavia, 20 days after emergence, were planted into Styrofoam plant boxes along the central line, spaced 25 cm apart (16 plants/m^2). The boxes (100 cm long × 25 cm wide × 10 cm high) were filled with 14 L of each mix at an approximate height of 6.5 cm. In this plant boxes drainage occurs laterally around the entire perimeter of the boxes. Treatments were arranged in a randomized complete block design with five replicate boxes per treatment (Figure 5).

Figure 5. Appearance of the experimental setup and lettuce plants ten days before harvest.

The nutrient solution used contained 14 mmol L^{-1} NO_3-N, 6.3 mmol L^{-1} NH_4-N, 1.32 mmol L^{-1} P, 11 mmol L^{-1} K, 3.5 mmol L^{-1} Ca, 3.5 mmol L^{-1} Mg, 1.31 mmol L^{-1} S, 46 µmol L^{-1} B, 7.86 µmol L^{-1} Cu chelated by EDTA, 8.95 µmol L^{-1} Fe chelated by EDTA, 18.3 µmol $L-^{1}$ Mn chelated by EDTA, 1 µmol L^{-1} Mo, 2 µmol L^{-1} Zn chelated by EDTA, 2.1 mmol L^{-1} Cl and 0.7 mmol L^{-1} Na. The pH of the nutrient solution was 6.4 ±0.5.

The two salinity levels were obtained by varying the concentration of the standard nutrient solution through the injection rate of the nutrient solution. From transplanting to 4 days after planting (DAP), a nutrient solution with an electrical conductivity of 1.1 dS m^{-1} was applied to all plants in the different treatments. After 4 days of planting, the nutrient solution was used with an electrical conductivity (EC) of 1.5 ± 0.2 and 2.5 ± 0.2 dS m^{-1} until one day before harvest. The irrigation schedule was optimized for the coir + compost + perlite mix. It was based on the substrate's volumetric water content at the Styrofoam box control, measured using a soil moisture probe (SM105T, Delta Devices, Cambridge, UK), and the volume of water drained. The nutrient solution was applied three to eight times daily, with an average drainage (leaching fraction) of 10 to 25% for each application.

4.2. Measurements

Before planting, pH, electrical conductivity (EC), mass wetness, moisture content, total porosity, and bulk density of the mixes were measured. pH and EC were determined in the aqueous extract, which was prepared using a 1:5 substrate-to-water ratio (*w*/*v*). Moisture content, total porosity, and bulk density were measured using the porometer procedure outlined by [50], with four replications for each mix. Leachate from each planting box was collected five times during the crop cycle, and its pH and electrical conductivity (ECW) were measured using a potentiometer (pH Micro 2000, Crison Instruments, Barcelona, Spain) and a conductivity meter (LF 330 WTW, Weilheim, Germany).

The substrate solution was extracted using hydrophilic porous polymer Rhizon soil moisture samplers (Eijkelkamp Agrisearch Equipment, Giesbeek, The Netherlands), following the procedures described by [33,51], with adjustments made based on the specific conditions of this study. Three samplers were installed vertically in each Styrofoam box at a depth of 5 cm and positioned 8 cm away from the crop row, perpendicular to the plants. The samples were placed in substrate mixes and collected 24 h later. Soil solution was collected at 9, 17, and 29 days after transplanting (DAT), pooled by treatment, and analyzed for pH and electrical conductivity using a potentiometer (pH Micro 2000, Crison Instruments, Barcelona, Spain) and a conductivity meter (LF 330, WTW, Weilheim, Germany), respectively.

The harvest of lettuce heads took place on 20 April 2022, corresponding to 38 days after transplanting (DAT). The heads of the plants were cut off above the surface of the media. One head from each box was washed, oven-dried at 70 °C for 2–3 days, weighed, ground to pass through a 40-mesh sieve, and then analyzed for N, P, K, Ca, Mg, B, Fe, Mn, Zn, and Na. See [42] for a description of the procedures used for each analysis. Head samples were taken by slicing a 2 cm thick section at 6 cm above the base, ensuring that inner, middle, and outer leaves were included. The contents of photosynthetic pigments—chlorophyll a (Ch a), chlorophyll b (Ch b), and carotenoids (Cc)—were measured following the procedure outlined in Machado et al. [42].

After harvesting the lettuces, two substrate samples were collected from each box. One sample was taken near the stem, perpendicular to the lateral wall of the box, and the other 6 cm away from the perpendicular line of the plant stem. To collect the samples, a 3 cm square cut was made from the surface to the bottom of the box. The samples were homogenized, and pH and electrical conductivity were measured in an aqueous extract prepared at a 1:5 (*w*/*v*) ratio.

A 1 g portion of plant material from each of the four treatments and their replicates was weighed for the preparation of the methanolic extract (MW80). The material was ground in a mortar and homogenized in 8 mL of a methanol-water mixture (80:10, *v*/*v*) [52]. Extracts were clarified by centrifugation at 6500 *g* for 15 min at 4 °C. Aliquots were stored at −20 °C for later analysis of total phenolic compounds (TPC), ascorbate (AsA), and proline (Pro) content.

The aqueous extract in buffered solution (BE) was obtained by grinding 1 g of plant sample (from all four treatments and replicates) in a mortar with liquid nitrogen, followed by homogenization in 5 mL phosphate buffer (0.12 mM, pH 7.2). The homogenate was centrifuged at 10,000 *g* for 20 min at 4 °C. Aliquots of the supernatant were stored at −20 °C for later determination of protein and glutathione (GSH) content, as well as proline dehydrogenase (PDH) enzymatic activity [53,54].

Total phenolics in the MW80 extract were quantified following the procedure of Bouayed et al. [55]. For the reaction, the extract was combined with Folin–Ciocalteu reagent diluted 1:10 and a 7.5% sodium carbonate solution in a 1:5:4 ratio. After mixing with a vortex, the samples were left in darkness for 90 min before analysis.

Absorbance was measured at 760 nm. Phenolic content was calculated via graphical interpolation from a standard curve (GAE, 6 standards ranging from 0 to 50 mg L^{-1}) and expressed as mg gallic acid equivalents (GAE) per 100 g fresh weight (FW).

Ascorbate quantification in the MW80 extract was performed as described by Cai and Tang [56]. A reaction mixture was prepared with MW80 extract, 5% TCA, absolute ethanol, 4% phosphoric acid in ethanol, 5% β-phenanthroline in ethanol, and 0.03% ferric chloride in ethanol (1:1:1:0.5:1:0.5). After vortexing, the mixture was incubated in a water bath at 30 °C for 90 min in the dark. Absorbance was measured at 534 nm. Ascorbate content was determined by interpolation from a calibration curve (AsA, 6 standards from 0 to 30 mg L^{-1}) and expressed in mg AsA per 100 g FW.

Proline content was determined by molecular absorption spectrometry, as described by Gruda and Schnitzler [57] and Bates [58]. MW80 extract was reacted with glacial acetic acid and ninhydrin (1:1:1). After vortexing, the mixture was incubated in a 75 °C water bath for 1 h in the dark. Absorbance was measured at 546 nm. Proline content was interpolated from a calibration curve (L-proline, 6 standards from 0 to 20 mg L^{-1}) and expressed in mg Pro per 100 g FW.

Glutathione levels were quantified using the protocol by Hissin and Hilf [59], which employed o-phthalaldehyde (OPT) to develop a measurable fluorophore. Diluted BE extracts or BE buffer (blank) were mixed with phosphate buffer (0.1 M, pH 8) containing 0.005 M EDTA and OPT. After vortexing, the mixture was left to stand for 10 min. Fluorescence was read at 25 °C (excitation: 350 nm; emission: 420 nm) using a Shimadzu RF-5001PC spectrofluorometer (Shimadzu Corpotation, Kyoto, Japan). Glutathione content was calculated from a standard curve (GSH, 6 standards from 0 to 100 mM) and expressed in mg GSH per g FW.

The catalytic activity of proline dehydrogenase (PDH, EC 1.5.5.2) was measured as described by Costilow [53]. A reaction mixture was prepared with sodium carbonate buffer (100 mM, pH 10.3), L-proline (2 mM), NAD^+ (10 mM), and an appropriate amount of BE extract. NADH production was monitored by recording absorbance at 340 nm for 180 s at 30 °C using Double-beam spectrophotometer Hitachi U2001 with temperature control (Hitachi, Ltd., Tokyo, Japan). Enzymatic activity was calculated from the slope of the absorbance curve using the molar extinction coefficient of NADH (6.22 $mM^{-1}\cdot cm^{-1}$) and expressed in nmol min^{-1} mg^{-1} protein.

The water-soluble protein content in the BE extract was calculated using the Lowry method [60]. A reaction mixture was prepared using 1.0 mL of Lowry reagent, 0.16 mL of 0.5 M NaOH, and 0.04 mL of BE extract. After vortexing and a 10 min rest period, 0.1 mL of diluted (1:2) Folin–Ciocalteu reagent was added. After vortexing again, the mixture was left to rest for 30 min. Protein content was determined from a calibration curve (bovine serum albumin, BSA, 6 standards from 0 to 200 mg mL^{-1}).

All spectrophotometric measurements were performed using a Thermo Scientific Genesys 10S UV/V spectrophotometer (Thermo Fisher Scientific, Waltham, MA, USA) except for the measurement of PDH enzyme activity.

Data were analysed using analysis of variance in SPSS Statistics 29.0.1.0 (171) software (Chicago, IL, USA). Means were separated at the 5% level using Duncan's new multiple range test.

5. Conclusions

This study shows that pine bark, an eco-friendly alternative, has the potential to replace perlite—a material with environmental and economic drawbacks—in coir-based substrates for lettuce cultivation, without affecting the yield and quality of lettuce. Replacing perlite with pine bark did not influence the effect of the nutrient solution electrical conductivity on substrate pH or the electrical conductivity of the substrate solution, which in combination with the nutrient solution, did not affect photosynthetic pigment content, leaf dry weight, fresh head yield, total phenolic compounds, ascorbic acid, or proline content in the leaves. The lack of response in electrical conductivity to the nutrient solution may be related to the irrigation methodology and the type of container used. Future research is advised to validate these results in crops with longer growing periods.

Author Contributions: R.M.A.M. conceived and designed the experiments; performed the experiments; analyzed and interpreted the data; contributed reagents, materials, analysis tools, or data; and wrote the paper. I.A.-P. and R.A.F. conceived, designed, and performed the enzyme assay and other biochemical experiments; analyzed and interpreted the data; contributed reagents, materials, analysis tools, or data; and wrote the paper. G.C.D. performed the experiments, analyzed, and interpreted the data. N.S.G. reviewing and editing. All authors have read and agreed to the published version of the manuscript.

Funding: This work was funded in Portugal by the FCT—Fundação para a Ciência e a Tecnologia—as part of Project UIDB/05183/2020.

Data Availability Statement: The data supporting the findings of this study are contained in this article.

Conflicts of Interest: The authors declare no conflicts of interest.

References

1. Gruda, N.S. Increasing sustainability of growing media constituents and stand-alone substrates in soilless culture systems. *Agronomy* **2019**, *9*, 298. [CrossRef]
2. Blok, C.; Eveleens, B.; Van Winkel, A. Growing media for food and quality of life in the period 2020–2050. *Acta Hortic.* **2021**, *1305*, 369–376. [CrossRef]
3. Defra. Towards Sustainable Growing Media: A Roadmap for the Horticultural Sector. Department for Environment, Food and Rural Affairs. UK 2011. Available online: https://assets.publishing.service.gov.uk/media/5a7c9843e5274a7b7e3217eb/pb13867-towards-sustainable-growing-media.pdf (accessed on 15 March 2025).
4. Gruda, N.S. Advances in Soilless Culture and Growing Media in Today's Horticulture—An Editorial. *Agronomy* **2022**, *12*, 2773. [CrossRef]
5. Sdao, A.E.; Gruda, N.S.; De Lucia, B. Beyond Peat: Wood Fiber and Two Novel Organic Byproducts as Growing Media—A Systematic Review. *Plants* **2025**, *14*, 1945. [CrossRef]
6. Savvas, D.; Gruda, N. Application of soilless culture technologies in the modern greenhouse industry-A review. *Eur. J. Hortic. Sci.* **2018**, *83*, 280–293. [CrossRef]
7. United States Department of Agriculture (USDA). Technical Evaluation Report: Perlite Handling and Production. Agricultural Marketing Service. 2024. Available online: https://www.ams.usda.gov/sites/default/files/media/2024TechnicalReportPerliteHandling.pdf (accessed on 2 April 2025).
8. Domeño, I.; Irigoyen, N.; Muro, J. Evolution of organic matter and drainages in wood fibre and coconut fibre substrates. *Scientia Hortic.* **2009**, *122*, 269–274. [CrossRef]
9. Rodriguez, J.C.; Cantliffe, D.J.; Shaw, N.L.; Karchi, Z. Soilless media and containers for greenhouse production of 'Galia' type muskmelon. *HortScience* **2006**, *41*, 1200–1205. [CrossRef]
10. Vinci, G.; Mattia, R. Hydroponic cultivation: Life cycle assessment of substrate choice. *Brit. Food J.* **2019**, *121*, 1801–1812. [CrossRef]
11. Thompson, R.; Delcour, I.; Berckmoes, E.; Stavridou, E. The Fertigation Bible; FERTINNOWA, EU, 2018. pp. 9–23. Available online: https://projectbluearchive.blob.core.windows.net/media/Default/Imported%20Publication%20Docs/The%20Fertigation%20Bible.pdf (accessed on 25 March 2025).
12. Grillas, S.; Lucas, M.; Bardopoulou, E.; Sarafopoulos, S.; Voulgari, M. Perlite-based soilless culture systems: Current commercial aplications and prospects. *Acta Hortic.* **2001**, *548*, 105–114. [CrossRef]
13. Londra. P.A. Simultaneous determination of water retention curve and unsaturated hydraulic conductivity of substrates using a steady-state laboratory method. *HortScience* **2010**, *45*, 1106–1112. [CrossRef]
14. Martínez, P.F.; Roca, D. Sustratos para el cultivo sin suelo. Materiales, propiedades y manejo. In *Sustratos, Manejo del Clima, Automatización y Control en Sistemas de Cultivo sin Suelo*; Flórez, R.V.J., Ed.; Editorial Universidad Nacional de Colombia: Bogotá, Colombia, 2011; pp. 37–77. Available online: http://hdl.handle.net/20.500.11939/3894 (accessed on 6 March 2025).

15. Altland, J.E.; Locke, J.C.; Krause, C.R. Influence of pine bark particle size and pH on cation exchange capacity. *Hortic. Technol.* **2014**, *24*, 554–559. [CrossRef]
16. Huang, J.; Fisher, P.R.; Argo, W.R.; Jeong, K.Y.; Altland, J. Macronutrient solubility in response to the pH of soilless container substrates. *J. Soil Sci. Plant Nutr.* **2025**, *25*, 3224–3240. [CrossRef]
17. Jackson, B.E.; Wright, R.D.; Alley, M.M. Comparison of fertilizer nitrogen availability, nitrogen immobilization, substrate carbon dioxide efflux, and nutrient leaching in peat-lite, pine bark, and pine tree substrates. *HortScience* **2009**, *44*, 781–790. [CrossRef]
18. Gruda, N.S.; Hirschler, O.; Stuart, J. Peat reduction in horticulture—An overview of Europe. *Acta Hortic.* **2024**, *1391*, 545–560. [CrossRef]
19. Gruda, N.; Rau, B.; Wright, R.D. Laboratory bioassay and greenhouse evaluation of a pine tree substrate used as a container substrate. *Europ. J. Hortic. Sci.* **2009**, *74*, 73–78. [CrossRef]
20. Gruda, N.S.; Gallegos-Cedillo, V.M.; Nájera, C.; Catalina, E.G.; Ochoa, J.; Fernández, J.A. Advancing Protected Cultivation: A Pathway for Nutrient-Rich Vegetables. *Crit. Rev. Plant Sci.* **2025**, *44*, 88–116. [CrossRef]
21. Gruda, N.S.; Dong, J.; Li, X. From salinity to nutrient-rich vegetables: Strategies for quality enhancement in protected cultiva-tion. *Crit. Rev. Plant Sci.* **2024**, *43*, 327–347. [CrossRef]
22. Sgherri, C.; Peirez-Loìpez, U.; Micaelli, F.; Miranda-Apodaca, J.; Mena-Petite, A.; MunÞoz-Rueda, A.; Quartacci, M.F. Elevated CO_2 and salinity are responsible for phenolics-enrichment in two differently pigmented lettuces. *Plant Physiol. Biochem.* **2017**, *115*, 269–278. [CrossRef]
23. Santander, C.; Vidal, G.; Ruiz, A.; Vidal, C.; Cornejo, P. Salinity eustress increases the biosynthesis and accumulation of phenolic compounds that improve the functional and antioxidant quality of red lettuce. *Agronomy* **2022**, *12*, 598. [CrossRef]
24. Adams, P. Aspectos de la nutrici6n mineral en cultivos sin suelo en relaci6n al suelo. In *Tratado de Cultivo sin Suelo*; Urrestarazu, G.M., Ed.; Editorial Mundi-Prensa: Madrid, Spain, 2004; pp. 81–111.
25. Machado, R.M.A.; Serralheiro, R.P. Soil salinity: Effect on vegetable crop growth. Management practices to prevent and mitigate soil salinization. *Horticulturae* **2017**, *3*, 30. [CrossRef]
26. Bunt, B.R. *Media and Mixes for Container-Grown Plants, A Manual on the Preparation and Use of Growing Media for Pot Plants*, 2nd ed.; Springer Science & Business Media: London, UK, 2012; pp. 86–93. [CrossRef]
27. Alarcón, A.L. *Tecnología Para Cultivos de Alto Rendimiento*; Novedades: Agrícolas, Spain, 2000; p. 459.
28. Marschner, H. *Marschner's Mineral Nutrition of Higher Plants*, 2nd ed.; Academic Press: New York, NY, USA, 2012.
29. Barcelos, C.; Machado, R.M.; Alves Pereira, I.; Ferreira, R.; Bryla, D.R. Effects of substrate type on plant growth and nitrogen and nitrate concentration in spinach. *Int. J. Plant Biol.* **2016**, *7*, 44–47. [CrossRef]
30. Singh, H.; Dunn, B. *Electrical Conductivity and pH Guide for Hydroponics*; Oklahoma Cooperative Extension Service HLA 6722: Stillwater, OK, USA, 2016; pp. 1–4.
31. Silber, A.; Xu, G.; Wallach, R.; Yermiyahu, U. High fertigation frequency: The effects on uptake of nutrients, water and plant growth. *Plant Soil* **2003**, *253*, 467–477. [CrossRef]
32. Scoggings, H.L.; Van Iersel, M.W. In situ probes for measurement of electrical conductivity of soilless substrates: Effects of temperature and substrate moisture content. *HortScience* **2006**, *41*, 210–214. [CrossRef]
33. Jeong, K.; Nelson, P.V.; Frantz, J.M. Comparison of rhizon soil moisture sampler pour-through and saturated media extract extractions of container root substrate. *Acta Hortic.* **2012**, *927*, 981–987. [CrossRef]
34. Sonneveld, C.; Voogt, W. *Plant Nutrition of Greenhouse Crops*; Springer: Dordrecht, The Netherlands, 2009.
35. Ondrasek, G.; Romic, D.; Romic, M.; Tomic, F.; Mustac, I. Salt distribution in peat substrate grown with melon (*Cucumis melo* L.). *Acta Hortic.* **2008**, *779*, 307–312. [CrossRef]
36. Valdés, R.; Miralles, J.; Ochoa, J.; Banón, S.; Sánchez-Blanco, M.J. The number of emitters alters salt distribution and root growth in potted gerbera. *HortScience* **2014**, *49*, 160–165. [CrossRef]
37. De Rijck, G.; Schrevens, E. Distribution of nutrients and water in rockwool slabs. *Sci. Hortic.* **1998**, *72*, 277–285. [CrossRef]
38. Awulachew, S.B.; Lemperiere, P.; Tulu, T. Training manual on agricultural water management. In *Module 5: Irrigation Methods*; ILRI Publications Unit: Addis Abeba, Ethiopia, 2009.
39. Altland, J.E.; Buamscha, M.G. Nutrient availability from Douglas fir bark in response to substrate pH. *HortScience* **2008**, *43*, 478–483. [CrossRef]
40. Kingston, P.H.; Scagel, C.F.; Bryla, D.R.; Strik, B. Suitability of sphagnum moss, coir, and Douglas fir bark as soilless substrates for container production of highbush blueberry. *HortScience* **2017**, *52*, 1692–1699. [CrossRef]
41. Hartz, T.K.; Johnstone, P.R.; Williams, E.; Smith, R.F. Establishing lettuce leaf nutrient optimum ranges through DRIS analysis. *HortScience* **2007**, *42*, 143–146. [CrossRef]
42. Machado, R.M.; Alves-Pereira, I.; Alves, I.; Ferreira, R.M.; Gruda, N.S. Reusing coir-based substrates for lettuce growth, Nutrient content and phytonutrients accumulation. *Horticulturae* **2023**, *9*, 1080. [CrossRef]
43. Ozgen, S.; Sekerci, S. Effect of leaf position on the distribution of phytochemicals and antioxidant capacity among green and red lettuce cultivars. *Span. J. Agric. Res.* **2011**, *9*, 801–809. [CrossRef]

44. Kleinhenz, M.D.; Gazula, A.; Scheerens, J.C.; French, D.G. Variety, shading, and growth stage effects on pigment concentrations in lettuce grown under contrasting temperature regimens. *HortTechnology* **2003**, *13*, 677–683. [CrossRef]
45. Sapkota, S.; Sapkota, S.; Liu, Z. Effects of nutrient composition and lettuce cultivar on crop production in hydroponic culture. *Horticulturae* **2019**, *5*, 72. [CrossRef]
46. Katsoulas, N.; Voogt, W. Recent trends in salinity control for soilless growing systems management. *Acta Hortic.* **2014**, *1034*, 433–442. [CrossRef]
47. Llorach, R.; Martínez-Sánchez, A.; Tomás-Barberán, F.A.; Gil, M.I.; Ferreres, F. Characterisation of polyphenols and antioxidant properties of five lettuce varieties and escarole. *Food Chem.* **2008**, *108*, 1028–1038. [CrossRef]
48. Gruda, N.S.; Samuolienė, G.; Dong, J.; Li, X. Environmental conditions and nutritional quality of vegetables in protected cultivation. *Compr. Rev. Food Sci. Food Saf.* **2025**, *24*, e70139. [CrossRef]
49. Marin, A.; Ferreres, F.; Barberá, G.G.; Gil, M.I. Weather variability influences color and phenolic content of pigmented baby leaf lettuces throughout the season. *J. Agric. Food Chem.* **2015**, *63*, 1673–168181. [CrossRef]
50. Fonteno, W.C.; Harden, C.T. Procedures for determining physical properties of horticultural substrates using the NCSU porometer. In *Horticultural Substrates Laboratory*; North Carolina State University: Raleigh, NC, USA, 2003.
51. Spangenberg, A.; Gecchini, G.; Lamersdorf, N. Analyzing the performance of a micro soil solution sampling device in a laboratory examination and a field experiment. *Plant Soil* **1997**, *196*, 59–70. [CrossRef]
52. Lichtenthaler, H.K.; Buschmann, C. Chlorophylls and carotenoids: Measurement and characterization by UV-VIS spectroscopy. *Curr. Protoc. Food Anal. Chem.* **1987**, *4*, 3.1–3.8. [CrossRef]
53. Costilow, R.N.; Cooper, D. Identity of proline dehydrogenase and D1-pyrroline-5-carboxylic acid reductase in *Clostridium sporogenes*. *J. Bacteriol.* **1978**, *134*, 139–146. [CrossRef]
54. Lake, B.G. Preparation and characterization of microsomal fractions for studies on xenobiotic metabolism. In *Biochemical Toxicology: A Practical Approach*, 1st ed.; Snell, K., Mullock, B., Eds.; IRL Press: Oxford, UK, 1987; pp. 183–215.
55. Bouayed, J.; Hoffmann, L.; Bohn, T. Total phenolics, flavonoids, anthocyanins and antioxidant activity following simulated gastro-intestinal digestion and dialysis of apple varieties: Bioaccessibility and potential uptake. *Food Chem.* **2011**, *128*, 14–21. [CrossRef] [PubMed]
56. Cai, W.M.; Tang, Z.C. Plant tolerance physiology. In *Experimental Guide for Modern Plant Physiology*, 1st ed.; Tang, Z.C., Ed.; Science Press: Beijing, China, 1999; pp. 315–316.
57. Gruda, N.; Schnitzler, W.H. The effect of the water supply of seedlings cultivated in different substrates and raising systems on the bio-morphological and plant-physiological parameters of lettuce. (In German: Einfluss der Wasserversorgung von Jungpflanzen angezogen in verschiedenen Substraten und Anzuchtsystemen auf biomorphologische und physiologische Merkmale von Kopfsalat). *J. Appl. Bot.* **2000**, *74*, 240–247.
58. Bates, L.S. Rapid determination of free proline for water stress studies. *Plant Soil.* **1973**, *39*, 205–207. [CrossRef]
59. Hissin, A.; Hilf, P.A. Fluorometric method for determination of oxidized and reduced glutathione in tissues. *Anal. Biochem.* **1976**, *74*, 214–226. [CrossRef] [PubMed]
60. Lowry, O.H.; Rosebroug, N.J.; Farr, A.L.; Randell, R.J. Protein measurement with the folin phenol reagent. *J. Biol. Chem.* **1951**, *193*, 265–275. [CrossRef] [PubMed]

plants

MDPI

Article

Effects of Seedling Substrate and Hydroponic Versus Aquaponic Nutrient Solution on Growth, Nutrient Uptake, and Eco-Physiological Response of Lemon Basil (*Ocimum × citriodorum*)

Linda Signorini [1], Giuseppe Carlo Modarelli [2,*], Prospero Di Pierro [1,3], Antonio Luca Langellotti [3], Chiara Cirillo [1,*], Stefania De Pascale [1] and Paolo Masi [1,3]

1 Department of Agricultural Sciences, University of Naples Federico II, Piazza Carlo di Borbone 1, 80055 Portici, Italy; linda.signorini@unina.it (L.S.); prospero.dipierro@unina.it (P.D.P.); depascal@unina.it (S.D.P.); pmasi@unina.it (P.M.)

2 Leibniz Institute of Vegetables and Ornamental Crops (IGZ), 14979 Großbeeren, Germany

3 Center for Innovation and Development of the Food Industry (CAISIAL), University of Naples Federico II, Via Università 100, 80055 Portici, Italy; langello@unina.it

* Correspondence: modarelli@igzev.de (G.C.M.); chiara.cirillo@unina.it (C.C.)

Abstract

Lemon basil (*Ocimum × citriodorum*) is a highly valued aromatic plant renowned for its distinct citrus aroma. This study aimed to evaluate sustainable substrates and cultivation systems for its production. Two complementary and sequential experiments were conducted: an initial experiment designed to compare coconut fiber mixed in varying proportions with perlite to rock wool, evaluating their effectiveness during germination and early growth (experiment 1), and a subsequent experiment aimed at assessing plant performance in a decoupled aquaponic system relative to hydroponics utilizing the best-performing coconut fiber-perlite mixture from the first phase along with rock wool as substrates (experiment 2). The substrate with 70% coconut fiber and 30% perlite (F70:P30) significantly improved seed germination, leaf number, and total leaf area of seedlings. The decoupled aquaponic cultivation system resulted in a 52.5% increase in flavonoid content, accompanied by higher calcium and magnesium uptake in stems and roots compared to hydroponics. These findings clearly underscore the potential of coconut fiber substrates mixed with perlite as sustainable alternatives to rock wool, reducing environmental impact, disposal costs, and health risks. Similarly, aquaponic cultivation emerges as a valuable strategy for sustainable lemon basil (*Ocimum × citriodorum*) production, offering comparable yields to hydroponics while improving plant nutritional and phytochemical quality through beneficial plant-microbe interactions. These results provide practical evidence supporting the adoption of environmentally friendly substrates and cultivation practices, thus contributing significantly toward sustainable intensive vegetable production systems.

Keywords: lemon basil (*Ocimum × citriodorum*); soilless cultivation; aquaponics; flavonoids; coconut fiber

1. Introduction

Basil (*Ocimum basilicum* L.) is a highly appreciated aromatic herb worldwide, renowned for its culinary versatility and distinctive fragrance [1]. Among its varieties, lemon basil (*Ocimum × citriodorum*), a hybrid between basil (*Ocimum basilicum*) and African basil

(*Ocimum americanum*), has recently gained increasing global popularity due to its unique, fresh citrus aroma which enhances various dishes, particularly seafood recipes, by adding freshness and vibrancy [1,2].

Aromatic herbs and leafy vegetables are widely cultivated worldwide. In Italy, one of the main production areas is the "Piana del Sele" in Campania, an intensive agricultural area spanning around 250 km^2 that has become a major hub for greenhouse vegetable production, especially fresh, ready-to-eat ("fourth range") leafy vegetables exported throughout Europe [3]. Over the past two decades, the rapid expansion and industrialization of agriculture in this area have led to significant environmental challenges, including soil degradation, salinization, soil compaction, coastal erosion, and the accumulation of potentially toxic substances due to the intensive use of agrochemicals [4,5].

Recently, many farms in the area have been confronted with increased contamination risks, particularly related to heavy metals such as cadmium and lead accumulating in leafy vegetables, posing threats to food safety, human health, and economic sustainability [6].

To address these challenges, soilless cultivation systems, such as hydroponics or aquaponics, have emerged as a promising alternative to traditional soil cultivation.

These methods can mitigate environmental impacts by significantly reducing agrochemical use, shortening crop cycles, minimizing water and nutrient losses, and avoiding soil-related contaminant issues [7,8]. Hydroponics allows precise control of nutrient supply tailored to plant needs; however, it still heavily relies on chemical fertilizers, leading to environmental concerns and associated economic costs [8,9]. In contrast, aquaponics integrates aquaculture and plant cultivation within a closed-loop system, using nutrient-rich fish wastewater as fertilizer for plant growth. Beneficial microorganisms convert fish waste into bioavailable nutrients, creating a synergistic and sustainable nutrient cycle. This reduces chemical fertilizer inputs, optimizes resource use efficiency, and minimizes environmental pollution [9,10].

Another significant sustainability issue in greenhouse vegetable production involves substrate selection. Generally, rock wool is still the most widespread substrate used in soilless cultivation despite its environmental drawbacks, high disposal costs, and potential toxicity risks during handling [11]. Therefore, identifying and validating alternative substrates with lower environmental impact, safer handling, and similar or improved agronomic performance is critical for achieving greater sustainability in greenhouse vegetable production. Furthermore, substrate composition during nursery-level production influences early plant growth and root system development, thereby affecting crop productivity in subsequent cultivation stages [8].

Given these pressing environmental and agronomic challenges, this study aimed to (experiment 1) evaluate coconut fiber mixed with different proportions of perlite during seed germination and early growth as an environmentally sustainable alternative to rock wool and subsequently, (experiment 2) assess eco-physiological responses and nutrient uptake of lemon basil cultivated in hydroponic versus decoupled aquaponic systems, utilizing the best-performing coconut fiber-perlite mixture identified in the first phase along with rock wool as substrates.

By implementing soilless cultivation systems that rely on environmentally sustainable substrates and nutrient solutions, this research aims to developing more sustainable techniques for intensive vegetable production, thereby reducing synthetic fertilizer use and substrate-related environmental impacts.

2. Results

2.1. Experiment 1

2.1.1. Seed Germination and Seedling Growth

Basil seed germination percentage was significantly higher in the treatment F70:P30 and the control (RW) (Table 1). Germination in the F70:P30% substrate was 26.1% points higher than in the F50:P50 treatment and 20.8% points higher than in the F90:P10 treatment.

Table 1. Percentage germination of seeds, total shoot Dry Matter (DM), total root Dry Matter (DM), total plant height, total root length of seedlings of Lemon basil (*Ocimum* × *citriodorum*) cultivated on different substrates under controlled growth chamber conditions. The substrates compared were: 50% coconut fiber mixed with 50% perlite (F50:P50), 70% coconut fiber mixed with 30% perlite (F70:P30), 90% coconut fiber mixed with 10% perlite (F90:P10), and 100% rock wool (RW; control).

Treatment	Germination	Shoot DM	Root DM	Plant Height	Root Length
Substrate		%		cm Plant^{-1}	
F50:P50	63.89 ± 3.67 b	8.41 ± 0.32 b	4.43 ± 0.08 b	7.12 ± 0.28 a	13.67 ± 0.91
F70:P30	80.56 ± 5.55 a	8.86 ± 0.23 b	4.30 ± 0.1 b	7.47 ± 0.31 a	14.71 ± 1.51
F90:P10	66.67 ± 2.41 b	9.79 ± 0.15 ab	4.75 ± 0.05 b	7.13 ± 0.35 a	13.08 ± 0.9
RW 100%	81.94 ± 5.55 a	11.09 ± 0.46 a	6.57 ± 0.73 a	5.83 ± 0.06 b	16.17 ± 1.18
Significance	0.04 *	0.00 **	0.01 **	0.01 *	0.32 ^{ns}

Data represent means ± standard error (n = 12). Different letters indicate significant differences among treatments according to Tukey's HSD post hoc test. Significance levels: ns = not significant; * $p < 0.05$, ** $p < 0.01$, *** $p < 0.001$.

The number of leaves per seedling was also significantly higher in all three treatments compared to the RW (Table 2). The F70:P30 treatment had the highest number of leaves, exceeding the control by 26.9%, followed by the F50:P50% and F90:P10 treatments, with an average leaf number of 22.9% higher than the control.

Table 2. Total shoot Fresh Weight (FW), total shoot Dry Weight (DW), total leaf number, and total leaf area of Lemon basil (*Ocimum* × *citriodorum*) seedlings cultivated on different substrates under controlled growth chamber conditions. The substrates compared were: 50% coconut fiber mixed with 50% perlite (F50:P50), 70% coconut fiber mixed with 30% perlite (F70:P30), 90% coconut fiber mixed with 10% perlite (F90:P10), and 100% rock wool (RW; control).

Treatment	Shoot FW	Shoot DW	Leaf Number	Total Leaf Area
Substrate	g Plant^{-1}		n Plant^{-1}	cm^2 Plant^{-1}
F50:P50	1.09 ± 0.08 a	0.10 ± 0.01	7.58 ± 0.22 a	24.78 ± 2.10 a
F70:P30	1.13 ± 0.07 a	0.10 ± 0.01	7.83 ± 0.2 a	24.55 ± 0.81 a
F90:P10	0.97 ± 0.05 ab	0.10 ± 0.01	7.58 ± 0.17 a	20.89 ± 0.66 ab
RW 100%	0.78 ± 0.02 b	0.09 ± 0.00	6.17 ± 0.17 b	16.98 ± 0.39 b
Significance	0.01 *	0.50 ^{ns}	0.00 **	0.01 *

Data represent means ± standard error (n = 12). Different letters indicate significant differences among treatments according to Tukey's HSD test. Significance levels: ns = not significant; * $p < 0.05$, ** $p < 0.01$, *** $p < 0.001$.

Plant height was also greater in all three treatments, averaging 7.24 cm, which was 24.2% taller than the control.

The total leaf area showed significant differences among the treatments. In particular, compared to the control, it was 44.6% greater in the F70:P30 treatment, 45.9% greater in the F50:P50 treatment, and 23% greater in the F90-P10 treatment.

The dry matter percentage in stems was significantly higher in the control treatment (RW), showing an increase of 22.9% compared to the substrate mix treatments. Similarly, the dry matter percentage in roots was higher in the RW, with an increment of 46.0% compared to the substrate mix treatments.

2.1.2. Cluster Heatmap Analysis

From the observation of the heatmap (x-axis, Figure 1), the clustering of the treatments reveals that RW formed a distinct cluster. Plants grown with RW exhibited a higher germination percentage, as well as increased dry matter content in both shoots and roots. In contrast, the Coconut-Perlite treatment formed a separate cluster, with the F70:P30 treatment showing, in order of proximity, the highest germination percentage compared to the other two treatments, along with longer roots and greater dry weight. Additionally, the F50:P50 and F90:P10 treatments generated a subcluster primarily characterized by lower germination activity and reduced root development.

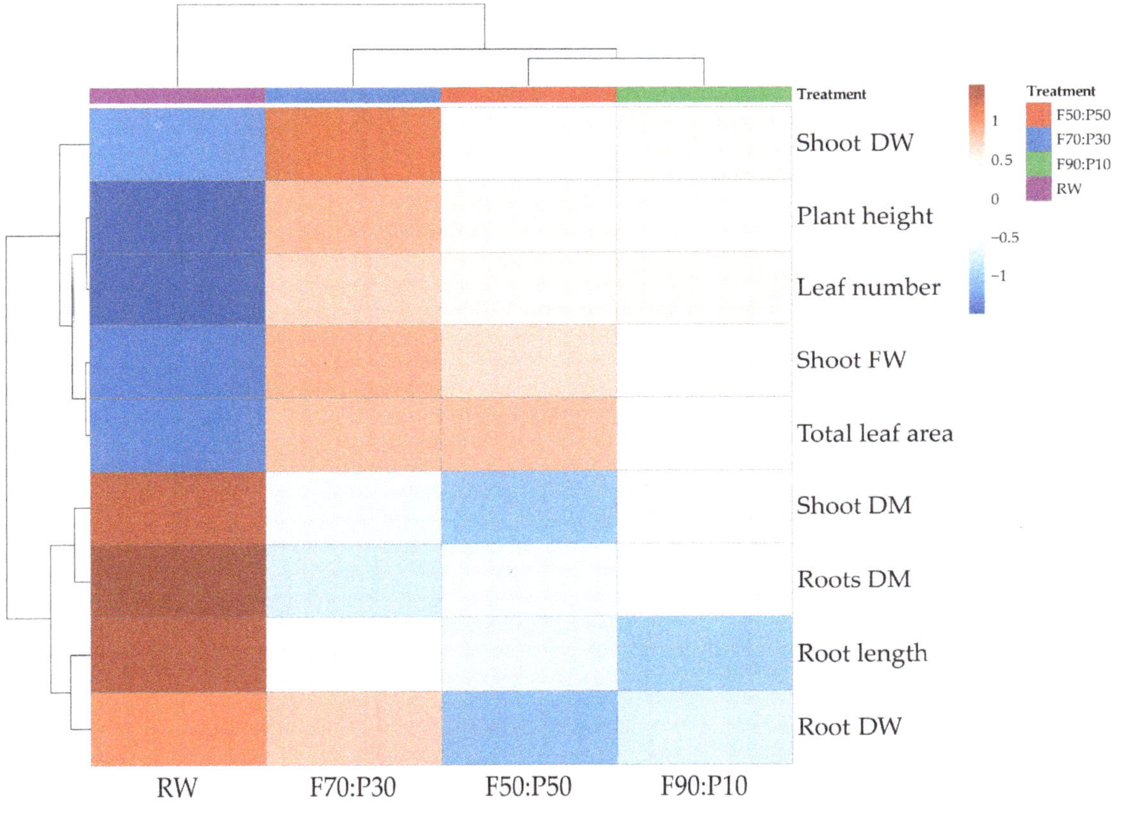

Figure 1. Cluster heatmap analysis displaying the relationships among growth parameters of lemon basil (*Ocimum × citriodorum*) seedlings grown on four different substrates: coconut fiber mixed with perlite at three ratios (F50:P50, F70:P30, F90:P10) and 100% rock wool (RW), in controlled growth chamber conditions. Each row represents a specific growth parameter (e.g., germination percentage, leaf number, root length), while each column represents an individual substrate treatment. Colors represent standardized values (ln(x + 1) transformed): higher values in red indicate above-average performance, whereas lower values in blue represent below-average performance. Hierarchical clustering was performed using Euclidean distances and complete linkage to group similar treatments and parameters together.

2.2. Experiment 2

2.2.1. Plant Growth

Basil plants exhibited similar leaf numbers, total leaf area, and total fresh biomass across all substrate treatments and cultivation systems (Table 3). The average leaf fresh weight was 50.8 g, and the mean number of leaves per plant was 129.6 n plant^{-1}. The dry weight of the roots was 22.8% higher in plants grown on RW than in those grown on F. Compared to the hydroponic system, the percentage of dry matter in the leaves was 10% higher in the aquaponic cultivation system. Furthermore, compared to hydroponic cultivation, plants grown in aquaponic systems exhibited a 19% reduction in root length (Table 4).

Table 3. Total leaf number, total leaf area, total Fresh Weight (FW), total Dry Weight (DW), total root Fresh Weight (FW) of Lemon basil (*Ocimum* × *citriodorum*) plants cultivated on two different substrates 70% coconut fiber mixed with 30% perlite (F) and 100% rock wool (RW) and in two different cultivation systems: aquaponics (A) and hydroponics (H).

Treatments		Leaf Number	Leaf FW	Leaf DW	Root FW	Total Leaf Area
Cultivation System (CS)	Substrate (S)	(n Plant^{-1})		(g Plant^{-1})		(cm^2 Plant^{-1})
A	F	127.55 ± 7.29	40.56 ± 4.36	3.81 ± 0.38	26.33 ± 2.79	1098.02 ± 78.09
	RW	143.77 ± 13.66	53.22 ± 4.86	5.02 ± 0.49	33.56 ± 2.94	1307.38 ± 103.30
Mean		135.67	46.89	4.42	29.94	1202.70
H	F	125.44 ± 12.10	53.67 ± 8.06	4.64 ± 0.76	31.11 ± 3.66	1060.47 ± 135.46
	RW	121.77 ± 6.59	55.78 ± 5.21	4.85 ± 0.48	34.67 ± 3.57	1147.14 ± 92.41
Mean		123.61	54.72	4.75	32.89	1103.81
Significance						
Cultivation system (CS)		0.25 ^{ns}	0.18 ^{ns}	0.56 ^{ns}	0.37 ^{ns}	0.35 ^{ns}
Substrates (S)		0.55 ^{ns}	0.21 ^{ns}	0.20 ^{ns}	0.11 ^{ns}	0.17 ^{ns}
CS × S		0.46 ^{ns}	0.25 ^{ns}	0.42 ^{ns}	0.29 ^{ns}	0.37 ^{ns}

Data represent means ± standard error (n = 9). Different letters indicate significant differences among treatments according to Tukey's HSD test. Significance levels: ns = not significant; * $p < 0.05$, ** $p < 0.01$, *** $p < 0.001$.

Table 4. Root dry weight (DW), leaf dry matter (DM) percentage, stem length, and root length of Lemon basil (*Ocimum* × *citriodorum*) plants cultivated on two different substrates: 70% coconut fiber mixed with 30% perlite (F) and 100% rock wool (RW) and in two different cultivation systems: aquaponics (A) and hydroponics (H).

Treatments		Root DW	Leaf DM	Stem Length	Root Length
Cultivation System (CS)	Substrate (S)	(g Plant^{-1})	(%)		(cm Plant^{-1})
A	F	1.71 ± 0.14	9.49 ± 0.22 a	34.36 ± 2.57	13.704 ± 2.16
	RW	2.31 ± 0.18	9.40 ± 0.28 a	31.73 ± 2.45	15.46 ± 0.88
Mean		2.01	9.45	33.05	14.58
H	F	2.06 ± 0.18	8.51 ± 0.31 b	30.80 ± 2.08	17.42 ± 1.68
	RW	2.32 ± 0.19	8.68 ± 0.25 b	32.93 ± 1.62	18.57 ± 1.33
Mean		2.19	8.59	31.87	18.00
Significance					
Cultivation system (CS)		0.31 ^{ns}	0.03 *	0.59 ^{ns}	0.04 *
Substrates (S)		0.02 *	0.88 ^{ns}	0.91 ^{ns}	0.37 ^{ns}
CS × S		0.06 ^{ns}	0.03 *	0.69 ^{ns}	0.16 ^{ns}

Data represent means ± standard error (n = 9). Different letters indicate significant differences among treatments according to Tukey's HSD test. Significance levels: ns = not significant; * $p < 0.05$, ** $p < 0.01$, *** $p < 0.001$.

2.2.2. Gas Exchanges and Chl a Fluorescence Emission

Gas exchanges and *chlorophyll* fluorescence measurement showed no significant differences between the cultivation system and the growth on different substrates (Table 5). The mean net photosynthesis rate (A) was 8.51 μmol CO_2 m^{-2} s^{-1} in hydroponic and 8.97 μmol CO_2 m^{-2} s^{-1} in aquaponics systems. The maximum photochemical efficiency of the PSII (F_v/F_m) was 0.83 in the hydroponic system and 0.84 μmol m^{-2} s^{-1} in the aquaponics system. The actual yield of PSII (Φ_{PSII}) and the non-photochemical quenching (NPQ) had mean values of 0.54 and 1.31, respectively, across all treatments. Similarly, the intrinsic water use efficiency (WUE_i) showed no significant variation between treatments, with a mean value of 105.1 and 117.6 μmol CO_2 m^{-2} s^{-1}/mol H_2O m^{-2} s^{-1} in hydroponic and aquaponic systems, respectively.

Table 5. Gas exchange and chlorophyll fluorescence parameters of lemon basil (*Ocimum × citriodorum*) plants grown on two different substrates: 70% coconut fiber mixed with 30% perlite (F) and 100% rock wool (RW) and in two different cultivation systems hydroponic (H) or aquaponic (A). Parameters measured include leaf net photosynthesis rate (A), intrinsic water use efficiency (WUE_i), quantum yield of photosystem II (Φ_{PSII}), maximal photochemical efficiency of photosystem II (F_v/F_m), and non-photochemical quenching (NPQ).

Treatments		**A**	**WUE_i**	**Φ_{PSII}**	**F_v/F_m**	**NPQ**
Cultivation System (CS)	**Substrate (S)**	**(μmol CO_2 m^{-2} s^{-1})**	**(μmol CO_2 m^{-2} s^{-1}/ mol H_2O m^{-2} s^{-1})**			
A	F	9.27 ± 0.89	107.67 ± 10.97	0.62 ± 0.04	0.84 ± 0.00	1.07 ± 0.09
	RW	8.67 ± 0.86	127.53 ± 14.44	0.49 ± 0.07	0.83 ± 0.01	1.48 ± 0.21
Mean		8.97	117.6	0.55	0.84	1.28
H	F	8.24 ± 0.48	112.56 ± 8.55	0.49 ± 0.06	0.83 ± 0.01	1.31 ± 0.14
	RW	8.78 ± 1.02	97.56 ± 12.05	0.57 ± 0.05	0.83 ± 0.01	1.36 ± 0.12
Mean		8.51	105.06	0.53	0.83	1.34
Significance						
Cultivation system (CS)		0.58 ^{ns}	0.30 ^{ns}	0.72 ^{ns}	0.82 ^{ns}	0.68 ^{ns}
Substrates (S)		0.97 ^{ns}	0.84 ^{ns}	0.67 ^{ns}	0.87 ^{ns}	0.14 ^{ns}
CS × S		0.50 ^{ns}	0.15 ^{ns}	0.07 ^{ns}	0.96 ^{ns}	0.24 ^{ns}

Data represent means ± standard error (n = 9). Different letters indicate significant differences among treatments according to Tukey's HSD test. Significance levels: ns = not significant; * $p < 0.05$, ** $p < 0.01$, *** $p < 0.001$.

2.2.3. Leaf Photosynthetic Pigments

The flavonoid index (FlvM) exhibited a significant difference among the cultivation systems, with a 52.5% increase observed in the aquaponic system compared to the hydroponic system (Table 6). In contrast, no significant differences were detected in the chlorophyll content index (CCla), which exhibited a mean index value of 23.8 for hydroponic and 21.6 for aquaponics systems.

2.2.4. Mineral Content (Leaf, Stem, and Root)

The cultivation system and substrate did not influence nitrate (NO_3) concentrations and mineral uptake in the leaves (Table 7). For the stems, compared to hydroponic cultivation, the aquaponic system resulted in a 21.9% higher calcium uptake. Similarly, with hydroponic cultivation, the magnesium uptake in the stems of plants grown in the aquaponic system was 37.5% higher. Regarding the interaction between the cultivation system and substrate, compared to both hydroponic cultivation and RW, the highest magnesium uptake in the stems was recorded in plants grown in the aquaponic system on coconut fiber substrate (AF) (Table 8). For the roots, compared to RW, calcium uptake was 20.8% higher in plants grown on F. Additionally, compared to the hydroponic system, magnesium uptake in the roots was 37.9% higher in plants grown in the aquaponic system (Table 9).

Table 6. Flavonoid index (FlvM) and chlorophyll content index (CCla) of lemon basil (*Ocimum × citriodorum*) plants grown on two different substrates: 70% coconut fiber mixed with 30% perlite (F) and 100% rock wool (RW) and in two different cultivation systems hydroponic (H) or aquaponic (A).

Treatments		FlvM	CCla
Cultivation System (CS)	**Substrate (S)**	**Index**	
A	F	0.92 ± 0.08	21.54 ± 1.33
	RW	0.93 ± 0.03	21.64 ± 2.23
Mean		0.93	21.60
H	F	0.66 ± 0.09	22.89 ± 2.00
	RW	0.56 ± 0.03	24.6 ± 200
Mean		0.61	23.75
Significance			
Cultivation system (CS)		0.00 ***	0.27 ^{ns}
Substrates (S)		0.51 ^{ns}	0.64 ^{ns}
CS × S		0.4 ^{ns}	0.68 ^{ns}

Data represent means ± standard error (n = 9). Different letters indicate significant differences among treatments according to Tukey's HSD test. Significance levels: ns = not significant; * $p < 0.05$, ** $p < 0.01$, *** $p < 0.001$.

Table 7. Leaf nitrate (NO_3^-) concentrations and phosphorus (P), potassium (K), calcium (Ca), magnesium (Mg), and sulfur (S) uptake of lemon basil (*Ocimum × citriodorum*) plants grown on two different substrates: 70% coconut fiber mixed with 30% perlite (F) and 100% rock wool (RW) and in two different cultivation systems hydroponic (H) or aquaponic (A). Minerals are expressed as follows: nitrate (NO_3^-) in g kg^{-1} fresh weight (f.w.), all other nutrients as g plant^{-1} dry weight (d.w.).

Treatments		NO_3	P	K	Ca	Mg	S
Cultivation System (CS)	**Substrate (S)**	**mg kg^{-1} f.w.**			**g Plant^{-1} d.w.**		
A	F	2451.67 ± 631.40	17.67 ± 1.67	153.67 ± 13.22	63.67 ± 3.84	18.67 ± 1.45	7.33 ± 1.20
	RW	5457.67 ± 208.73	23.00 ± 2.00	181.67 ± 24.18	82.00 ± 7.00	25.00 ± 1.53	18.67 ± 1.76
Mean		3954.67	20.34	167.67	72.83	21.83	13.00
H	F	3402.00 ± 1480.88	18.33 ± 9.24	211.00 ± 48.40	59.00 ± 12.74	18.00 ± 4.04	17.33 ± 5.04
	RW	1643.33 ± 815.76	16.67 ± 6.98	224.67 ± 10.81	66.67 ± 2.85	19.33 ± 0.67	20.33 ± 14.44
Mean		2522.67	17.50	217.84	62.83	18.67	18.83
Significance							
Cultivation system (CS)		0.154 ^{ns}	0.646 ^{ns}	0.115 ^{ns}	0.228 ^{ns}	0.206 ^{ns}	0.472 ^{ns}
Substrates (S)		0.512 ^{ns}	0.765 ^{ns}	0.484 ^{ns}	0.128 ^{ns}	0.135 ^{ns}	0.380 ^{ns}
CS × S		0.08 ^{ns}	0.877 ^{ns}	0.356 ^{ns}	0.245 ^{ns}	0.201 ^{ns}	0.648 ^{ns}

Data represent means ± standard error (n = 9). Different letters indicate significant differences among treatments according to Tukey's HSD test. Significance levels: ns = not significant; * $p < 0.05$, ** $p < 0.01$, *** $p < 0.001$.

2.2.5. Cluster Heatmap Analysis

To obtain an in-depth overview of the morpho-anatomical and eco-physiological variations induced by the two cultivation systems and growth substrates, a clustered heatmap was created for all the parameters to highlight these differences better (Figure 2). As a result, the primary factor influencing clustering was the cultivation system rather than the substrates. Plants grown in the aquaponic system, particularly when combined with F substrate, demonstrated significantly increased stem length and leaf dry matter. Additionally, the aquaponic system showed a markedly higher leaf number and total leaf area, especially when associated with RW substrate. Furthermore, the fresh and dry weights of leaves and roots were notably higher in the aquaponic system when associated with RW substrate. In contrast, the hydroponic cultivation system, regardless of the substrate on which the plants were grown, yielded lower values for leaf number, total leaf area, and leaf dry matter than the aquaponic system. Nevertheless, the hydroponic system produced results comparable to those of the aquaponic system concerning both fresh and dry weights of leaves and roots, as well as root length, particularly when plants were grown on RW substrate.

Table 8. Stem nitrate (NO_3^-) concentrations and phosphorus (P), potassium (K), calcium (Ca), and magnesium (Mg) uptake of lemon basil (*Ocimum* × *citriodorum*) plants grown on two different substrates: 70% coconut fiber mixed with 30% perlite (F) and 100% rock wool (RW) and in two different cultivation systems hydroponic (H) or aquaponic (A). Minerals are expressed as follows: nitrate (NO_3^-) in g kg^{-1} fresh weight (f.w.), all other nutrients as g $plant^{-1}$ dry weight (d.w.).

Treatments		NO_3	P	K	Ca	Mg
Cultivation System (CS)	**Substrate(S)**	**mg kg^{-1} f.w.**	**g $plant^{-1}$ d.w.**			
A	F	2485.00 ± 694.26	2.00 ± 0.00	94.00 ± 3.06	7.00 ± 0.00	4.00 ± 0.00 a
	RW	5783.00 ± 3230.14	1.67 ± 0.33	102.00 ± 8.72	6.00 ± 0.58	3.33 ± 0.33 a
Mean		4134.00	1.83	98.00	6.50	3.67
H	F	3712.00 ± 452.27	1.67 ± 0.33	91.00 ± 12.12	5.33 ± 0.67	2.67 ± 0.33 b
	RW	2020.67 ± 397.76	1.67 ± 0.33	87.33 ± 3.18	5.33 ± 0.33	2.67 ± 0.33 b
Mean		2866.33	1.67	89.17	5.33	2.67
Significance						
Cultivation system (CS)		0.472 ns	0.580 ns	0.289 ns	0.038 *	0.009 **
Substrates (S)		0.645 ns	0.580 ns	0.788 ns	0.320 ns	0.282 ns
CS × S		0.44 ns	0.802 ns	0.609 ns	0.109 ns	0.032 *

Data represent means ± standard error (n = 9). Different letters indicate significant differences among treatments according to Tukey's HSD test. Significance levels: ns = not significant; * $p < 0.05$, ** $p < 0.01$, *** $p < 0.001$.

Table 9. Root nitrate (NO_3^-) concentrations and phosphorus (P), potassium (K), calcium (Ca), magnesium (Mg) uptake of lemon basil (*Ocimum* × *citriodorum*) plants grown on two different substrates: 70% coconut fiber mixed with 30% perlite (F) and 100% rock wool (RW) and in two different cultivation systems hydroponic (H) or aquaponic (A). Minerals are expressed as follows: nitrate (NO_3^-) in g kg^{-1} fresh weight (f.w.), all other nutrients as g $plant^{-1}$ dry weight (d.w.).

Treatments		NO_3	P	K	Ca	Mg
Cultivation System (CS)	**Substrate (S)**	**mg kg^{-1} f.w.**	**g $Plant^{-1}$ d.w.**			
A	F	2998.70 ± 63.19	0.58 ± 0.12	75.94 ± 6.77	4.28 ± 0.22	10.25 ± 0.63
	RW	2476.03 ± 158.60	0.68 ± 0.25	81.95 ± 10.90	3.17 ± 0.24	11.43 ± 1.69
Mean		2737.36	0.63	78.95	3.73	10.84
H	F	2356.53 ± 125.05	0.68 ± 0.10	77.77 ± 13.13	3.73 ± 0.07	7.70 ± 1.33
	RW	2491.60 ± 178.48	0.54 ± 0.18	85.25 ± 4.71	3.46 ± 0.37	8.04 ± 0.49
Mean		2424.07	0.61	81.51	3.60	7.86
Significance						
Cultivation system (CS)		0.053 ns	0.915 ns	0.793 ns	0.620 ns	0.032 *
Substrates (S)		0.199 ns	0.920 ns	0.497 ns	0.020 *	0.527 ns
CS × S		0.050 ns	0.920 ns	0.897 ns	0.066 ns	0.141 ns

Data represent means ± standard error (n = 9). Different letters indicate significant differences among treatments according to Tukey's HSD test. Significance levels: ns = not significant; * $p < 0.05$, ** $p < 0.01$, *** $p < 0.001$.

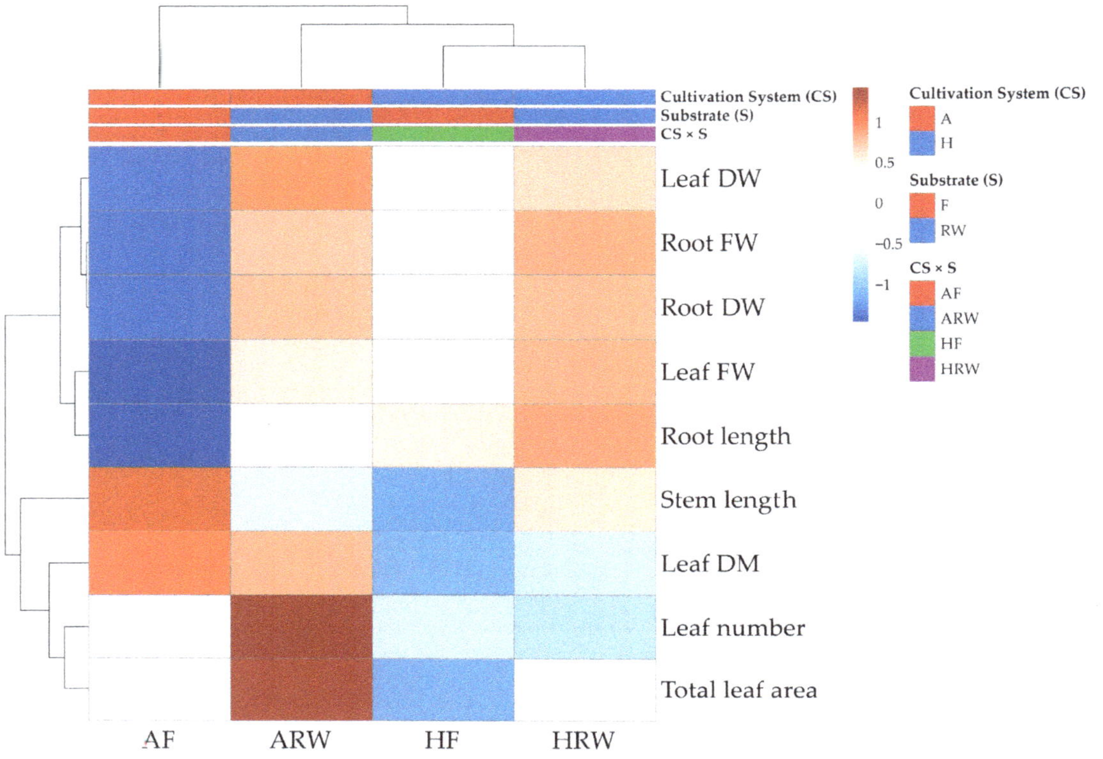

Figure 2. Cluster heatmap analysis illustrating morpho-anatomical parameters of lemon basil (*Ocimum* × *citriodorum*) plants grown on two different substrates: 70% coconut fiber mixed with 30% perlite (F) and 100% rock wool (RW) and in two different cultivation systems, hydroponic (H) or aquaponic (A). Rows represent the measured parameters (e.g., leaf number, total leaf area, root length, flavonoid index), and columns represent each combination of substrate and cultivation system. Color intensity reflects standardized values (ln(x + 1) transformed), with red indicating higher-than-average values and blue indicating lower-than-average values. Hierarchical clustering was conducted using Euclidean distances and complete linkage.

3. Discussion

This study evaluated the potential of coconut fiber substrates combined with perlite as sustainable alternatives to rock wool (experiment 1) and compared plant growth, nutrient uptake, and eco-physiological responses under hydroponic and aquaponic conditions (experiment 2) of lemon basil (*Ocimum* × *citriodorum*). Our key findings from experiment 1 revealed that a substrate mix composed of 70% coconut fiber and 30% perlite (F70:P30) promoted optimal seed germination and seedling growth compared to other tested substrates and was comparable or superior to rock wool. Furthermore, in experiment 2, while total leaf biomass did not differ significantly between hydroponic and aquaponic systems, aquaponics increased leaf dry matter and flavonoid content and improved calcium and magnesium uptake compared to hydroponics, indicating potential nutritional and phytochemical advantages. These findings align with previous research indicating the potential of coconut fiber substrates and aquaponics to enhance plant health and sustainability [12,13].

Environmental challenges linked to intensive agriculture practices in soil-grown leafy vegetables, such as heavy metal contamination, have become increasingly concerning in southern Italy, especially in the Campania region [3]. A more sustainable and safer alternative is represented by the adoption of closed-loop soilless cultivation systems. Soilless cultivation, such as hydroponics and aquaponics, represents viable solutions because it reduces plant exposure to contaminated soils, minimizes agrochemical use, and significantly improves resource-use efficiency compared to traditional soil-based cultivation practices [14,15].

Traditionally, rock wool (RW) has been widely used in commercial greenhouse production due to its favorable physical properties for root development [16]. However, RW is environmentally problematic, expensive to dispose of, and hazardous to human health during handling [11].

Our results highlight coconut fiber substrates, particularly the F70:P30 blend, as a promising alternative substrate. In alignment with studies, the improved water retention and high cation exchange capacity (CEC) of coconut fiber likely facilitated superior germination and early seedling growth [17]. Additionally, the enhanced growth (leaf number, plant height, leaf area) observed in seedlings grown on F70:P30 indicates that this substrate might have provided optimal conditions for early establishment, potentially influencing subsequent plant productivity positively.

In Experiment 2, which focuses on the comparison of aquaponic and hydroponic systems, despite the initial differences observed during germination and seedling stages (experiment 1), the total fresh biomass and leaf number did not significantly differ among treatments. This outcome supports previous research that reported comparable yields between hydroponic and aquaponic systems [12]. Interestingly, root dry biomass was greater in RW-grown plants, likely due to RW's stable physical structure enhancing root growth [18]. Nevertheless, from an environmental sustainability perspective, coconut fiber substrates offer clear advantages over RW, with lower environmental impact, safer handling, and reduced disposal costs [11,16].

While hydroponics remains a widely adopted soilless technique, its dependence on synthetic fertilizers raises concerns regarding environmental sustainability [15]. In contrast, decoupled aquaponics offers a promising alternative by utilizing nutrient-rich aquaculture water, thereby significantly reducing the need for supplementary fertilization [19,20]. Aquaponic cultivation resulted in a 10% higher leaf dry matter content compared to hydroponics. This finding aligns with recent hypotheses suggesting beneficial interactions between aquaponic solution microbial communities and plant roots, potentially increasing nutrient availability and promoting physiological adaptations [21,22]. Moreover, the significantly higher flavonoid content (52.5% increase) observed in aquaponically grown lemon basil plants highlights aquaponics' potential to enhance phytochemical quality. Previous studies have also reported increased flavonoid concentrations in aromatic plants grown aquaponically [23,24]. This increase in flavonoids may be linked to osmotic stress resulting from higher sodium concentrations in aquaponic solutions, prompting plants to activate secondary metabolic pathways as a protective mechanism against oxidative stress [25–27]. Lemon basil (*Ocimum* × *citriodorum*), specifically, appears particularly responsive to mild salinity stress, producing elevated flavonoid concentrations under such conditions [28]. Root length was significantly shorter in aquaponic-grown plants compared to hydroponics, likely due to increased sodium and salinity stress [25]. However, this reduced root growth did not negatively affect overall biomass production or photosynthetic performance. Indeed, gas exchange and chlorophyll fluorescence measurements (net photosynthesis, PSII efficiency) showed no significant differences between systems or substrates, in line with previous findings indicating stable photosynthetic performance under both cultivation methods [12,29]. Regarding mineral nutrition, aquaponically cultivated plants exhibited significantly higher calcium and magnesium uptake in stems and roots compared to hydroponically grown plants. This result supports previous findings demonstrating distinct nutrient uptake patterns influenced by microbial activities inherent to aquaponic nutrient solutions [30]. The microbial diversity within aquaponic systems, particularly nitrifying bacteria, can enhance nutrient solubility and uptake efficiency, explaining these observed differences [31,32]. Additionally, the chemical properties of coconut fiber substrates, particularly their high cation exchange capacity (CEC), likely enhanced nutrient availability to the plants [33]. Probably, RW, due to its favorable physical properties, promoted root growth; however, as an inert substrate, it did not enhance nutrient uptake to the same as coconut fiber. Although plants grown in RW developed more root biomass, this did not translate into significant differences in nutrient composition compared to those grown on F, as the high CEC allowed plants with less root system to absorb comparable amounts of nutrients as those grown on RW. However, further research is needed to clarify the precise interactions between coconut fiber substrates, aquaponic microbial communities, and plant nutrient uptake dynamics. Despite being a decoupled system, the aquaponic setup still contains residues of fish feed and fish feces, which inevitably influence the nutrient solution composition over time [26]. Although pH, CEC, and other parameters were carefully managed to remain similar between aquaponic and hydroponic solutions, the presence of these organic inputs leads to ongoing changes in nutrient availability and solution dynamics. This aspect represents a key difference from hydroponic systems and highlights the complex interactions between microbial communities and nutrient cycling in aquaponics. Such considerations suggest valuable directions for future research on optimizing nutrient solution characteristics concerning fish feeding regimes in aquaponic cultivation.

4. Materials and Methods

4.1. *Experiment 1: Seed Germination and Seedling Growth*

4.1.1. Experimental Site and Duration

The germination experiment was conducted from 15 November to 6 December 2022 at the University Centre for Innovation and Development in the Food Industry (CAISIAL) of the University of Naples Federico II—Department of Agriculture, University of Naples Federico II, Portici, Italy (40°48′57.9″ N 14°21′01.6″ E, 29 m a.s.l.), within a climate-controlled chamber.

4.1.2. Plant Material and Experimental Conditions

Seeds of lemon basil (*Ocimum × citriodorum*) (Bingenheimer Saatgut, Echzell, Germany) were used. The experiment was conducted under controlled conditions: ambient temperature was set at 26 °C, while inside the germination containers, temperature was maintained at 29 °C, with relative humidity (R.H.) kept constant at 89%. Seeds were initially hydrated with a half-strength Hoagland nutrient solution, with a pH of 6.2 and electrical conductivity (EC) of 1600 µS/cm.

4.1.3. Substrates and Sowing Procedure

Four substrate treatments were tested: (1) 100% rock wool (RW; control), (2) 50% coconut fiber (F) mixed with 50% perlite (P) (F50:P50), (3) 70% coconut fiber mixed with 30% perlite (F70:P30), and (4) 90% coconut fiber mixed with 10% perlite (F90:P10). A completely randomized design was adopted, with three replicate containers per substrate type. Each container contained 12 sowing holes, with two seeds sown per hole (24 seeds per container, 72 seeds per treatment). The sowing holes measured 3.81 × 3.81 cm with a depth of 0.5 cm, resulting in a soil volume of 7.25 cm^3 per hole for both substrates.

4.1.4. Germination and Seedling Management

Seeds underwent an initial 48-h dark period, followed by illumination at 243 μmol m^{-2} s^{-1} photosynthetic photon flux density (PPFD) provided by full-spectrum fluorescent lamps (ATI Powermodule, Viareggio, LU, Italy) positioned 50 cm above seedling containers. After the first 7 days, light intensity was increased to 343 μmol m^{-2} s^{-1} with a 16/8 h photoperiod (day/night). Initially, substrates were hydrated every two days with osmotic water until the emergence of the first radicle. Subsequently, substrates were irrigated at the same frequency with half-strength Hoagland nutrient solution (EC 1100 μS/cm; pH 5.7), ensuring consistent moisture without water stagnation.

4.1.5. Seedling Growth Measurements

Seed germination was monitored daily over 7 consecutive days. A seed was considered germinated upon emergence of the two cotyledons above the substrate surface. The germination percentage (% GP) was calculated on day 7 as (number of germinated seeds/total number of seeds sown) × 100. At twenty days after sowing, seedling height (collar base to vegetative apex), leaf number, total leaf area, and root length were measured by analyzing digital images using ImageJ software, version 1.50i (Wayne Rasband National Institute of Health, Bethesda, MD, USA). The fresh weight of the shoot and root was measured using an electronic balance; dry weights were determined after oven drying at 70 °C for 48 h.

4.2. Experiment 2: Comparison of Aquaponic and Hydroponic Systems

4.2.1. Experimental Site, Duration, and Setup

This greenhouse experiment was conducted from 6 December 2022 to 13 January 2023 at the same university facility (40°48′57.9″ N, 14°21′01.6″ E). Twenty-day-old lemon basil seedlings grown on 70% coconut fiber mixed with 30% perlite (F) and on rock wool (RW) substrates (Experiment 1) were transplanted into floating raft systems of decoupled aquaponic (A) and hydroponic (H) units at a density of 20 plants m^{-2}. These two treatments were selected based on the results of the germination experiment: RW served as the control substrate, while F70:P30 achieved the highest germination percentage, outperforming F50:P50 and F90:P10 by 26.1 and 20.8 percentage points, respectively.

4.2.2. Greenhouse and Cultivation Systems Description

The experiment utilized a recirculating aquaponic system (RAS) prototype comprising four aquaculture tanks (2800 L each). Nile tilapia (*Oreochromis niloticus*) was stocked at an average density of 8.7 ± 5.4 kg m^{-3}. The RAS included an 800-L mechanical and biological filtration unit (Superbead, Air-aqua, Staphorst, The Netherlands), a 400-L trickling filter (Scubla srl, Udine, Italy), and a 40-W UV sterilizer (Air-aqua, Staphorst, The Netherlands). Air was supplied at 0.05 v v^{-1} min^{-1}. Floating raft units (2 m^2 each) were independently managed as decoupled systems for aquaponic and hydroponic treatments.

4.2.3. Decoupled Aquaponic and Hydroponic Nutrient Solution Management

Initially, the aquaponic tank was filled with RAS water and subsequently decoupled. Both systems were maintained at a constant water temperature of 23 °C, with daily monitoring of pH and EC using Thermo Scientific Expert Testers (Segrate, Italy). Nutrient solutions were analyzed biweekly, and the pH was adjusted to a target of 6.2–6.3 using nitric acid (HNO_3) and potassium hydroxide (KOH). The electrical conductivity (EC) was maintained at approximately 1600 μS cm^{-1}. The half-strength Hoagland hydroponic nutrient solution was prepared by dissolving the following salts per liter of solution: KNO_3 = 0.75 g L^{-1}; $CaCl_2$ = 0.28 g L^{-1}; $MgSO^4 * 7H_2O$ = 0.25 g L^{-1}; H_3PO^4= 0.04 g L^{-1};

$FeSO^4$ * $7H_2O$ = 0.02 g L^{-1}; Fe-EDTA = 0.04 g L^{-1}, H_3BO_3 (0.0014 g L^{-1}), $MnCl_2{\cdot}4H_2O$ (0.0011 g L^{-1}), $ZnSO_4{\cdot}7H_2O$ (0.0001 g L^{-1}), $CuSO_4{\cdot}5H_2O$ (0.00004 g L^{-1}), and $H_2MoO_4{\cdot}H_2O$ (0.0001 g L^{-1})". Nutrient concentrations for both systems are detailed in Table 10.

Table 10. Nutrient concentration (mg L^{-1}) in hydroponic (H) and aquaponic (A) systems.

System	NO_3	PO_4	SO_4	K	Ca	Mg	Na	Cl
H	390.28	45.70	80.43	95.77	113.64	35.98	10.35	7.63
A	469.93	50.04	60.04	40.65	107.46	31.77	41.23	59.08

4.2.4. Plant Growth Measurements

At 38 days after transplanting (DAT), corresponding to 59 days after sowing, plant height, leaf number, total leaf area, and root length were measured using ImageJ software, version 1.50i (Wayne Rasband, National Institute of Health, Bethesda, MD, USA). Fresh and dry weights of shoots and roots were recorded after drying samples at 70° C for 48 h.

4.2.5. Leaf Vegetation Indices

Leaf anthocyanin (AnthM), flavonoid (FlvM), and chlorophyll indices (CCla) were measured at 38 DAT using a portable MPM-100, Multi Pigment Meter (ADC BioScientific Ltd., Hoddesdon, UK). The anthocyanin index was calculated at 660 and 525 nm as AnthM = log (f_{660}/f_{525}). The flavonoid index was calculated at 660 and 375 nm as FlvM = log (f_{660}/f_{375}). The chlorophyll content index (CCIa) was calculated using the formula CCIa = (T_{850}/T_{720}) × 0.86532 + 0.39, where T_{850} and T_{720} are the transmittance measurements at 850 nm and 720 nm, respectively.

4.2.6. Gas Exchanges and Chlorophyll Fluorescence Measurements

Leaf gas exchange was measured at 38 DAT on one fully expanded leaf from each plant on 3 plants × 3 replicates × cultivation system × substrate, using an Infrared Gas Analyser (LCi T, ADC Bioscientific Ltd., Hoddesdon, UK). The measurements were conducted at noon under ambient CO_2 (on average 513 ± 25 ppm), mean temperature 28 °C, R.H. 33%, and PPFD 486 μmol m^{-2} s^{-1}. Intrinsic water use efficiency (WUE_i) was calculated as leaf net photosynthesis (A) divided by stomatal conductance (g_s).

On the same leaves used for gas exchange measurements, the evaluation of chlorophyll fluorescence was performed. A portable Plant Stress Kit (Opti-Sciences Inc., Hudson, NY, USA) was utilized, and light-induced measurements were conducted using a ΦPSII meter, applying a saturating pulse of 4286 μmol m^{-2} s^{-1} for 1.1 s to obtain the maximum light-adapted fluorescence (F_m) and steady-state fluorescence (F_s). For dark-adapted measurements, the leaves were dark-adapted for 30 min with a dark leaf clip (Opti-Sciences Inc., Hudson, NY, USA). An F_v/F_m meter (Opti-Sciences Inc., Hudson, NY, USA), capable of delivering a 1.0-s saturating light pulse (3429 μmol m^{-2} s^{-1}), was employed to obtain the F_m and F_0 values. The PSII maximal photochemical efficiency (F_v/F_m) was determined using $F_v/F_m = (F_m - F_0)/F_m$. The quantum yield of PSII electron transport (Φ_{PSII}) was computed as $\Phi_{PSII} = (F_{m'} - F_s)/F_{m'}$ [34]. The non-photochemical quenching (NPQ) was calculated as $(F_m/F_{m'}) - 1$ [35].

4.2.7. Leaf, Stem, and Root Mineral Content Analyses

Mineral content (NO_3^-, P, K, Ca, Mg) of leaves, stems, and roots were analyzed at 38 DAT using 38 DAT on samples from each of 3 plants × 3 replicates × substrate × cultivation system. A 250 mg aliquot of a ground dry tissue sample was utilized for the determination of leaf mineral composition according to the procedures outlined by Pannico et al. [36]. Subsequently, mineral analysis was conducted following 0.45-micrometer filtration using an ion chromatography system (model ICS-3000, Dionex, Sunnyvale, CA, USA), with quantification facilitated by an electrical conductivity detector equipped with IonPac CS12A and IonPac AS11-HC analytical columns for cationic and anionic content analysis, respectively (Dionex, Sunnyvale, CA, USA). Accounting for the dry matter content of the leaves, all mineral (leaf, stem, and root) uptake values were calculated by multiplying the concentrations expressed in grams per kilogram (g kg^{-1}) on a dry weight (DW) basis by the dry weight of the analyzed organ and expressed as gram per plant (g $plant^{-1}$). Only nitrate was expressed as concentration (g kg^{-1}) on a fresh weight (FW) basis.

4.2.8. Statistical Analysis

Experiment 1

Measurements and statistical analysis were carried out on 4 plants × 3 replicates × 4 substrates, resulting in 48 plants. Data were analyzed using one-way ANOVA. The analysis of the variance was performed using SPSS software version 27 (IBM, Chicago, IL, USA). Means were compared by Tukey's HSD post hoc test at a significance level of $p < 0.05$. A cluster heatmap was generated with ClustVis online software [37], using Euclidean distance and complete linkage clustering; data were log(x + 1) transformed.

Experiment 2

Measurements and statistical analysis were carried out on 3 plants × 3 replicates × 2 substrates × 2 cultivation systems, resulting in 36 plants. Data were analyzed using two-way ANOVA. The analysis of the variance was performed using SPSS software version 27 (IBM, Chicago, IL, USA). Means were compared by Tukey's HSD post hoc test at a significance level of $p < 0.005$. A cluster heatmap was generated with ClustVis online software [37], using Euclidean distance and complete linkage clustering; data were log(x + 1) transformed.

5. Conclusions

This research highlights the effectiveness of the F70:P30 coconut fiber-perlite substrate mix in optimizing seed germination and early plant growth, providing a sustainable and environmentally friendly alternative to rock wool substrate. This alternative reduces disposal costs and minimizes potential health risks associated with rock wool handling.

Additionally, decoupled aquaponics proved to be a viable cultivation method, achieving yields comparable to traditional hydroponics while significantly decreasing dependence on chemical fertilizers.

The observed increase in flavonoid content in aquaponics-grown lemon basil suggests additional nutritional and nutraceutical benefits, potentially attributable to mild osmotic stress conditions.

This enhancement offers growers a marketing advantage and the potential for higher economic returns from health-conscious consumers.

Collectively, these findings offer direct practical implications for commercial greenhouse production, supporting the adoption of more sustainable, cost-effective, and resource-efficient agricultural practices. Future research should further investigate the integration and optimization of sustainable substrates and aquaponic systems, particularly for commercial vegetable production in regions affected by soil contamination.

Author Contributions: Conceptualisation, L.S., and G.C.M.; methodology, L.S., G.C.M., and C.C.; validation, G.C.M., L.S., C.C., and A.L.L.; formal analysis, L.S., G.C.M.; investigation, G.C.M., and L.S.; resources, C.C., A.L.L., P.M., and S.D.P.; data curation, L.S., and G.C.M.; writing—original draft preparation, L.S., G.C.M., and C.C.; writing—review and editing, L.S., G.C.M., CC., and S.D.P.; visualization, G.C.M., L.S., C.C., S.D.P., and P.M.; supervision, C.C., G.C.M., and P.D.P.; project administration, C.C., and P.M.; funding acquisition, C.C., S.D.P., and P.M. All authors have read and agreed to the published version of the manuscript.

Funding: This research was funded by Rago Group, an agricultural company based in Battipaglia, Salerno. This project was carried out under the title "Innovation in Cultivation Techniques for Leafy Vegetables to Reduce Environmental Impact through the Reduction of Chemical Fertilizers." The work was also supported by the National Biodiversity Future Center—NBFC funded by the European Union—NextGenerationEU under the NRRP, Mission 4 Component 2 Investment 1.4. Code CN_00000033, CUP E63C22000990007, and by the Italian Ministry of Education and Research (MUR) within the call for Research Projects of National Interest (PRIN), within the project "VFARM-Sustainable Vertical Farming" (Project code: 2020ELWM82, CUP: J33C20002350001).

Data Availability Statement: The data are contained within the article.

Acknowledgments: The authors would like to thank Antonio Matteo, Lucia Vanacore, Giovanni Luca Russo, Francesco Angelone, and Gabriele Buono for their technical assistance during the experiments.

Conflicts of Interest: The authors declare no conflict of interest.

References

1. Spence, C. Sweet Basil: An Increasingly Popular Culinary Herb. *Int. J. Gastron. Food Sci.* **2024**, *36*, 100927. [CrossRef]
2. Boronat, Ò.; Mora, M.; Romeo-Arroyo, E.; Vázquez-Araújo, L. Unraveling the Mediterranean Cuisine. What Ingredients Drive the Flavor of the Gastronomies Included in the Mediterranean Diet? *Int. J. Gastron. Food Sci.* **2023**, *34*, 100802. [CrossRef]
3. Matarazzo, N.; Imbrenda, V.; Coluzzi, R.; Pace, L.; Samela, C.; Lanfredi, M. 'Industrializzazione Dell' Agricoltura Nella Piana Del Sele: Una Prospettiva Geografica Basata Sull' Urban Atlas Copernicus. *Bull. Dell'assoc. Ital. Cartogr.* **2022**, *175*, 101–113. [CrossRef]
4. Ferreira, C.S.S.; Seifollahi-Aghmiuni, S.; Destouni, G.; Ghajarnia, N.; Kalantari, Z. Soil Degradation in the European Mediterranean Region: Processes, Status and Consequences. *Sci. Total Environ.* **2022**, *805*, 150106. [CrossRef] [PubMed]
5. Stolte, J.; Tesfai, M.; Øygarden, L.; Hessel, R.; Panagos, P.; Kværnø, S.; Verheijen, F.; Ballabio, C.; Stolte, J.; Keizer, J. *Soil Threats in Europe*; Publications Office: Luxemburg, 2015. [CrossRef]
6. Baldantoni, D.; Morra, L.; Zaccardelli, M.; Alfani, A. Cadmium Accumulation in Leaves of Leafy Vegetables. *Ecotoxicol. Environ. Saf.* **2016**, *123*, 89–94. [CrossRef]
7. Junge, R.; Schmautz, Z.; Milliken, S. Toward Nutrient Cycling from Organic Waste Streams for Soilless Cultivation. *Curr. Opin. Food Sci.* **2025**, *61*, 101257. [CrossRef]
8. Manimozhi, R.; Krishnamoorthy, G. Innovative Techniques in Agriculture: Transitioning From Traditional Farming to Precision and Hydroponic Agriculture. *Environ. Qual. Manag.* **2025**, *34*, e70047. [CrossRef]
9. Suhl, J.; Dannehl, D.; Baganz, D.; Schmidt, U.; Kloas, W. An Innovative Suction Filter Device Reduces Nitrogen Loss in Double Recirculating Aquaponic Systems. *Aquac. Eng.* **2018**, *82*, 63–72. [CrossRef]
10. Yen, H.Y.; Chou, J.H. Water Purification by Oyster Shell Bio-Medium in a Recirculating Aquaponic System. *Ecol. Eng.* **2016**, *95*, 229–236. [CrossRef]
11. Mansour, E.; Loxton, C.; Elias, R.M.; Ormondroyd, G.A. Assessment of Health Implications Related to Processing and Use of Natural Wool Insulation Products. *Environ. Int.* **2014**, *73*, 402–412. [CrossRef]
12. Modarelli, G.C.; Vanacore, L.; Rouphael, Y.; Langellotti, A.L.; Masi, P.; De Pascale, S.; Cirillo, C. Hydroponic and Aquaponic Floating Raft Systems Elicit Differential Growth and Quality Responses to Consecutive Cuts of Basil Crop. *Plants* **2023**, *12*, 1355. [CrossRef]
13. Körner, O.; Bisbis, M.B.; Baganz, G.F.; Baganz, D.; Staaks, G.B.; Monsees, H.; Goddek, S.; Keesman, K.J. Environmental impact assessment of local decoupled multi-loop aquaponics in an urban context. *J. Clean. Prod.* **2021**, *313*, 127735. [CrossRef]
14. Fussy, A.; Papenbrock, J. Techniques—Chances, Challenges and the Neglected Question of Sustainability. *Plants* **2022**, *11*, 1153. [CrossRef]
15. Vanacore, L.; El-nakhel, C.; Modarelli, G.C.; Rouphael, Y. Growth, Ecophysiological Responses, and Leaf Mineral Composition of Lettuce and Curly Endive in Hydroponic and Aquaponic Systems. *Plants* **2024**, *13*, 2852. [CrossRef] [PubMed]
16. Barrett, G.E.; Alexander, P.D.; Robinson, J.S.; Bragg, N.C. Achieving Environmentally Sustainable Growing Media for Soilless Plant Cultivation Systems—A Review. *Sci. Hortic.* **2016**, *212*, 220–234. [CrossRef]
17. He, L.; Ding, X.; Jin, H.; Zhang, H.; Cui, J.; Chu, J.; Li, R.; Zhou, Q.; Yu, J. Comparison of Rockwool and Coir for Greenhouse Cucumber Production: Chemical Element, Plant Growth, and Fruit Quality. *Heliyon* **2022**, *8*, e10930. [CrossRef]
18. Inden, H.; Torres, A. Comparison of Four Substrates on the Growth and Quality of Tomatoes. *Acta Hortic.* **2004**, *644*, 205–210. [CrossRef]

19. Kloas, W.; Groß, R.; Baganz, D.; Graupner, J.; Monsees, H.; Schmidt, U.; Staaks, G.; Suhl, J.; Tschirner, M.; Wittstock, B.; et al. A New Concept for Aquaponic Systems to Improve Sustainability, Increase Productivity, and Reduce Environmental Impacts. *Aquac. Environ. Interact.* **2015**, *7*, 179–192. [CrossRef]
20. Francisco, E.C.; Freato, T.A.; Piolli, A.L.; Poz, M.E.S.D. Analysis of the Aquaponic System Sustainability via System Dynamics Modelling—FEW Nexus Approach. *Circ. Econ. Sustain.* **2024**. [CrossRef]
21. Chiquito-Contreras, R.G.; Hernandez-Adame, L.; Alvarado-Castillo, G.; Martínez-Hernández, M.d.J.; Sánchez-Viveros, G.; Chiquito-Contreras, C.J.; Hernandez-Montiel, L.G. Aquaculture—Production System and Waste Management for Agriculture Fertilization—A Review. *Sustainability* **2022**, *14*, 7257. [CrossRef]
22. Lobanov, V.; Keesman, K.J.; Joyce, A. Plants Dictate Root Microbial Composition in Hydroponics and Aquaponics. *Front. Microbiol.* **2022**, *13*, 848057. [CrossRef]
23. Braglia, R.; Costa, P.; Di Marco, G.; D'Agostino, A.; Redi, E.L.; Scuderi, F.; Gismondi, A.; Canini, A. Phytochemicals and Quality Level of Food Plants Grown in an Aquaponics System. *J. Sci. Food Agric.* **2022**, *102*, 844–850. [CrossRef]
24. Flores-Aguilar, P.S.; Rico-Chávez, A.K.; Rodriguez-deLeón, E.; Aguirre-Becerra, H.; Zamora-Castro, S.A.; Soto-Zarazúa, G.M. Bioactive Compounds of Endemic Medicinal Plants (*Cuphea* spp.) Cultured in Aquaponic Systems: A Short Study. *Agriculture* **2023**, *13*, 2018. [CrossRef]
25. Tarchoune, I.; Sgherri, C.; Izzo, R.; Lachaal, M.; Ouerghi, Z.; Navari-Izzo, F. Antioxidative Responses of *Ocimum basilicum* to Sodium Chloride or Sodium Sulphate Salinization. *Plant Physiol. Biochem.* **2010**, *48*, 772–777. [CrossRef] [PubMed]
26. Goddek, S.; Joyce, A.; Kotzen, B.; Burnell, G.M. *Aquaponics Food Production Systems: Combined Aquaculture and Hydroponic Production Technologies for the Future*; Springer: New York, NY, USA, 2019; pp. 1–620. [CrossRef]
27. Cramer, G.R.; Läuchli, A.; Epstein, E. Effects of NaCl and $CaCl_2$ on Ion Activities in Complex Nutrient Solutions and Root Growth of Cotton. *Plant Physiol.* **1986**, *81*, 792–797. [CrossRef] [PubMed]
28. Ciriello, M.; Formisano, L.; Kyriacou, M.C.; Carillo, P.; Scognamiglio, L.; De Pascale, S.; Rouphael, Y. Morpho-Physiological and Biochemical Responses of Hydroponically Grown Basil Cultivars to Salt Stress. *Antioxidants* **2022**, *11*, 2207. [CrossRef] [PubMed]
29. Mourantian, A.; Aslanidou, M.; Mente, E.; Katsoulas, N.; Levizou, E. Basil Functional and Growth Responses When Cultivated via Different Aquaponic and Hydroponics Systems. *PeerJ* **2023**, *11*, e15664. [CrossRef]
30. Aslanidou, M.; Elvanidi, A.; Mourantian, A.; Levizou, E.; Mente, E.; Katsoulas, N. Nutrients Use Efficiency in Coupled and Decoupled Aquaponic Systems. *Horticulturae* **2023**, *9*, 1077. [CrossRef]
31. Eck, M.; Szekely, I.; Massart, S.; Jijakli, M.H. Microorganisms in Aquaponics: Insights on the Composition of the Root Microbiome of Lettuces of Varying Age. *Acta Hortic.* **2021**, *1321*, 213–219. [CrossRef]
32. Kasozi, N.; Abraham, B.; Kaiser, H.; Wilhelmi, B. The Complex Microbiome in Aquaponics: Significance of the Bacterial Ecosystem. *Ann. Microbiol.* **2021**, *71*, 1. [CrossRef]
33. Tuckeldoe, R.B.; Maluleke, M.K.; Adriaanse, P. The Effect of Coconut Coir Substrate on the Yield and Nutritional Quality of Sweet Peppers (*Capsicum annuum*) Varieties. *Sci. Rep.* **2023**, *13*, 2742. [CrossRef]
34. Genty, B.; Briantais, J.-M.M.; Baker, N.R. The Relationship between the Quantum Yield of Photosynthetic Electron Transport and Quenching of Chlorophyll Fluorescence. *Biochim. Biophys. Acta—Gen. Subj.* **1989**, *990*, 87–92. [CrossRef]
35. Maxwell, K.; Johnson, G.N. Chlorophyll Fluorescence—A Practical Guide. *J. Exp. Bot.* **2000**, *51*, 659–668. [CrossRef] [PubMed]
36. Pannico, A.; El-Nakhel, C.; Kyriacou, M.C.; Giordano, M.; Stazi, S.R.; De Pascale, S.; Rouphael, Y. Combating Micronutrient Deficiency and Enhancing Food Functional Quality Through Selenium Fortification of Select Lettuce Genotypes Grown in a Closed Soilless System. *Front. Plant Sci.* **2019**, *10*, 1495. [CrossRef] [PubMed]
37. Metsalu, T.; Vilo, J. ClustVis: A web tool for visualizing clustering of multivariate data using Principal Component Analysis and heatmap. *Nucleic Acids Res.* **2015**, *43*, 566–570. [CrossRef] [PubMed]

plants

MDPI

Article

Using Recycled Construction Waste Amended with Pine Bark as a Substrate for Urban Plantings

Claire Kenefick *, Stephen J. Livesley, John P. Rayner and Claire Farrell

School of Agriculture, Food and Ecosystem Sciences, Burnley Campus, The University of Melbourne, 500 Yarra Boulevard, Richmond, VIC 3121, Australia; sjlive@unimelb.edu.au (S.J.L.); jrayner@unimelb.edu.au (J.P.R.); c.farrell@unimelb.edu.au (C.F.)

* Correspondence: c.kenefick@student.unimelb.edu.au

Abstract

In urban plantings, mined sand and scoria are commonly used as low-nutrient substrates to improve plant establishment and growth. However, reliance on mined materials conflicts with sustainability policies promoting resource circularity and waste reuse. Construction wastes are readily available, and while their high alkalinity and density may limit plant growth, incorporating organic matter, like pine bark, can ameliorate these issues. We investigated whether construction waste amended with pine bark can support plant growth. We evaluated physical and chemical properties of 12 substrates combining four mineral components—scoria (mined), sand (recycled), crushed concrete (recycled), and crushed rock (recycled)—with pine bark (10%, 20%, and 50% v/v). We then tested eight of these substrates in a container experiment, evaluating the growth of two woody shrubs: *Alyogyne huegelii* and *Goodenia ovata*. All mineral components were alkaline (pH 9.2–12.3), with crushed concrete remaining hyper-alkaline despite pine bark addition. Greater pine bark volumes improved air-filled porosity but reduced water retention. Substrates with 50% v/v pine bark had lower plant biomass compared to those with 10% v/v. However, plant biomass was similar across all mineral components. This demonstrates that construction waste–pine bark substrates can support plant growth in urban plantings, supporting broader sustainability goals in cities.

Keywords: alkalinity; crushed concrete; crushed rock; growing media; recycled substrate; sand; scoria; substrate properties; urban greening; woody plants

1. Introduction

Naturalistic plantings can be a low-maintenance approach to increase the amenity and biodiversity of informal green spaces, such as streetscapes and railway easements [1]. They are designed to mimic the structure of natural vegetation communities, layering diverse understorey plant species to create a complex structure with abundant and prolonged flowering [2]. A key design feature of naturalistic plantings is including species that can establish and persist with low inputs [3,4]. However, establishing naturalistic plantings on urban soils can be challenging, as they are often degraded and compacted [5,6].

Urban soils can be highly fertile due to excessive or long-term fertiliser application and runoff [7], which can lead to high weed competition [8]. Urban soils can be improved via (i.) amendments (e.g., gypsum, lime, compost) to improve chemical and physical structure [9,10], (ii.) mechanical ripping or tillage to reduce dominant weed cover, reduce compaction and improve physical structure [6,11,12], or (iii.) a combination of tillage and deep soil amendments [12–14]. However, these treatments can be impractical at larger scales and may not address the issues of persistent weed seed banks [6]. In such instances, the options are to remove upper soil layers or to add new upper soil layers [4].

Topsoil removal (or "scalping") removes the upper 10–50 cm of soil layers containing excess nutrients and weed propagules [15] and has been used for grassland restoration in Australia [16,17], Europe [18–20], and North America [21]. However, in urban landscapes, scalping can be limited by the high cost of topsoil removal and disposal [22] or by contaminants that prohibit removal or disturbance of soil [23]. In these cases, new soil layers can be added on top of existing site soil as a "capping substrate", providing an uncompacted, low-nutrient, weed-free growing medium, and acting as a mulch to suppress weeds [9,24].

Capping substrates are commonly used in naturalistic plantings to mimic free-draining, low-nutrient natural soils, favouring the establishment of target species over nutrient-demanding weeds [8]. To support plant establishment and growth, capping substrates must provide adequate air, water, and structure [25–27]. There are currently no guidelines for capping substrates for naturalistic plantings, but acceptable ranges for substrate physical and chemical properties (Table 1) can be drawn from guidelines for horticultural container media [28–31], green roof substrates [32–35], and landscape soils [9,36,37].

Table 1. Key substrate properties and their recommended ranges for healthy plant growth in urban landscape plantings based on guidelines for container media, green roof substrates, and landscape soils.

Property	Unit	Definition	Acceptable Range	References
Air-filled porosity (AFP)	% *v*/*v*	The percentage of substrate that is composed of air-filled pores	7–50% depending on plant species	[29–31,34,35]
Dry bulk density (BD)	g cm^{-3}	Mass in grams of a dried cubic centimetre of substrate	≤1.7 g cm^{-3}	[29,32,36,38]
Coarseness index (CI)	% *w*/*w*	Summary of particle size distribution: cumulative percentage by weight of particle sizes < 1 mm	<40%	[28]
Water-holding capacity (WHC)	% *v*/*v*	The total amount of water able to be held by a substrate after gravity drainage (~−10 kPa)	20–65%	[30–33,35]
Plant available water (PAW)	% *w*/*w*	The amount of water available to a plant between field capacity (−10 kPa) and permanent wilting point (−1500 kPa)	>20% depending on plant species water use	[29]
pH	unitless	The concentration of hydrogen ions (H^+) in a medium. Indicates if the substrate is acidic, neutral, or alkaline	4.5–8.5	[29,30,32,36,39]
Electrical conductivity (EC)	dS m^{-1}	The sum of all salt ions in a solution which indicates the salinity of the substrate	≤2.2 dS m^{-1}	[29,30,37]

Capping substrates are typically composed of mined or quarried materials such as sand, gravel, or scoria, due to their low-nutrient content, minimal weed seed load, and well-characterised physical properties [8,40]. However, with a growing desire to reduce the environmental impact of cities, more sustainable materials are required—particularly those that reuse waste generated in cities [41].

Mineral materials recycled from construction and demolition waste could be used as capping substrates for naturalistic plantings. Construction wastes include materials such as crushed concrete, crushed rock, recycled sand, crushed brick, crushed roof tile, and subsoil [39]. They can have lower environmental and economic cost compared to mined materials, so they could be sustainable replacements for existing mineral substrates [42,43]. Previous studies have evaluated construction wastes as the inorganic substrate component for green roofs [44–47] and in technosols for tree or garden plantings, e.g., [48–52]. However, their use as capping substrates for naturalistic plantings is limited [2,8,22,24]. Construction wastes are generally low in nutrients and may contain fewer weed seeds [22], which would

https://doi.org/10.3390/plants15030403

make them ideal capping substrates. However, they can have high bulk density (e.g., recycled sands or subsoils; [39,53]), poor water-holding capacity (e.g., crushed rock; [22]), and extreme alkalinity (e.g., crushed concrete; [47,54]).

Adding organic matter to construction wastes could ameliorate unfavourable physical and chemical properties [55], potentially bringing them into the acceptable ranges for plant growth. Organic matter can increase air and water capacity, decrease density, improve the structure [46,56], and reduce the pH of alkaline construction wastes [54]. There are many types of organic matter, and their origin and pre-treatment (e.g., ageing, composting) can determine the maturity, cellulose-to-lignin ratio, nutrient content, and stability (i.e., rate of decomposition and longevity; [57,58]). For capping substrates, organic matter should be stable with low nutrients and weed seed load. Aged pine bark is a stable organic matter commonly used in soil-less container media and as landscape mulch, due to its favourable physical and chemical properties, low weed seed content, and slow decomposition [25,57,59]. Pine bark is a softwood material produced as a waste by-product from the timber industry and is carbon-rich with low nitrogen availability [57]. Pine bark has been shown to increase air-filled porosity (AFP), water-holding capacity (WHC), and plant available water (PAW), as well as reduce bulk density (BD) in sand container media [60]; it is also acidic [61], which could help ameliorate high alkalinity in construction waste substrates. Pine bark can ameliorate highly alkaline substrates by replacing alkaline mineral particles on a volume-to-volume basis (i.e., physical dilution) as well as through chemical reactions.

We aimed to evaluate the use of construction and demolition mineral wastes with and without pine bark addition as capping substrates for naturalistic plantings. We evaluated the physical, hydrological, and chemical properties of four mineral components—scoria, recycled sand, crushed concrete, and crushed rock—pine bark, and 12 substrate mixes made of the four mineral components with three rates of pine bark addition (10, 20, or 50% *v*/*v*). Based on their properties, we then evaluated plant growth in eight substrates with 10% or 50% pine bark in a nursery-based container experiment. To compare plant response across a range of substrate alkalinity, we selected two drought-tolerant species from low-nutrient Australian shrublands that have demonstrated vigorous growth in woody naturalistic plantings with scoria capping substrates [62] and have contrasting alkalinity tolerance. *Alyogyne huegelii* is found naturally across a range of soil types from acidic to alkaline [63], whereas *Goodenia ovata* is found naturally on neutral to acidic soils [64]. We hypothesised that

1. Greater rates of pine bark addition to mineral substrates will improve physical and hydrological properties and result in greater plant growth as compared to mixes with lower rates of pine bark addition.
2. Greater rates of pine bark addition to alkaline substrates will improve chemical properties and result in greater growth of alkaline-sensitive plant species (*G. ovata*) than alkaline-tolerant species (*A. huegelii*).

2. Results

2.1. Effect of Pine Bark Addition on Substrate Properties

Coarseness index (CI) exceeded the acceptable threshold for plant growth in pine bark, all mineral components, and substrate mixes, except for recycled sand with no pine bark and recycled sand with 10% pine bark (Figure 1i–iv). The interaction between pine bark addition rate and mineral component significantly affected CI ($\chi^2_{(11)} = 994.0$, $p < 0.001$). Pine bark addition significantly increased the CI of all mineral components, except crushed rock. For recycled sand, the addition of 50% pine bark increased CI beyond the acceptable threshold for plant growth (45%, Figure 1i). Particle size distributions show a clear grouping

https://doi.org/10.3390/plants15030403

of the substrate components and mixes into three statistically distinct categories (Figure 2): "dominantly fine" (CI < 40%), comprising recycled sand with no pine bark and recycled sand with 10% pine bark; "intermediate" (CI 45–55%), comprising scoria with no pine bark and recycled sand with 50% pine bark; and "dominantly coarse" (CI < 55%), comprising all other substrate components and substrate mixes.

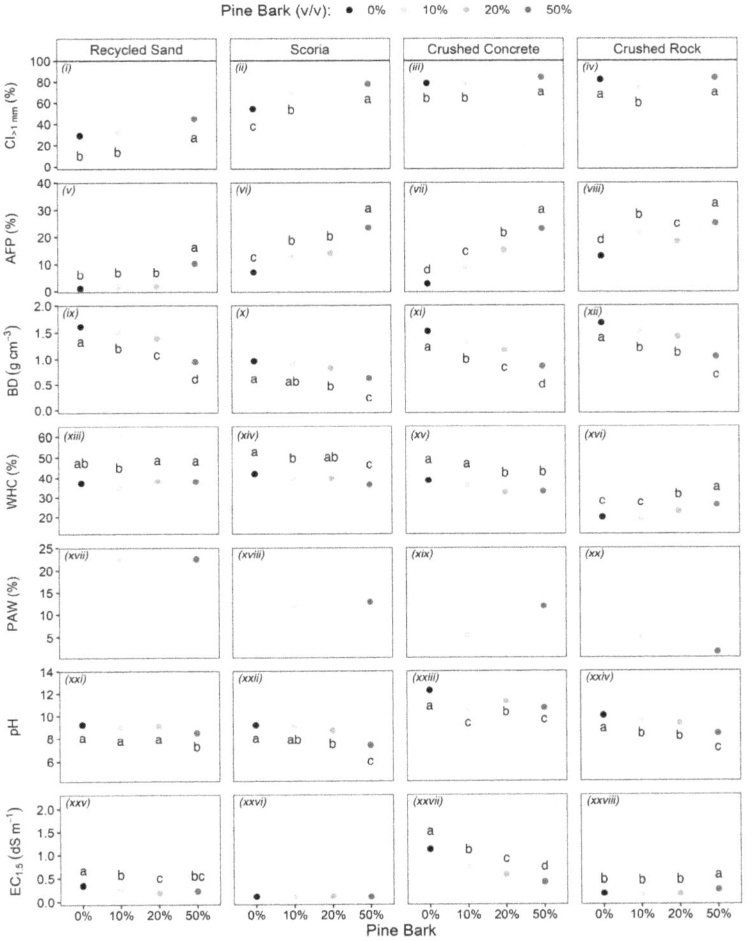

Figure 1. Effects of pine bark addition on physical properties: (**i–iv**) coarseness index, ($CI_{>1\ mm}$), (**v–viii**) air-filled porosity (AFP), (**ix–xii**) dry bulk density (BD); hydrological properties: (**xiii–xvi**) water-holding capacity (WHC), and (**xvii–xx**) plant available water (PAW); and chemical properties: (**xxi–xxiv**) pH, and (**xxv–xxviii**) electrical conductivity ($EC_{1:5}$). Columns represent the four mineral components: recycled sand, scoria, crushed concrete, and crushed rock. The X-axis and colours represent pine bark rate v/v (0% = black, 10% = light blue, 20% = grey-blue, and 50% = purple). Values represent means (n = 3) for AFP, BD, WHC, pH, and EC; (n = 2) for CI; and (n = 1) for PAW. Green shading indicates the acceptable range for plant growth: CI < 40%, AFP = 7–50%, BD < 1.7 g cm^{-3}, WHC = 20–65%, PAW < 20%, pH = 4.5–8.5, and EC < 2.2 dS m^{-1}. Different letters denote significant differences among property means across pine bark rates within each mineral component (for all $p < 0.001$, except WHC $p = 0.002$, and no analysis for PAW). CI was measured on 3 pine bark rates (0, 10, and 50%). PAW was measured on 2 pine bark rates (10 and 50%).

https://doi.org/10.3390/plants15030403

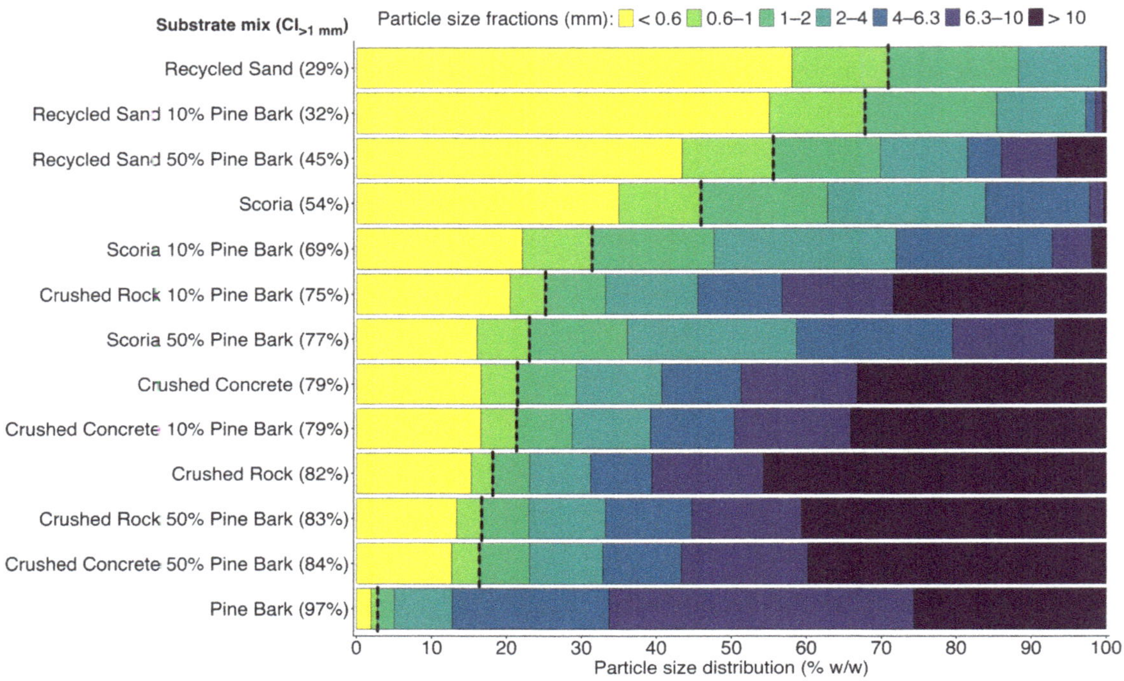

Figure 2. Particle size distribution of the four mineral components (recycled sand, scoria, crushed concrete, and crushed rock), pine bark, and the eight substrate mixes (10% and 50% pine bark within each mineral component) from 2 L samples. The X-axis represents % weight/weight of particles in each size fraction (coloured bars; <0.6 mm, 0.6–1 mm, 1–2 mm, 2–4 mm, 4–6.3 mm, 6.3–10 mm, and >10 mm). The Y-axis represents substrate mix or component with coarseness index ($CI_{>1\ mm}$); the cumulative percent of particles < 1 mm. Vertical black dashed lines indicate a 1 mm threshold used for CI.

Mineral components scoria and crushed rock had air-filled porosity (AFP) within the acceptable range for plant growth, while the AFP of recycled sand and crushed concrete was below the threshold (Figure 1v–viii). A significant interaction between pine bark addition and mineral component affected AFP ($\chi^2_{(9)} = 132.2$, $p < 0.001$), as shown by the variable response of AFP to increasing pine bark addition across substrate mixes. Recycled sand required the highest pine bark addition rate (50%) to achieve acceptable AFP, whereas a 10% addition was sufficient to bring crushed concrete into the acceptable range (>7%, Figure 1v,vii).

The dry bulk density (BD) of all mineral components was within the threshold for unrestricted root growth, although crushed rock was near the upper limit (1.7 g cm^{-3}; Figure 1ix–xii). Increasing pine bark addition rates generally reduced BD across all mineral components, with the 50% rate producing the lowest BD values and bringing crushed rock closer in line with the other substrate mixes (Figure 1ix–xii). A significant interaction between pine bark addition rate and mineral component affected the BD of the substrate mixes ($\chi^2_{(9)} = 66.4$, $p < 0.001$).

All mineral components had a water-holding capacity (WHC) within the acceptable range for plant growth, although crushed rock was at the lower limit (20%, Figure 1xiii–xvi). A significant interaction between pine bark addition rate and mineral component affected WHC ($\chi^2_{(9)} = 160.8$, $p < 0.001$). In scoria and crushed concrete, increasing rates of pine bark

https://doi.org/10.3390/plants15030403

addition significantly lowered WHC compared to 0% pine bark, though values remained within the acceptable range. In contrast, rates of 20% and 50% pine bark addition significantly increased WHC in crushed rock, bringing it further into the acceptable range (23% and 26%, respectively).

Plant available water (PAW) was measured only in substrate mixes with 10% and 50% pine bark addition. Adding 50% pine bark had no effect on PAW in recycled sand and scoria, increased PAW in crushed concrete, and decreased it in crushed rock (Figure 1xvii–xx). Only the recycled sand mixes had acceptable PAW for plant growth (22% and 23% *w*/*w*). All other substrate mixes had unacceptable PAW, with the lowest values observed in crushed concrete with 10% pine bark addition (5% *w*/*w*) and in both crushed rock mixes (5% and 2% *w*/*w*). Notably, 50% pine bark addition improved PAW in crushed concrete by 7% *w*/*w*, making it equivalent to both scoria mixes (12% *w*/*w*). Though not formally measured, we observed that mineral components had drastically different water retention and drainage times after irrigation, regardless of pine bark addition rates. Recycled sand drained much more slowly than the coarser components, especially crushed rock, which drained rapidly. All mineral components were very strongly or hyper-alkaline (pH 9.2–12.3) and unsuitable for plant growth (acceptable threshold: pH < 8.5; Figure 1xxi–xxiv), but crushed concrete was the most alkaline (pH 12.3). A significant interaction between pine bark addition rate and mineral component affected pH ($\chi^2_{(9)} = 171.9$ and $p < 0.001$). Except for crushed concrete, adding 50% pine bark reduced pH to acceptable levels across all other mineral components, with reductions most pronounced in scoria, followed by crushed rock and recycled sand (Figure 1xxi–xxiv).

Electrical conductivity (EC) followed a similar pattern to alkalinity, with crushed concrete being the most saline (1.1 dS m^{-1}, $p < 0.001$). However, all mineral components and substrate mixes had EC in the acceptable range for plant growth (<2.2 dS m^{-1}; Figure 1xxv–xxviii). In crushed concrete, 50% pine bark addition significantly lowered EC, aligning it with recycled sand with no pine bark (0.4 dS m^{-1}, $p < 0.001$).

2.2. Plant Biomass and Allocation in the Container Experiment

There was no plant mortality in the container experiment. Shoot and root biomass were highly correlated with total biomass (R = 0.97 and 0.93, $p < 0.001$); therefore, only total biomass results are presented. Both species shared similar patterns of total biomass response across substrate mixes (Figure 3). A significant interaction between pine bark addition rate and mineral component affected total biomass ($\chi^2_{(3)} = 15.1$, $p = 0.002$). In most substrate mixes, 50% pine bark addition resulted in lower total biomass compared to 10% addition, with the largest decrease observed in scoria and crushed rock substrates. However, in crushed concrete, the negative effect of 50% pine bark was less pronounced, with no significant difference in mean total biomass between 10% and 50% pine bark mixes ($p = 0.2$; Figure 3).

Biomass allocation—root mass fraction (RMF) and leaf mass fraction (LMF)—was not significantly affected by pine bark addition rate, mineral component, or their interaction ($p > 0.05$, Table S1).

Overall, *Goodenia ovata* was the more vigorous species, exhibiting significantly greater total, shoot, and root biomass, and LMF than *Alyogyne huegelii* when comparing within pine bark rates (all $p < 0.001$; Table S2). However, *Alyogyne huegelii* had significantly greater RMF than *Goodenia ovata* ($p < 0.005$; Table S1).

https://doi.org/10.3390/plants15030403

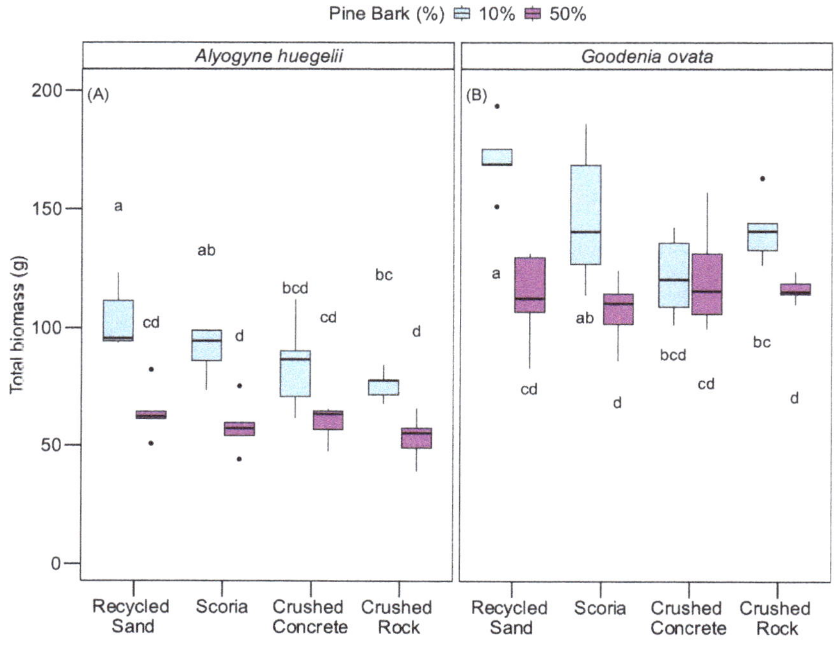

Figure 3. Effect of substrate mix (pine bark rate x mineral component) on total biomass (g) for (**A**) *Alyogyne huegelii* and (**B**) *Goodenia ovata*. The X-axis represents mineral component (recycled sand, scoria, crushed concrete, and crushed rock). Colours indicate pine bark rate (10% = blue, 50% = purple). Black dots indicate outlier values. Different letters above or below boxes denote significant differences among mean total biomass across the interaction of pine bark addition rate and mineral component within each species.

2.3. Relationships Between Substrate Physical and Hydrological Properties and Plant Biomass

We assessed relationships between substrate physical (CI, AFP, BD) and hydrological (WHC, PAW) properties and total biomass. Air-filled porosity and CI were strongly positively correlated with each other ($R = 0.8$, $p = 0.02$) and exhibited similar slopes in their relationships with total biomass (Table S3). Therefore, only CI is presented here and in Figure 4 to represent physical properties. Coarseness index and PAW were strongly negatively correlated with each other ($R = -0.86$, $p = 0.006$). For *Alyogyne huegelii*, total biomass was significantly negatively related to CI in the 10% pine bark substrate mixes ($p = 0.02$; Figure 4A), but this relationship was not significant in the 50% mixes ($p = 0.22$). In contrast, total biomass showed a significant positive relationship with PAW in the 10% pine bark substrate mixes ($p = 0.003$; Figure 4C) and a marginally positive relationship in 50% mixes ($p = 0.04$). For *Goodenia ovata*, a similar pattern was observed; total biomass was significantly negatively related to CI in the 10% pine bark substrate mixes ($p = 0.02$; Figure 4B), but this relationship was marginal for 50% mixes ($p = 0.04$). Total biomass was significantly positively related to PAW in the 10% mixes ($p = 0.02$; Figure 4D), but this was marginal for 50% mixes ($p = 0.05$). No significant relationships were found between total biomass and either BD or WHC in either species, across both pine bark addition rates ($p < 0.05$; Table S3).

https://doi.org/10.3390/plants15030403

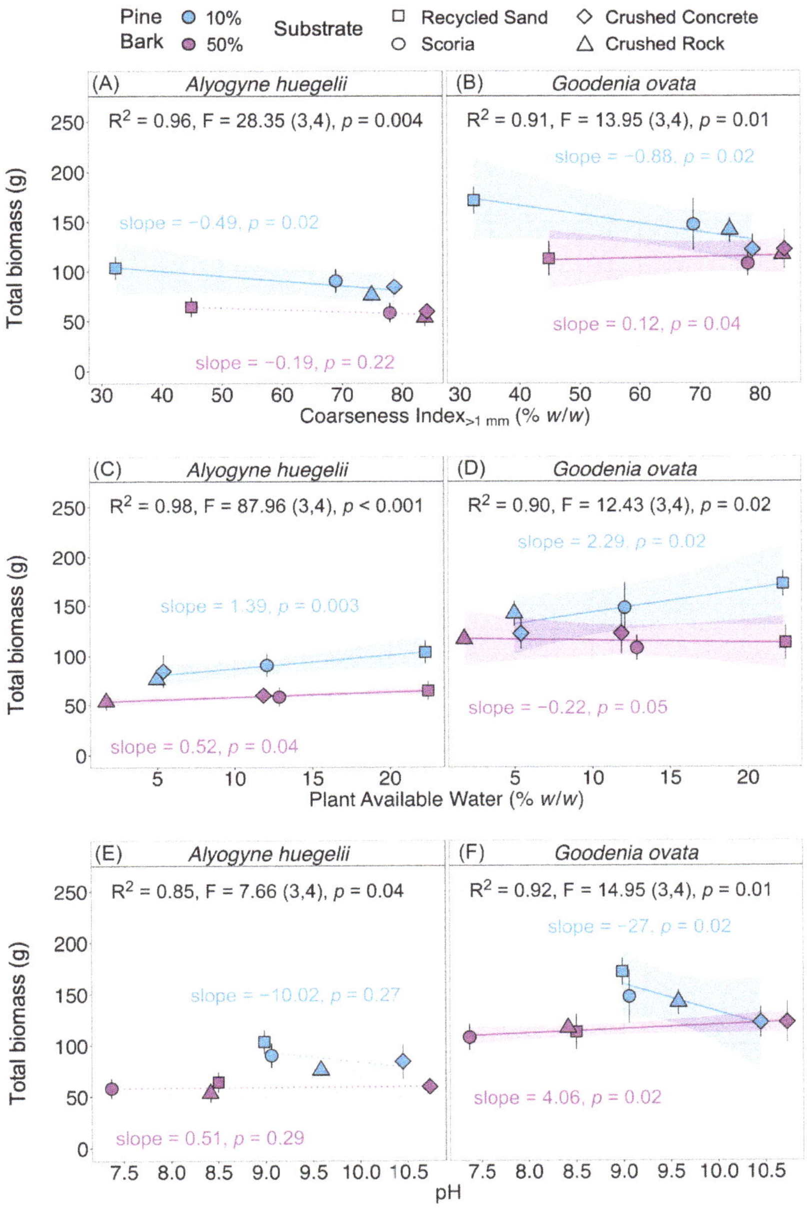

Figure 4. Relationships of mean total biomass (g), against (**A**,**B**) coarseness index$_{>1\ mm}$ (% *w*/*w*), (**C**,**D**) plant available water (% *w*/*w*), and (**E**,**F**) pH. Columns indicate species: *Alyogyne huegelii* (**A**,**C**,**E**) and *Goodenia ovata* (**B**,**D**,**F**). Colours indicate pine bark volume (10% = blue, 50% = purple). Symbols represent mineral components (recycled sand = squares, scoria = circles, crushed concrete = diamonds, and crushed rock = triangles). Linear relationship significance is represented by lines (solid = significant and dotted = insignificant), with shading representing standard error. Property–pine bark interaction is summarised as slope and *p*-value (*p*). Linear model fit for each property within each species is presented as model coefficient (R^2), F-statistic (degrees of freedom), and *p*-value (*p*).

https://doi.org/10.3390/plants15030403

2.4. Relationships Between Substrate Chemical Properties and Total Biomass

We assessed the relationships between substrate chemical (pH and EC) properties and total biomass. For *Alyogyne huegelii*, there were no significant relationships between total biomass and pH in either the 10% or 50% pine bark substrate mixes ($p = 0.2$; Figure 4E). In *Goodenia ovata*, total biomass had a significant negative relationship with pH in the 10% pine bark substrate mixes ($p = 0.02$; Figure 4F), with a marginal effect of pH observed in the 50% mixes ($p = 0.02$). No significant relationships were found between total biomass and EC in either species ($p < 0.05$; Table S3).

3. Discussion

Results from this study demonstrate that pine bark addition improved the physical and chemical properties of alkaline construction wastes so that they may be used as capping substrates for naturalistic plantings. This study demonstrates that even at low rates of pine bark addition, construction wastes can be a sustainable substitute to quarried scoria in capping substrates, though high coarseness and low water retention may limit biomass growth, and hyper-alkaline crushed concrete may limit biomass in alkali-sensitive species.

3.1. Effect of Pine Bark Addition on Substrate Properties

Overall, greater rates of pine bark improved air-filled porosity (AFP) and dry bulk density (BD) across all mineral components. However, these benefits were consistently accompanied by increased coarseness index (CI) and often by reductions in water-holding capacity (WHC) or plant available water (PAW). These results generally support our hypothesis that greater rates of pine bark will improve physical properties, except in the case of CI. However, greater rates of pine bark had inconsistent effects on hydrological properties, so we cannot confirm that aspect of the hypothesis. In scoria and recycled sand, pine bark increased AFP and reduced BD, but also raised CI, which, in combination, may explain the lower WHC observed in scoria. However, PAW remained unchanged between 10% and 50% rates of pine bark in both substrates. In crushed concrete, greater rates of pine bark improved AFP, BD, and PAW, but also increased coarseness and reduced WHC. In crushed rock—the coarsest mineral component—pine bark improved AFP and BD but reduced both WHC and PAW.

These variable effects on hydrological properties can be attributed to the particle size distributions (summarised as CI) of both the mineral components and pine bark. The pine bark used was "dominantly coarse" (6–10 mm), likely increasing macro-porosity in already coarse substrates—scoria, crushed concrete, and crushed rock—leading to reduced water retention, expressed as either lower WHC or PAW. Previous studies have shown that pine bark's particle size strongly influences its effect on substrate physical and hydrological properties [60,65]. Coarse particles (>6.4 mm) can increase AFP and drainage but lower water availability [66], whereas fine particles (<0.5 mm) can decrease AFP and increase WHC and PAW [60].

The trade-off between CI and water retention was most extreme in crushed rock substrates at both pine bark rates, where high CI and low PAW indicated faster drainage rates. The crushed rock particles had the most angular particles out of all the recycled mineral components (Figure S1). Interlocking of angular particles can create larger pores and atypical capillary action [22], which has been reported to increase drainage and decrease water retention in recycled substrate studies using crushed porcelain, demolition aggregate, and crushed brick [24,45,67]. However, crushed concrete substrates, which had similar CI to crushed rock but greater WHC and PAW, had more rounded particles. Concrete is more reactive to weathering than basalt rock as it has faster dissolution rates [68,69], which

https://doi.org/10.3390/plants15030403

may produce less angular particles over a shorter time (concrete: 1–2 years vs. basalt: >10 years; [70,71].

In our study, pine bark addition increased AFP and decreased BD, which aligns with physical and hydrological property trade-offs observed in the literature. For example, Blythe and Merhaut [72] observed a negative relationship between total porosity (AFP and WHC) and density (BD) across 127 container media blends, including pine bark, peat, and sand. Similarly, Graceson [46] found that organic matter additions decreased BD and increased WHC, while AFP increased with coarse compost (10–25 mm) addition and decreased with finer compost (0–10 mm). However, unlike these studies, we did not observe a consistent relationship between BD and WHC or AFP and WHC. This may be due to our limited sample of mineral components, with two dominantly coarse materials and only one dominantly fine material, as well as one intermediate material, which limited the sample of CI to more extreme values on the scale. Future work should assess a greater range of materials sampling at regular intervals along the CI range to fully characterise the trade-offs between physical and hydrological properties of construction wastes.

Collectively, these results emphasise the importance of balancing particle size distribution when designing capping substrates. Overly coarse mixtures may compromise water retention, while blends with appropriate proportions of fine and coarse particles can support favourable physical and hydrological properties for plant growth [26,58].

All mineral components were too alkaline for plant growth, although salinity was within the acceptable range. Increasing pine bark addition reduced pH non-linearly, and only the 50% rate lowered pH into the acceptable range for recycled sand, scoria, and crushed rock. Crushed concrete remained hyper-alkaline even at the highest pine bark rate. These results generally support our second hypothesis that greater rates of pine bark addition improve chemical properties, except in the case of crushed concrete.

In contrast, previous green roof studies have reported that even low rates of organic matter can significantly reduce pH in strongly to hyper-alkaline mineral substrates [73,74]. For example, Molineux [54] found that a slightly acidic organic mix of 1:1 conifer-bark compost and medium clay soil, added at 15% and 25% v/v to alkaline "Carbon8" pellet green roof substrates, significantly reduced the pH from 11.8 to 9.1. Although not explicitly stated, the inclusion of medium clay soil likely contributed high proportions of mineral and organic fine particles, which may have enhanced the acidifying capacity of the organic matter, leading to the observed pH reduction [9,54].

As with physical properties, fine particles (<0.6 mm) strongly influence chemical properties such as pH by altering the composition of the substrate solution (also "soil solution"; [9,36]). In our study, the pine bark was moderately acidic (pH 5.8) but dominantly coarse (CI = 97%), with only 2% of particles < 0.6 mm. At lower addition rates (10% and 20% v/v), this equated to just 0.2% and 0.4% of total substrate volume composed of fine pine bark particles, making it unlikely that they could meaningfully alter the substrate solution.

It is therefore more likely that the pH reductions observed were driven by physical dilution, with coarse pine bark particles displacing alkaline mineral fines, rather than chemical acidification. This would explain why a high pine bark addition rate (50%) was required to reduce pH in the strongly alkaline substrates (recycled sand, scoria, and crushed rock), and why even at this rate, crushed concrete remained hyper-alkaline. Our findings highlight the limited acidifying effect of coarse pine bark, particularly in hyper-alkaline substrates like crushed concrete, where substantial additions were still insufficient to bring pH into the suitable range for plant growth.

Overall, these results reinforce the importance of particle size in shaping physical, hydrological, and chemical substrate properties. Achieving pH levels suitable for plant growth in alkaline mineral components may require not only organic amendments, but

https://doi.org/10.3390/plants15030403

also adequate quantities of fine mineral or organic particles capable of chemically altering the substrate solution to reduce alkalinity.

3.2. Plant Response to Physical and Hydrological Properties Altered by Pine Bark Rate

Contrary to our first hypothesis, substrates with 10% pine bark addition had greater total biomass than substrates with 50% addition, except in crushed concrete. Within substrates with 10% pine bark, plant biomass was greatest in recycled sand and lowest in crushed rock and crushed concrete, whereas within substrates with 50% pine bark, plant biomass was similar among all mineral components.

We found that biomass was related to AFP, CI, and PAW, which were generally more suitable for plant growth in 10% mixes compared to 50% mixes, except for AFP. Recycled sand substrates were "dominantly fine" and had the highest PAW and lowest AFP, which, according to guidelines, would limit plant growth. However, we observed the opposite effect, like in the study by Ekşi and Rowe [45], where a green roof substrate with 3% AFP had the greatest biomass. As the recycled sand is a washed soil material screened to $\leq$3 mm, it likely contained silt and clay particles < 0.6 mm in size. As discussed, mineral and organic fines are known to increase WHC and PAW of substrates, while reducing AFP and increasing BD [9,36]. This increase in water retention can result in greater plant biomass yield [75].

Ekşi and Rowe [45] found expanded shale substrates retained more water after irrigation than coarser recycled substrates using foamed glass or crushed porcelain, resulting in greater total biomass of sedum and basil plants. They also found that even if the recycled substrates had similar or higher WHC to expanded shale, the combination of their greater coarseness (CI < 40%), considerably higher AFP, and rapid drainage resulted in lower water retention, which reduced plant biomass [45]. These results are reflected in our study, where crushed concrete and crushed rock 10% pine bark substrates and all 50% pine bark substrates had high CI and AFP, indicating greater drainage [45,75], which resulted in lower biomass compared to recycled sand with 10% pine bark.

Graceson [76] found similar relationships among coarseness, water retention, and biomass of sedum, where they identified WHC as an important predictor for water retention, and thus biomass growth on green roofs. Biomass was greatest in crushed tile substrates with 30% v/v of composted green waste, which had the highest proportion of fine particles (41% < 1 mm) and the greatest WHC (33%) compared to crushed brick and Lytag® substrates [76]. However, when the rate of composted green waste was lowered to 20% v/v, all substrates had lower WHC and generally greater coarseness, which resulted in lower sedum biomass compared to 30% substrates, and similar biomass across all mineral components [76]. This aligns with the relationships we found among CI, PAW, and biomass, with increasing CI and lower PAW reducing biomass.

Our study demonstrates that substrate coarseness, which was influenced by mineral component and pine bark rate, is a key driver of available water and thus plant performance. The finest substrate mixes had the highest water availability and the best plant growth, which highlights the importance of supplying sufficient fine particles in capping substrates. We also found that while 50% pine bark addition did not affect plant survival, it reduced plant growth, which contrasts with previous green roof and agricultural studies that have demonstrated a strong positive effect of greater organic matter addition rates on plant biomass or yield [73,77]. In addition to the greater drainage and reduced water retention from 50% pine bark addition, it is possible that nitrogen availability was reduced due to microbial immobilisation (i.e., nitrogen drawdown; [78]) in response to the large carbon source provided by pine bark. This limitation may have added to the unfavourable physical and hydrological conditions, resulting in lower plant biomass in substrates with 50% pine

https://doi.org/10.3390/plants15030403

bark. However, pine bark at 50% also reduced the variation in physical and hydrological properties among mineral components, which aligned more extreme crushed concrete and crushed rock with scoria and recycled sand, making them comparable candidates for capping substrates.

3.3. Alkaline-Sensitive Plant Response to pH Altered by Coarseness

Despite greater rates of pine bark improving chemical properties in recycled sand, scoria, and crushed rock, this did not translate to greater biomass in either species, as 50% mixes generally had worse growth. For alkaline-sensitive *Goodenia ovata*, biomass was similar between pine bark rates in crushed concrete, where the pH was extreme and unaffected by increased pine bark addition. The biomass of both species in 50% mixes seemed to be influenced more by physical properties, rather than pH.

However, in 10% mixes, there was clear evidence of an alkalinity effect on *G. ovata*, with hyper-alkaline crushed concrete resulting in the lowest biomass of all the mineral components. Poor growth of alkaline-sensitive plants in concrete-based substrates was documented in a green roof study where meadow species with a strong preference for acid soils did not survive, but species with broad pH tolerance or alkaline tolerance had high survival and abundance [47]. Despite the high alkalinity of the substrates in our study, there was no mortality of either species in any of the substrate mixes, and though the biomass of sensitive species was reduced in crushed concrete substrates, it was no different from substrates with 50% pine bark, where all mineral components had similar biomass. This is a promising outcome which shows construction wastes such as recycled sand, crushed rock, and even the hyper-alkaline crushed concrete, may be used as capping substrates in place of scoria, the current preferred mineral material for capping substrates in woody naturalistic plantings. However, construction wastes must be amended with pine bark as their properties are too extreme for plant growth.

3.4. Limitations

We focused on only two plant species to understand how alkalinity may affect plant growth using waste substrates. However, to assess the broader application of construction wastes as capping substrates in urban naturalistic plantings, experiments with a larger plant palette are required. In addition, container studies of monocultures are not always representative of true field conditions in the urban landscape [24]. Future work should assess these capping substrates overlying urban soil and the interacting effect of substrate depth on plant performance. Further, to comprehensively compare construction wastes to industry standards like scoria and sand, we must understand how they can function as mulch to suppress weeds. Future work should assess the viability of construction wastes functioning as mulch in comparison to sand or scoria. The results obtained in this study show that pine bark-amended construction wastes can support plant growth in containers, but similar experiments in a landscape context are required to fully validate their function as capping substrates in urban naturalistic plantings.

4. Materials and Methods

4.1. Substrate Components

Four types of mineral components were evaluated, scoria—as a reference mineral component—and three recycled waste materials: recycled sand, crushed concrete, and crushed rock. The scoria was sourced from Aerolite Quarries Pty Ltd. (Anakie, VIC, Australia) and is a quarried volcanic material, crushed and screened to $\leq$8 mm, including fine particles. The recycled materials were sourced from Repurpose It® Pty Ltd. (Epping, VIC, Australia), a construction and demolition waste recycling company based in Melbourne,

https://doi.org/10.3390/plants15030403

Australia. The crushed concrete and crushed rock materials are crushed and screened to $\leq$20 mm minus, and can have variable origins, depending on supply from construction projects. The crushed concrete used in this experiment was sourced from construction and demolition mixed waste from the Greater Melbourne region and included 1.7% metal, glass, brick, ceramics, or slag. The crushed rock used in this experiment is composed of Tertiary Olivine basalt, which is naturally alkaline, excavated from the Greater Melbourne region. The recycled sand is a washed and screened soil material derived from EPA category C and D contaminated soils [79]. The recycled sand is processed through a wash plant to remove contaminants and then screened to $\leq$3 mm, including fines.

The organic component—aged *Pinus radiata* bark mulch (hereafter "pine bark")—was sourced from Repurpose It® Pty Ltd. (Epping, VIC, Australia) and is a screened waste material from the timber industry (6–10 mm). The pine bark had the following properties: high coarseness index (CI = 97% w/w), high air-filled porosity (AFP = 49% v/v), low dry bulk density (BD = 0.2 g cm^{-3}), moderate water-holding capacity (WHC = 25% v/v), moderate acidity (pH = 5.8), and low salinity (EC = 0.2 dS m^{-1}). Pine bark typically has low nitrogen content, between 0.2 and 0.76% dry mass [9,80]. Figure S1 shows the appearance of each substrate component.

4.2. Physical and Chemical Properties of Substrate Components and Mixes

Substrate mineral and organic components and substrate mixes were tested for physical, hydrological, and chemical properties. The four mineral components were mixed with three rates of pine bark (10%, 20%, and 50% v/v) to create 12 different substrate mixes for testing physical (AFP and BD), hydrological (WHC), and chemical properties (pH and EC). Particle size distribution, summarised as coarseness index (CI), was tested on the individual substrate components and on substrate mixes with 10% or 50% v/v pine bark addition. Plant available water was evaluated on mixes with 10% or 50% v/v pine bark addition. Table 1 describes the acceptable ranges for each substrate property.

Physical properties and WHC (all water held in substrate below −10 kPa) were measured on three replicates of each substrate component and substrate mix. Air-filled porosity, WHC, and BD were measured using methods described in the Australian Standard for potting mixes (AS 3743-2003, [31]). In summary, using a special apparatus (described in Conn [33]), approximately 720 mL of each substrate mix was immersed in water to soak for 30 min and then drained for 5 min, followed by two more soak (10 min)-and-drain (5 min) cycles. Then, the top part of the apparatus (as described in Conn [33]) was removed, and excess substrate was scraped flush with the top of the apparatus base. The apparatus was then immersed a final time for 2 min to achieve full saturation, then drained into a measuring container for 30 min. Before the last drain period, drainage holes at the bottom of the apparatus were sealed before the samples were removed from the water to ensure all water held within the apparatus was kept and released into the measuring container. The volume of the final drained water (V_{water}) was measured in mL and used with the substrate volume (V_{sub}) to calculate AFP:

$$\text{AFP } (\%\ v/v) = \left(\frac{V_{water}}{V_{sub}}\right) \times 100 \tag{1}$$

The ~720 mL substrate was then oven-dried (105 °C for 5 days) and weighed. The substrate oven-dry weight (W_{dry} in g), saturated weight after 30 min of drainage (W_{sat} in g), and its volume were used to calculate WHC and dry BD.

$$\text{WHC } (\%\ w/w) = \left(\frac{W_{sat} - W_{dry}}{V_{sub}}\right) \times 100 \tag{2}$$

https://doi.org/10.3390/plants15030403

$$BD_{dry}(\text{g cm}^3) = \frac{W_{dry}}{V_{sub}} \quad (3)$$

Particle size distribution was measured following the methods described in Cao [32] and Werdin [58]. Substrates were oven-dried at 105 °C to achieve constant weight, and two replicates of 1 L were passed through a nest of sieves in descending order from 10 mm, 6.3 mm, 4 mm, 2 mm, 1 mm to 0.6 mm aperture. The coarseness index [28] was calculated as the weight of all particles < 1 mm ($W_{>1\,mm}$ in g) divided by the total sample weight (W_{Total} in g):

$$CI_{>1mm}\%(w/w) = \left(\frac{W_{>1\,mm}}{W_{Total}}\right) \times 100 \quad (4)$$

Plant available water was calculated as moisture content (%) measured at container capacity (−10 kPa) minus moisture content at the permanent wilting point (−1500 kPa). To measure moisture content at container capacity, we adapted the methods of Nelson [81] for use with coarse substrates < 2 mm in particle size. Duplicate samples were placed into 100 mm diameter funnels with Whatman no. 1 filter paper and saturated with deionised water, and then left to drain for 24 h. Wet substrate (W_{wet} in g) samples were weighed, then oven-dried (105 °C for 4 days) and reweighed (W_{dry} in g) to determine moisture content at container capacity as percentage weight (% w/w):

$$\textit{Moisture content}\ \%(w/w) = \left(W_{wet} - W_{dry}\right) \times 100 \quad (5)$$

To measure moisture content at permanent wilting point, we adapted the pressure plate method described in Yeates [82] for use with coarse substrates < 2 mm in particle size. Approximately 70 mL of each substrate mix was placed into 50 mm diameter × 30 mm deep metal rings on a ceramic pressure plate and then saturated with deionised water overnight. The ceramic pressure plate with the saturated samples was then placed into the pressure plate apparatus (Soilmoisture Equipment Corp., CA, USA), sealed, and the upper chamber was slowly brought up to 1500 kPa, and the sample moisture content was left to reach equilibrium over seven days. The wet substrate samples were weighed after removal from the pressure plate apparatus, then oven-dried (105 °C for 5 days) and reweighed to determine moisture content at permanent wilting point as % w/w using Equation (5).

Chemical properties were measured on three replicates of each substrate component and substrate mix. We measured EC and pH in a 1:5 substrate to water solution, according to the Australian standard for soils for landscaping and garden use (Appendix D; AS 4419-2018, [38]) using a lab-bench pH and conductivity meter (labCHEM, TPS International Pty Ltd., Brisbane, QLD, Australia). To account for the larger particle sizes of the substrate components, we adjusted the method to use 80 g of air-dried substrate and 400 mL of deionised water in a 500 mL bottle for each sample. Substrate–water solutions were shaken for one hour in an end-to-end shaker, and pH was measured immediately. The solution was then left to settle overnight and then filtered using Whatman No. 1 filter papers before measuring EC.

4.3. Plant Growth Experiment

We evaluated the plant response of two species in a container experiment undertaken at the Burnley Campus, University of Melbourne, Australia (latitude 37°47′ S; longitude 144°58′ E). Eight substrate mixes, composed of the four mineral components (recycled sand, scoria, crushed concrete, crushed rock) and pine bark at two rates of addition (10% and 50% v/v), were evaluated. Substrate components were mixed for two minutes in a clean 99 L concrete mixer (Lightburn & Co., Ltd., Adelaide, SA, Australia) to create a total of 126 L of each substrate mix for the experiment. After mixing, substrates were dumped on a concrete

https://doi.org/10.3390/plants15030403

pad and turned manually with a shovel for one minute to incorporate coarse particles. We filled 10 replicates of 10.5 L "Citrus" containers (d 250 mm × h 275 mm, P250C, Garden City Plastics Pty Ltd., Dandenong South, VIC, Australia) with nine litres of each substrate mix for a total of 80 containers. Substrate was lightly shaken to level but not compact it in each container.

We selected two vigorous fast-growing woody plant species that have been shown to perform well in previous installations of woody naturalistic plantings using a scoria capping substrate [62]: *Alyogyne huegelii* and *Goodenia ovata*. *A. huegelii* is a medium-sized ornamental shrub up to 2.5 m high with dull green, hairy, deeply three- to five-lobed leaves and large, lilac or mauve flowers [83] (Figure 5A). *G. ovata* is a small–medium-sized ornamental shrub up to 1 m high with glossy green oval leaves that have toothed margins, and small bright yellow fan-like flowers [84] (Figure 5B). These species have contrasting tolerance of alkalinity; *A. huegelii* tolerates acidic to alkaline conditions [63], whereas *G. ovata* tolerates acidic to neutral conditions [64].

Figure 5. The two woody plant species used in the experiment are (**A**) *Alyogyne huegelii* and (**B**) *Goodenia ovata*. Photos show the whole plant, leaf, and flower morphology. Species occurrences in Australia are displayed in the bottom panel for each species, adapted from Atlas of Living Australia [83,84], red dots indicate occurrences.

Mature plants were purchased in 14 cm pots from a wholesale nursery (Plantmark Pty Ltd., Melbourne, VIC, Australia). Planting into containers filled with substrate mixes was completed on 6 March 2023 (late summer), with each plant receiving a surface application of controlled-release fertiliser (2.7 g per container of Osmocote® Pro Low P, Scotts Australia Pty. Ltd., Sydney, NSW, Australia; NPK of 14:1.3:14.9), equivalent to 0.38 g nitrogen per plant. One week after planting, all plants were pruned to 40 cm above the substrate surface to promote multi-stemmed growth. All plants also received additional liquid fertiliser applications (1 L per container) of Peters Professional® All Rounder (NPK of 20:8.7:16.6, rate: 0.5 g L^{-1}), equivalent to 0.1 g nitrogen per plant on four occasions; once in May and October, and twice in November 2023. Plants were irrigated two to four times daily with overhead rotary sprinklers (12 L min^{-1}) except for three months over the winter (May–August 2023). Additional hand-watering (approximately 1 L per container) occurred once (18 March 2023) when the maximum daily temperature exceeded 35 °C. Although soil moisture was not directly monitored, it is unlikely that containers remained at saturated

https://doi.org/10.3390/plants15030403

WHC following irrigation events, given the inherent water loss through large container drainage holes.

4.4. Plant Biomass and Allocation

Plants were destructively harvested after 10 months of growth (March 2023 to January 2024) to measure shoot, leaf, and root biomass. Above-ground biomass was removed and separated into leaves and stems. Roots were removed from the containers and carefully washed to remove the substrate mix. Stems, leaves, and roots were oven-dried at 60 °C until a constant dry weight was achieved. Dry mass was used to calculate total biomass (g), and plant allocation metrics; root mass fraction (g g^{-1}; RMF), and leaf mass fraction (g g^{-1}; LMF).

4.5. Statistical Analyses

All statistical analyses were performed in R version 4.1.2 [85]. All response variables (substrate properties and plant biomass or allocation) were tested for normality using the Shapiro–Wilk test.

To test whether greater rates of pine bark addition improved physical, hydrological, and chemical properties and resulted in greater plant growth, we used linear regression to analyse the normally distributed response variables. These included substrate properties (AFP, BD, pH, and WHC) and plant traits (mean total, shoot, and root biomass; RMF; and LMF). Response variables were analysed against treatments of pine bark addition rate, mineral component, and their interaction (substrate mix). To test the significance of treatments, we used two-way ANOVA (type III) from the "car package" [86]. Where significant effects were found, the "emmeans" package [87] was used to determine estimates of the group means for each response variable and differences between group means were tested with post hoc pairwise comparisons using Tukey's HSD method at a significance level of 0.05. Mean pairwise comparisons were performed (1) across pine bark rates (0, 10, 20, and 50% v/v) within each mineral component, and (2) among substrate mixes (interaction term) for each response variable. Analyses of plant response were conducted separately for each species.

For right-skewed EC, transformations could not achieve a normal distribution, so generalised linear regression using the Gamma family with a log link was used to analyse the response against treatments of pine bark addition rate, mineral component, and substrate mix. For the left-skewed CI, values were bounded (0–100%) and so were rescaled to a proportion (0–1) and then analysed with beta regression using the "betareg" package [88]. Post hoc pairwise comparisons of estimated means among pine bark addition rates were performed within each mineral component and across substrate mixes, with Bonferroni-adjusted p-values to control for multiple comparisons of the four pine bark addition rates and eight substrate mixes.

Plant available water (PAW) was excluded as a response variable due to limited replication (n = 1 per substrate mix) and is summarised descriptively.

To test whether improved substrate properties led to greater plant growth, we used linear regression models with plant biomass and allocation as response variables and substrate properties as predictors, including interaction terms with pine bark rate. Analyses were conducted separately for each species. We report slope estimates, model fit (R^2, F-statistic, p-value), and significant interactions. Degrees of freedom for the F-statistic are reported as "F(df1, df2)". Collinearity among substrate properties was tested using Pearson's correlation from the "psych" package [89], with a significance threshold < 0.05. The correlation coefficient (R) among properties is reported in text as (R, p-value).

To test whether reduced alkalinity led to better growth of the alkaline-sensitive species (*G. ovata*) compared to the tolerant species (*A. huegelii*), we used linear regression models with plant traits as response variables and species as the predictor. Where significant effects were found, differences between group means were tested with post hoc pairwise comparisons using Tukey's HSD method at a significance level of p-value < 0.05.

Significant pairwise or multiple comparison differences were visualised using compact letter displays from the "multcomp" package [90]; groups sharing the same letter are not statistically different. All figures were generated using the ggplot2 package [91]. Data shown in figures and tables are on the original (non-transformed) scale.

5. Conclusions

The three recycled mineral components and scoria did not have acceptable properties for plant growth without pine bark addition. While greater pine bark rates generally improved the physical properties of the mineral substrates more than lower rates, this did not translate to greater plant growth. In the nursery experiment, *Goodenia ovata* had greater biomass than *Alyogyne huegelii*, but its growth was reduced by high alkalinity in 10% pine bark and high coarseness in 50% pine bark substrates. Across mineral components, 10% pine bark mixes had greater biomass than 50% pine bark mixes for both species, and recycled sand with 10% pine bark had the greatest biomass out of the construction wastes. However, biomass was similar in all substrates with 50% pine bark addition. We identified that dominantly fine substrates that had high PAW and lower alkalinity, such as recycled sand with 10% pine bark, were the best candidates for increasing plant biomass. This emphasises the importance of balancing particle size so that there are enough mineral and organic fine particles to increase water retention and lower pH. We have identified recycled sand, crushed concrete, and crushed rock as three new recycled materials that can be used with pine bark as capping substrates for urban woody naturalistic plantings. Replacing scoria with these recycled materials could enhance the sustainability of capping substrates for naturalistic plantings, while enhancing the biodiversity and amenity of informal green spaces.

Supplementary Materials: The following supporting information can be downloaded at: https://www.mdpi.com/article/10.3390/plants15030403/s1, Figure S1: Photographs illustrating the shape and size of particles in the five substrate components. Table S1: ANOVA output of pine bark addition rate and mineral component effects on plant response variables. Table S2: Pearson correlation of substrate physical, hydrological, and chemical properties. Table S3: Linear regression model output of substrate properties and pine bark addition rate effects on total plant biomass, per species.

Author Contributions: Conceptualization, C.F., C.K., J.P.R. and S.J.L.; methodology, C.F., C.K., J.P.R. and S.J.L.; validation, C.K.; formal analysis, C.K.; investigation, C.F., C.K., J.P.R. and S.J.L.; resources, C.F., C.K., J.P.R. and S.J.L.; data curation, C.K.; writing—original draft preparation, C.K.; writing—review and editing, C.F., C.K., J.P.R. and S.J.L.; visualisation, C.K.; supervision, C.F., J.P.R. and S.J.L.; project administration, C.K.; funding acquisition, C.F., C.K. and J.P.R. All authors have read and agreed to the published version of the manuscript.

Funding: This research was funded by the Australian Research Council (ARC) Linkage Partnership grant (LP190100536—Resilient and adaptable urban landscapes: low input Woody Meadows) and the partners of the Woody Meadow Project. C.K. was supported by the Australian Government Research Training Program, the Maurice and Helen Wood Memorial Scholarship (2022–2023), the Frank Keenan Trust Fund Scholarship (2023), and the Dr Betty Elliot Horticulture Scholarship (2023).

Data Availability Statement: The original data presented in the study are openly available in FigShare at https://doi.org/10.26188/30918680.

Acknowledgments: The authors acknowledge the Traditional Custodians of the land on which the project took place, the Wurundjeri and Bunurong people of the Kulin Nations, and we pay our respects to their Elders, past, present, and emerging. We would like to sincerely thank Tania Hall and the team at Repurpose IT for the supply of recycled sand, crushed rock, and crushed concrete for this study, and Sam Oddo for his technical advice. We would like to thank the University of Melbourne colleagues, Lavinia Chu, Bihan Guo, Chris Szota, Theresa Lam, Jennifer Fouad Khattar, Nicole Dowling, Claudia Helena Giraldo Escobar, Sophia Blosfelds, and Julia Schiller, for their help planting and harvesting the experiment; Sivashneel Sanjappa for help with lab work and data acquisition; and David J Price for statistical advice. We would also like to thank Burnley campus and the nursery staff, Lisa Wittick, Rowan Berry, and Brett Hough, for their technical assistance throughout the experiment. We also thank our funding partners for supporting our work. This research is part of the Australian Research Council (ARC) Linkage Partnership grant (LP190100536–Resilient and adaptable urban landscapes: low input woody meadows).

Conflicts of Interest: The authors declare no conflicts of interest. The funders had no role in the design of the study; in the collection, analyses, or interpretation of data; in the writing of the manuscript; or in the decision to publish the results.

Abbreviations

The following abbreviations are used in this manuscript:

AFP	Air-filled porosity
AS	Australian standard
BD	Dry bulk density
CI	Coarseness index
EC	Electrical conductivity
EPA	Environmental protection agency
LMF	Leaf mass fraction
PAW	Plant available water
RMF	Root mass fraction
WHC	Water-holding capacity

References

1. Babington, A.; Hughes, M.; Farrell, C.; Chambers, J.; Standish, R.J. Preference for Multi-Layered, Flowering, Woody Streetscape Plantings in a Mediterranean-Type Climate. *Urban For. Urban Green.* **2023**, *89*, 128094. [CrossRef]
2. Hitchmough, J.; Wagner, M.; Ahmad, H. Extended Flowering and High Weed Resistance within Two Layer Designed Perennial "Prairie-Meadow" Vegetation. *Urban For. Urban Green.* **2017**, *27*, 117–126. [CrossRef]
3. Bretzel, F.; Vannucchi, F.; Pezzarossa, B.; Paraskevopoulou, A.; Romano, D. Establishing Wildflower Meadows in Anthropogenic Landscapes. *Front. Hortic.* **2024**, *2*, 428–436. [CrossRef]
4. Horsfall, K.; Livesley, S.J.; Delpratt, J.; Hirst, M.J.; Williams, N.S.G. Interacting Effects of Sand, Slugs and Jute Drive Community Composition in Direct-seeded Urban Wildflower Meadows. *J. Appl. Ecol.* **2024**, *61*, 2762–2773. [CrossRef]
5. Craul, P.J. Urban Soil: Problems and Promise. *Arnoldia* **1991**, *51*, 23–32. [CrossRef]
6. Fischer, L.K.; von der Lippe, M.; Rillig, M.C.; Kowarik, I. Creating Novel Urban Grasslands by Reintroducing Native Species in Wasteland Vegetation. *Biol. Conserv.* **2013**, *159*, 119–126. [CrossRef]
7. Sloan, J.J.; Ampim, P.A.Y.; Basta, N.T.; Scott, R. Addressing the Need for Soil Blends and Amendments for the Highly Modified Urban Landscape. *SSSAJ* **2012**, *76*, 1133–1141. [CrossRef]
8. Horsfall, K.; Williams, N.S.G.; Michael, R.N.; Livesley, S.J. Rapid Root Development in Clay Subsoils Enhances the Early Growth of Native Grassland Species. *Plant Soil* **2024**, *511*, 1151–1172. [CrossRef]
9. Jim, C.Y. Edaphic Properties and Horticultural Applications of Some Common Growing Media. *Commun. Soil Sci. Plant Anal.* **1996**, *27*, 2049–2064. [CrossRef]
10. Rojas, J.A.; Dhar, A.; Naeth, M.A. Urban Green Spaces Restoration Using Native Forbs, Site Preparation and Soil Amendments—A Case Study. *Land* **2022**, *11*, 498. [CrossRef]
11. Bretzel, F.; Vannucchi, F.; Pini, R.; Scatena, M.; Marradi, A.; Cinelli, F. Use of Coarse Substrate to Increase the Rate of Water Infiltration and the Bearing Capacity in Tree Plantings. *Ecol. Eng.* **2020**, *148*, 105798. [CrossRef]
12. Somerville, P.D.; May, P.B.; Livesley, S.J. Effects of Deep Tillage and Municipal Green Waste Compost Amendments on Soil Properties and Tree Growth in Compacted Urban Soils. *J. Environ. Manag.* **2018**, *227*, 365–374. [CrossRef]
13. Schwartz, S.S.; Smith, B. Restoring Hydrologic Function in Urban Landscapes with Suburban Subsoiling. *J. Hydrol.* **2016**, *543*, 770–781. [CrossRef]

14. Somerville, P.D.; Farrell, C.; May, P.B.; Livesley, S.J. Biochar and Compost Equally Improve Urban Soil Physical and Biological Properties and Tree Growth, with No Added Benefit in Combination. *Sci. Total Environ.* **2020**, *706*, 135736. [CrossRef]
15. Valkó, O.; Kelemen, A.; Kiss, O.; Bátori, Z.; Kiss, R.; Deák, B. Grassland Restoration on Linear Landscape Elements—Comparing the Effects of Topsoil Removal and Topsoil Transfer. *BMC Ecol. Evol.* **2024**, *24*, 112. [CrossRef]
16. Gibson-Roy, P. Australian Grassy Community Restoration: Recognizing What Is Achievable and Charting a Way Forward. *Ecol. Manag. Restor* **2022**, *23*, 10–24. [CrossRef]
17. Morris, E.C.; Gibson-Roy, P. Comparison of Biomass Removal, Nutrient Manipulation and Native Seed Addition to Restore the Ground Layer of a Degraded Grassy Woodland. *Aust. J. Bot.* **2017**, *66*, 1–12. [CrossRef]
18. Hölzel, N.; Otte, A. Restoration of a Species-rich Flood Meadow by Topsoil Removal and Diaspore Transfer with Plant Material. *Appl. Veg. Sci.* **2003**, *6*, 131–140. [CrossRef]
19. Jaunatre, R.; Buisson, E.; Dutoit, T. Topsoil Removal Improves Various Restoration Treatments of a Mediterranean Steppe (La Crau, Southeast France). *Appl. Veg. Sci.* **2013**, *17*, 236–245. [CrossRef]
20. Resch, M.C.; Schütz, M.; Graf, U.; Wagenaar, R.; van der Putten, W.H.; Risch, A.C. Does Topsoil Removal in Grassland Restoration Benefit Both Soil Nematode and Plant Communities? *J. Appl. Ecol.* **2019**, *56*, 1782–1793. [CrossRef]
21. Buisson, E.; Holl, K.D.; Anderson, S.; Corcket, E.; Hayes, G.F.; Torre, F.; Peteers, A.; Dutoit, T. Effect of Seed Source, Topsoil Removal, and Plant Neighbor Removal on Restoring California Coastal Prairies. *Restor. Ecol.* **2006**, *14*, 569–577. [CrossRef]
22. Hitchmough, J.; Kendle, T.; Paraskevopoulou, A.T. Seedling Emergence, Survival and Initial Growth of Forbs and Grasses Native to Britain and Central/Southern Europe in Low Productivity Urban "Waste" Substrates. *Urban Ecosyst.* **2001**, *5*, 285–308. [CrossRef]
23. Egendorf, S.P.; Cheng, Z.; Deeb, M.; Flores, V.; Paltseva, A.; Walsh, D.; Groffman, P.; Mielke, H.W. Constructed Soils for Mitigating Lead (Pb) Exposure and Promoting Urban Community Gardening: The New York City Clean Soil Bank Pilot Study. *Landsc. Urban Plan.* **2018**, *175*, 184–194. [CrossRef]
24. Hitchmough, J.D.; Kendle, A.D.; Paraskevopoulou, A.T. Emergence, Survival and Initial Growth of North American Prairie Forbs and British Meadow Forbs and Grasses in Low-Productivity Urban "Waste" Soils. *J. Hortic. Sci.* **2003**, *78*, 89–99. [CrossRef]
25. Barrett, G.E.; Alexander, P.D.; Robinson, J.S.; Bragg, N.C. Achieving Environmentally Sustainable Growing Media for Soilless Plant Cultivation Systems—A Review. *Sci. Hortic.* **2016**, *212*, 220–234. [CrossRef]
26. Jim, C.Y. Urban Soil Characteristics and Limitations for Landscape Planting in Hong Kong. *Landsc. Urban Plan.* **1998**, *40*, 235–249. [CrossRef]
27. Séré, G.; Schwartz, C.; Ouvrard, S.; Sauvage, C.; Renat, J.-C.; Morel, J.L. Soil Construction: A Step for Ecological Reclamation of Derelict Lands. *J. Soils Sediments* **2008**, *8*, 130–136. [CrossRef]
28. Abad, M.; Fornes, F.; Carrión, C.; Noguera, V.; Noguera, P.; Maquieira, Á.; Puchades, R. Physical Properties of Various Coconut Coir Dusts Compared to Peat. *HortScience* **2005**, *40*, 2138–2144. [CrossRef]
29. Handreck, K.A.; Black, N.D. *Growing Media for Ornamental Plants and Turf*, 4th ed.; UNSW Press: Sydney, Australia, 2010; ISBN 978-1-74223-082-5.
30. Mathers, H.M.; Lowe, S.B.; Scagel, C.; Struve, D.K.; Case, L.T. Abiotic Factors Influencing Root Growth of Woody Nursery Plants in Containers. *HortTech* **2007**, *17*, 151–162. [CrossRef]
31. *AS 3743-2003*; Australian Standard for Potting Mixes. Standards Australia International: Sydney, Australia, 2003.
32. Cao, C.T.N.; Farrell, C.; Kristiansen, P.E.; Rayner, J.P. Biochar Makes Green Roof Substrates Lighter and Improves Water Supply to Plants. *Ecol. Eng.* **2014**, *71*, 368–374. [CrossRef]
33. Conn, R.; Werdin, J.; Rayner, J.P.; Farrell, C. Green Roof Substrate Physical Properties Differ between Standard Laboratory Tests Due to Differences in Compaction. *J. Environ. Manag.* **2020**, *261*, 110206. [CrossRef] [PubMed]
34. Farrell, C.; Cao, C.T.N.; Ang, X.Q.; Rayner, J.P. Use of Water-Retention Additives to Improve Performance of Green Roof Substrates. *Acta Hortic.* **2016**, *1108*, 271–278. [CrossRef]
35. Landscape Development and Landscaping Research Society e.V. (FLL). *Guidelines for the Planning, Construction and Maintenance of Green-Roofs*; Landscape Development and Landscaping Research Society e.V. (FLL): Bonn, Germany, 2018.
36. Craul, P.J. The Nature of Urban Soils: Their Problems and Future. *Arboric. J.* **1994**, *18*, 275–287. [CrossRef]
37. Hazelton, P.; Murphy, B. *Interpreting Soil Test Results*, 3rd ed.; CSIRO Publishing: Clayton South, VIC, Australia, 2016; ISBN 978-1-4863-0396-0.
38. *AS 4419-2018*; Australian Standard for Soils for Landscaping and Garden Use. Standards Australia International: Sydney, Australia, 2018.
39. Ampim, P.A.Y.; Sloan, J.J.; Cabrera, R.I.; Harp, D.A.; Jaber, F.H. Green Roof Growing Substrates: Types, Ingredients, Composition and Properties. *J. Environ. Hortic.* **2010**, *28*, 244–252. [CrossRef]
40. Kazemi, F.; Mohorko, R. Review on the Roles and Effects of Growing Media on Plant Performance in Green Roofs in World Climates. *Urban For. Urban Green.* **2017**, *23*, 13–26. [CrossRef]

41. United Nations DESA. *Transforming Our World: The 2030 Agenda for Sustainable Development*; United Nations DESA: New York, NY. USA, 2015.
42. Jiang, N.; Zou, W.; Lu, Y.; Liao, Z.; Wu, L. Using Recycled Construction Waste Materials with Varying Components and Particle Sizes for Extensive Green Roof Substrates: Assessment of Its Effects on Vegetation Development. *Sustainability* **2024**, *16*, 414. [CrossRef]
43. Kader, S.; Gratchev, I.; Michael, R.N. Recycled Waste Substrates: A Systematic Review. *Sci. Total Environ.* **2024**, *953*, 176029. [CrossRef]
44. Bates, A.J.; Sadler, J.P.; Greswell, R.B.; Mackay, R. Effects of Recycled Aggregate Growth Substrate on Green Roof Vegetation Development: A Six Year Experiment. *Landsc. Urban Plan.* **2015**, *135*, 22–31. [CrossRef]
45. Eksi, M.; Rowe, D.B. Green Roof Substrates: Effect of Recycled Crushed Porcelain and Foamed Glass on Plant Growth and Water Retention. *Urban For. Urban Green.* **2016**, *20*, 81–88. [CrossRef]
46. Graceson, A.; Hare, M.; Hall, N.; Monaghan, J. Use of Inorganic Substrates and Composted Green Waste in Growing Media for Green Roofs. *Biosyst. Eng.* **2014**, *124*, 1–7. [CrossRef]
47. Jauri, M.; Kuoppamäki, K.; Hagner, M.; Prass, M.; Suonio, T.; Fransson, A.-M.; Lehvävirta, S. Alkaline Habitat for Vegetated Roofs? Ecosystem Dynamics in a Vegetated Roof with Crushed Concrete-Based Substrate. *Ecol. Eng.* **2020**, *157*, 105970. [CrossRef]
48. Barredo, O.; Vilela, J.; Garbisu, C.; Besga, G.; Alkorta, I.; Epelde, L. Technosols Made from Urban and Industrial Wastes Are a Good Option for the Reclamation of Abandoned City Plots. *Geoderma* **2020**, *377*, 114563. [CrossRef]
49. Cannavo, P.; Guénon, R.; Galopin, G.; Vidal-Beaudet, L. Technosols Made with Various Urban Wastes Showed Contrasted Performance for Tree Development during a 3-Year Experiment. *Environ. Earth Sci.* **2018**, *77*, 650. [CrossRef]
50. Rokia, S.; Séré, G.; Schwartz, C.; Deeb, M.; Fournier, F.; Nehls, T.; Damas, O.; Vidal-Beaudet, L. Modelling Agronomic Properties of Technosols Constructed with Urban Wastes. *Waste Manag.* **2014**, *34*, 2155–2162. [CrossRef]
51. Weiler, J.; Firpo, B.A.; Schneider, I.A.H. Technosol as an Integrated Management Tool for Turning Urban and Coal Mining Waste into a Resource. *Miner. Eng.* **2020**, *147*, 106179. [CrossRef]
52. Pruvost, C.; Mathieu, J.; Vallet, J.; Dubs, F.; Gigon, A.; Lerch, T.; Blouin, M. Technosols Made of Urban Wastes Are Suitable Habitats for Flora and Soil Macrofauna. *Ecol. Eng.* **2025**, *211*, 107457. [CrossRef]
53. Mikajlo, I.; Pando, A.; Robain, H.; Lerch, T.Z. Reusing Asphalt Millings with Excavated Materials and Compost to Construct Technosols: Effects on Soil Properties and Plant Growth. *J. Soils Sediments* **2024**, *25*, 565–577. [CrossRef]
54. Molineux, C.J.; Fentiman, C.H.; Gange, A.C. Characterising Alternative Recycled Waste Materials for Use as Green Roof Growing Media in the U.K. *Ecol. Eng.* **2009**, *35*, 1507–1513. [CrossRef]
55. Bretzel, F.; Vannucchi, F.; Romano, D.; Malorgio, F.; Benvenuti, S.; Pezzarossa, B. Wildflowers: From Conserving Biodiversity to Urban Greening—A Review. *Urban For. Urban Green.* **2016**, *20*, 428–436. [CrossRef]
56. Graceson, A.; Hare, M.; Monaghan, J.; Hall, N. The Water Retention Capabilities of Growing Media for Green Roofs. *Ecol. Eng.* **2013**, *51*, 328–334. [CrossRef]
57. Gruda, N. Increasing Sustainability of Growing Media Constituents and Stand-Alone Substrates in Soilless Culture Systems. *Agronomy* **2019**, *9*, 298. [CrossRef]
58. Werdin, J.; Conn, R.; Fletcher, T.D.; Rayner, J.P.; Williams, N.S.G.; Farrell, C. Biochar Particle Size and Amendment Rate Are More Important for Water Retention and Weight of Green Roof Substrates than Differences in Feedstock Type. *Ecol. Eng.* **2021**, *171*, 106391. [CrossRef]
59. Duryea, M.L.; English, R.J.; Hermansen, L.A. A Comparison of Landscape Mulches: Chemical, Allelopathic, and Decomposition Properties. *J. Arboric.* **1999**, *25*, 88–97. [CrossRef]
60. Handreck, K.A. Particle Size and the Physical Properties of Growing Media for Containers. *Commun. Soil Sci. Plant Anal.* **1983**, *14*, 209–222. [CrossRef]
61. Altland, J.E.; Locke, J.C.; Krause, C.R. Influence of Pine Bark Particle Size and pH on Cation Exchange Capacity. *HortTech* **2014**, *24*, 554–559. [CrossRef]
62. Farrell, C.; Rayner, J.P.; Bathgate, R. *Woody Meadow Guidelines: Naturalistic Plantings of Australian Woody Plants for People and Nature*; The University of Melbourne: Richmond, VIC, Australia, 2025.
63. Botanic Gardens of South Australia. Plant Selector—*Alyogyne huegelii*. Available online: https://plantselector.botanicgardens.sa.gov.au/Plants/Details/863 (accessed on 16 May 2025).
64. Botanic Gardens of South Australia. Plant Selector—*Goodenia ovata*. Available online: https://plantselector.botanicgardens.sa.gov.au/Plants/Details/565 (accessed on 16 May 2025).
65. Nkongolo, N.V.; Caron, J. Bark Particle Sizes and the Modification of the Physical Properties of Peat Substrates. *Can. J. Soil Sci.* **1999**, *79*, 111–116. [CrossRef]
66. Gartner, J.B.; Still, S.M.; Klett, J.E. The Use of Bark Waste as a Substrate in Horticulture. *Acta Hortic.* **1974**, *37*, 2003–2012. [CrossRef]
67. Eksi, M.; Sevgi, O.; Akburak, S.; Yurtseven, H.; Esin, İ. Assessment of Recycled or Locally Available Materials as Green Roof Substrates. *Ecol. Eng.* **2020**, *156*, 105966. [CrossRef]

https://doi.org/10.3390/plants15030403

68. del Pilar León, H.; Martinez, S.; del Mar Delgado, M.; Gabriel, J.L.; Alvarez, S. Plant and Soil Responses to Concrete and Basalt Amendments Under Elevated CO_2: Implications for Plant Growth, Enhanced Weathering and Carbon Sequestration. *Agriculture* **2025**, *15*, 2435. [CrossRef]
69. Knapp, W.J.; Tipper, E.T. The Efficacy of Enhancing Carbonate Weathering for Carbon Dioxide Sequestration. *Front. Clim.* **2022**, *4*, 928215. [CrossRef]
70. Sanger, M.; Natarajan, B.M.; Wang, B.; Edil, T.; Ginder-Vogel, M. Recycled Concrete Aggregate in Base Course Applications: Review of Field and Laboratory Investigations of Leachate pH. *J. Hazard. Mater.* **2020**, *385*, 121562. [CrossRef] [PubMed]
71. Burke, T.M.; Kamber, B.S.; Rowlings, D. Microscopic Investigation of Incipient Basalt Breakdown in Soils: Implications for Selecting Products for Enhanced Rock Weathering. *Front. Clim.* **2025**, *7*, 1572341. [CrossRef]
72. Blythe, E.K.; Merhaut, D.J. Grouping and Comparison of Container Substrates Based on Physical Properties Using Exploratory Multivariate Statistical Methods. *HortScience* **2007**, *42*, 353–363. [CrossRef]
73. Nagase, A.; Dunnett, N. The Relationship between Percentage of Organic Matter in Substrate and Plant Growth in Extensive Green Roofs. *Landsc. Urban Plan.* **2011**, *103*, 230–236. [CrossRef]
74. Xue, M.; Farrell, C. Use of Organic Wastes to Create Lightweight Green Roof Substrates with Increased Plant-Available Water. *Urban For. Urban Green.* **2020**, *48*, 126569. [CrossRef]
75. Ekşi, M.; Rowe, D.B. Effect of Substrate Depth and Type on Plant Growth for Extensive Green Roofs in a Mediterranean Climate. *J. Green. Build.* **2019**, *14*, 29–44. [CrossRef]
76. Graceson, A.; Monaghan, J.; Hall, N.; Hare, M. Plant Growth Responses to Different Growing Media for Green Roofs. *Ecol. Eng.* **2014**, *69*, 196–200. [CrossRef]
77. Barzegar, A.R.; Yousefi, A.; Daryashenas, A. The Effect of Addition of Different Amounts and Types of Organic Materials on Soil Physical Properties and Yield of Wheat. *Plant Soil* **2002**, *247*, 295–301. [CrossRef]
78. Abbruzzini, T.F.; Mora, L.; Prado, B. Evaluation of Technosols Constructed with Construction and Excavation Debris for Greenhouse Production of Ornamental Plants. *J. Soils Sediments* **2021**, *22*, 745–756. [CrossRef]
79. Environment Protection Authority Victoria. *Waste Disposal Categories—Characteristics and Thresholds*; Environment Protection Authority Victoria: Melbourne, Australia, 2021; p. 10.
80. Webber, B.; Madgwick, H.A.I. Biomass and Nutrient Content of a 29-Year-Old. *N. Z. J. For. Sci.* **1983**, *13*, 222–228.
81. Nelson, J.T.; Adjuik, T.A.; Moore, E.B.; VanLoocke, A.D.; Ramirez Reyes, A.; McDaniel, M.D. A Simple, Affordable, Do-It-Yourself Method for Measuring Soil Maximum Water Holding Capacity. *Commun. Soil Sci. Plant Anal.* **2023**, *55*, 1190–1204. [CrossRef]
82. Yeates, G.W.; Dando, J.L.; Shepherd, T.G. Pressure Plate Studies to Determine How Moisture Affects Access of Bacterial-feeding Nematodes to Food in Soil. *Eur. J. Soil Sci.* **2002**, *53*, 355–365. [CrossRef]
83. Atlas of Living Australia. *Alyogyne huegelii*: Lilac Hibiscus. Available online: https://bie.ala.org.au/species/https://id.biodiversity.org.au/node/apni/2908199 (accessed on 14 January 2026).
84. Atlas of Living Australia. *Goodenia ovata*: Hop Goodenia. Available online: https://bie.ala.org.au/species/https://id.biodiversity.org.au/node/apni/2898281 (accessed on 14 January 2026).
85. R Core Team. *R, Version 4.1.2; A Language and Environment for Statistical Computing*; R Foundation for Statistical Computing: Vienna, Austria, 2021.
86. Fox, J.; Weisberg, S. *An R Companion to Applied Regression*; Sage: Thousand Oaks, CA, USA, 2019; ISBN 1-5443-3645-4.
87. Lenth, R.V. *Emmeans: Estimated Marginal Means, Aka Least-Squares Means, Version 1.8.5*; CRAN: Vienna, Austria, 2023.
88. Cribari-Neto, F.; Zeileis, A. Beta Regression in R. *J. Stat. Softw.* **2010**, *34*, 1–24. [CrossRef]
89. Revelle, W. *psych: Procedures for Psychological, Psychometric, and Personality Research, Version 2.4.3*; CRAN: Vienna, Austria, 2024.
90. Hothorn, T.; Bretz, F.; Westfall, P. Simultaneous Inference in General Parametric Models. *Biom. J.* **2008**, *50*, 346–363. [CrossRef] [PubMed]
91. Wickham, H. *ggplot2: Elegant Graphics for Data Analysis*; Springer: New York, NY, USA, 2016; ISBN 978-3-319-24277-4.

https://doi.org/10.3390/plants15030403

plants

MDPI

Article

Rapid Reduction of Phytotoxicity in Green Waste for Use as Peat Substitute: Optimization of Ammonium Incubation Process

Wenzhong Cui [1,2], Juncheng Liu [1,2], Qi Bai [1,2], Lingyi Wu [1,2], Zhiyong Qi [2] and Wanlai Zhou [2,*]

[1] School of Mechanical Engineering, Chengdu University, Chengdu 610100, China; cuiwenzhong@stu.cdu.edu.cn (W.C.); liujuncheng@stu.cdu.edu.cn (J.L.); baiq_99@163.com (Q.B.); 82101225393@caas.cn (L.W.)

[2] Chengdu National Agricultural Science and Technology Center, Institute of Urban Agriculture, Chinese Academy of Agricultural Sciences, Chengdu 610213, China; qizhiyong@caas.cn

* Correspondence: zhouwanlai@caas.cn

Abstract: The rapid growth of the horticultural industry has increased demand for soilless cultivation substrates. Peat, valued for its physical and chemical properties, is widely used in soilless cultivation. However, peat is non-renewable, and over-extraction poses serious ecological risks. Therefore, sustainable alternatives are urgently needed. Ammonium incubation, a novel method to reduce phytotoxicity, offers the potential for green waste, a significant organic solid waste resource, to substitute peat. This study optimized the ammonium incubation process to reduce green waste phytotoxicity. It systematically examined different nitrogen salts (type and amount) and environmental conditions (temperature, aeration, duration) affecting detoxification efficiency. Results show a significant reduction in phytotoxicity with ammonium bicarbonate, carbonate, and sulfate, especially carbonate, at 1.5%. Optimal conditions were 30 °C for 5 days with regular aeration. Under these conditions, ammonium salt-treated green waste significantly reduced total phenolic content and stabilized germination index (GI) at a non-phytotoxic level (127%). Using treated green waste as a partial peat substitute in lettuce cultivation showed promising results. This low-cost, low-energy method effectively converts green waste into sustainable peat alternatives, promoting eco-friendly horticulture and environmental conservation.

Keywords: ammonium incubation; phytotoxicity; green waste; peat substitute

1. Introduction

Due to the rapid growth of the horticultural industry, demand for soilless cultivation substrates has been increasing annually [1]. Peat is widely used as the most versatile plant growth medium due to its excellent aeration, water retention, buffering capacity, and nutrient supply capabilities [2,3]. However, peat is a non-renewable resource. Over-extraction can lead to biodiversity loss and excessive greenhouse gas emissions [4,5]. Therefore, finding sustainable alternatives to peat as plant growth media is urgently needed. Numerous studies have shown that green waste, after proper processing, exhibits favorable physicochemical properties and can serve as a substitute for peat in plant growth media compositions [6–8].

Green waste mainly refers to various plant residues and debris generated in landscaping, such as pruning branches and fallen leaves [9,10]. The volume of green waste is enormous; in China alone, it reached 40 million tons in 2018, with even higher quantities globally [11,12]. Currently, most green waste is disposed of through landfilling or incineration [13–15], causing environmental pollution and resource wastage [16,17]. Utilizing green waste as a substitute for peat in plant growth media not only reduces environmental pollution but also decreases the horticultural industry's reliance on peat. However, green waste often possesses phytotoxicity due to substances like phenols and organic acids, which can inhibit plant growth. Therefore, it is crucial to remove or reduce its phytotoxicity before incorporating green waste into plant growth media [18]. The chemical constituents of green waste can dissolve into water and migrate to the plant root zone during application, Therefore, the removal of material phytotoxicity can be achieved by removing water-soluble phytotoxic substances from green wastes.

Ammonium incubation is an emerging technique for reducing phytotoxicity by rapidly decreasing the activity or concentration of phytotoxic substances in green waste. Previous studies on ammonium incubation involved mixing green waste and ammonium salts in a mass ratio of 1% to 2%, adjusting moisture content to 60% to 70%, and leaving them at room temperature for 5 days. This method significantly reduced phytotoxic substances and toxicity in green waste materials [19]. Compared to commonly used techniques like composting and heat treatment, ammonium incubation offers advantages such as low energy consumption, minimal pollution, simplicity, and low cost [5,19–21]. Additionally, it preserves the original volume and weight of green waste as much as possible [18]. Despite its benefits, ammonium incubation is a relatively new technology lacking comprehensive research and widespread application. This study also found significant variability in the effectiveness of ammonium incubation when applied to different types of green waste. Moreover, the incubation process likely involves complex chemical reactions and microbial activities. To enhance the detoxification efficiency and effectiveness of ammonium incubation, optimal environmental conditions for incubation need to be clearly defined.

This study systematically investigates the effects of detoxifying agents (types and amounts) and environmental conditions, including incubation temperature, aeration, and duration, on the effectiveness of ammonium incubation. The aim is to optimize key parameters of the ammonium incubation process and improve its ability to remove phytotoxicity from green waste effectively and consistently.

2. Materials and Methods

2.1. Materials

The green waste used in this study originated from street tree pruning waste in Changsha City, Hunan Province, China. Initially, the green waste was preliminarily crushed to a particle size of 1–2 cm, followed by drying at 85 °C for 24 h. Subsequently, a cutting mill (ST-R200, Xuxinshengke, Beijing, China) was used for further cutting and grinding, and the material was sieved through a 20-mesh screen (aperture size 0.9 mm). All processed green waste had particle sizes smaller than 0.9 mm. The bulk density of the waste material was 284 g/L, with a macroporosity of 5.02%, microporosity of 57.26%, and total porosity of 62.28%. The cellulose content was 20.51%, hemicellulose content was 18.47%, and lignin content was 32.41%.

2.2. Experimental Design

2.2.1. The Effects of Detoxifiers

This study selected seven common nitrogen-containing compounds, including ammonium carbonate, ammonium sulfate, ammonium bicarbonate, ammonium chloride, ammonium dihydrogen phosphate, calcium nitrate, and potassium nitrate, to investigate their effects on the efficacy of detoxification treatments. Following the methodology of previous studies, green waste materials were mixed separately with $(NH_4)_2CO_3$, $(NH_4)_2SO_4$, NH_4HCO_3, NH_4Cl, $NH_4H_2PO_4$, $Ca(NO_3)_2$, and KNO_3. The nitrogen application rate for each treatment was 2.92 $mg \cdot g^{-1}$ based on the dry weight of green waste (nitrogen quantity in compound/mass of green waste). The mixture was adjusted to 66.7% moisture content, thoroughly mixed, and then incubated at room temperature (22–30 °C) for 5 days [19]. Each incubation treatment was replicated three times.

To assess the detoxification effects of ammonium carbonate, four levels of ammonium salts were applied (0.5%, 1%, 1.5%, and 2% of ammonium carbonate relative to the dry weight of green waste). Specifically, 25 mg, 50 mg, 75 mg, and 100 mg of $(NH_4)_2CO_3$ were added to 5 g of green waste, mixed thoroughly, adjusted to 66.7% moisture content, and then incubated at room temperature (22–30 °C) for 5 days. Each incubation treatment was replicated three times.

2.2.2. The Effects of Environmental Conditions

Environmental conditions were also evaluated using 1.5% $(NH_4)_2CO_3$ (relative to the dry weight of green waste) as the detoxifying agent. Different incubation durations (1 day, 3 days, 5 days, 7 days, 9 days, 11 days) and temperatures (20 °C, 25 °C, 30 °C, 35 °C) were tested to explore their impact on the detoxification efficacy of ammonium salts. Ventilation conditions were varied with two treatments: sealed incubation and once-daily ventilation. Each incubation treatment was replicated three times (Figure 1).

Figure 1. Ammonium incubation process and regulation of environmental conditions.

2.2.3. Cultivation Experiment of Green Waste as a Peat Substitute

For cultivation experiments replacing peat with detoxified green waste, a standard peat-based growing medium (CG) (peat:perlite:vermiculite = 9:3:1 *v*/*v*/*v*) was used as control. Two plant-growing media were prepared using shredded green waste to replace peat: the first using untreated green waste (OG) and the second using detoxified green waste under optimized ammonium incubation parameters (AG). In both media, green waste replaced 75% of peat by volume, while the proportions of perlite and vermiculite remained unchanged (green waste:peat:perlite:vermiculite = 6.75:2.25:3:1 *v*/*v*/*v*/*v*). The peat used in the study was moss peat (Pindstrup Sphagnum, Pindstrup, Denmark), size 10–30 mm. No fertilizers were added during the preparation of these growing media.

After seedlings were grown in seedling trays for 20 days, lettuce (Lactuca sativa, variety Rex Swan, obtained from Rijk Zwaan Export B.V., De Lier, The Netherlands) were transplanted into 500 cm^3 plastic pots containing the aforementioned plant-growing media. Fifteen lettuce pots were randomly placed in a growth chamber (Sefo Experimental Instruments Co., Ltd., Ningbo, China) set to alternate between nighttime temperatures of 23 °C and daytime temperatures of 25 °C, with 67% humidity. Each lettuce pot received 100 mL of water every two days, with 100 mL of 50% Hoagland's solution replacing water on the 6th, 12th, and 18th days after transplanting to supplement nutrients. Lettuce was harvested 28 days after transplanting.

2.3. Analytical Techniques

2.3.1. Phytotoxicity Assessment

In this study, seed germination tests were used to assess the phytotoxicity of green wastes. *Brassica rapa chinensis* (Chinese cabbage) seeds, sensitive to phytotoxicity [22], were used for the germination tests. Water extracts were obtained following the classic method at a ratio of 1:10 (*w*/*v*). Specifically, 5 g of green waste samples were mixed with 50 mL of deionized water, shaken for 2 h, and centrifuged at 8000 rpm for 10 min at 4 °C. Subsequently, 10 mL of supernatant was collected and filtered through a 0.45 μm filter paper [23]. A 2 mL aliquot of the extract (with deionized water as control, denoted as CK for all control groups) was added to Petri dishes containing germination paper (diameter 9.0 cm), followed by the placement of 10 seeds on the germination paper. Each treatment's extract underwent three replicate germination tests. Petri dishes were covered and sealed with parafilm to prevent moisture loss, then incubated in darkness at 25 °C for 72 h. The number of germinated seeds and the length of the radicle were recorded. Germination indices, including germination rate (GR), root length (RL), and germination index (GI), were calculated as follows [22]:

$$GR(\%) = \frac{seed\ germination\ in\ treatment}{seed\ germination\ in\ control} * 100\% \tag{1}$$

$$RL(\%) = \frac{root\ length\ in\ treatment}{root\ length\ in\ control} * 100\% \tag{2}$$

$$GI(\%) = GR * RL \tag{3}$$

If the germination index (GI) is 80% or higher, the phytotoxicity of the material is considered extremely low, meeting the standard of harmlessness.

2.3.2. pH, EC

The samples in this study were prepared using the same method as for the assessment of phytotoxicity experiments to obtain water extracts of green waste. pH and EC of the materials were measured using a multiparameter tester (Mettler-Toledo Instruments Ltd., Shanghai, China).

2.3.3. Determination of Total Phenol

Total phenol in green waste materials was determined using a Plant Total Phenol Assay Kit (Boxbio, Beijing Bokai Biotechnology Co., Ltd., Beijing, China). According to the manufacturer's instructions, the method is based on the reduction of tungsten molybdic acid to form a blue compound by phenolic substances under alkaline conditions. The product exhibits a characteristic absorption peak at 760 nm, and the total phenol content in plants is quantitatively detected by the change in absorbance. A 2 mL aliquot of water extract prepared before and after optimal ammonium incubation of green waste was tested following the kit's instructions. The extraction method followed the procedures outlined in Section 2.3.1. Results are expressed in terms of gallic acid equivalents.

2.3.4. Lettuce Biomass Measurement

Lettuce plants were harvested after 28 days of cultivation, and the above-ground parts were weighed fresh after clearing the growing medium. Subsequently, the samples are dried in an oven at 65 °C until a constant weight is achieved to determine the dry weight.

2.3.5. OJIP Parameters Measurement

Chlorophyll fluorescence parameters of lettuce leaves were measured using a handheld chlorophyll fluorometer (FluorPen FP110, Photon Systems Instruments, Brno, Czech Republic). Due to higher physiological activity at the top leaves, which better reflects plant responses to environmental stressors, measurements were taken from the first fully developed leaf from the top left side, avoiding the main vein. Before measurement, leaves were dark-adapted using leaf clips for 30 min until reaching stable states. The rapid fluorescence rise from minimal to maximal fluorescence after dark adaptation was detected, with the instrument's detector sensitivity exceeding 100,000 counts per second. Parameters interpretation and calculation formulas are detailed in Table A1 in Appendix A [24,25].

2.4. *Statistical Analyses*

Each treatment in the cultivation experiment was replicated 5 times, and other treatments and analyses were replicated three times each. Data are presented as mean ± standard deviation. Statistical analysis of the phytotoxicity assessment experiment was performed using SPSS 27. One-way ANOVA and Tukey's post hoc test were used to explain the differences in the measured parameters in the cultivation experiment.

3. Results and Discussion

3.1. Effects of Detoxifying Agent Types and Dosages on Detoxification Efficacy

Ammonium salts significantly reduce the phytotoxicity of materials, but their effectiveness varies greatly depending on the type of ammonium salt used. As shown in Figure 2, water extracts from raw garden waste materials markedly inhibited seed germination and root growth, indicating high phytotoxicity. Incubation with $(NH_4)_2CO_3$, NH_4HCO_3, and $(NH_4)_2SO_4$ significantly increased seed germination rates (GR) from 25% to 102%, 95%, and 98%, respectively, and ultimately elevated the seed germination index (GI) from 6.4% to 86.1%, 37.3% and 72.9%, demonstrating a significant reduction in the phytotoxicity of green waste. Additionally, compared to incubation with $(NH_4)_2CO_3$ and $(NH_4)_2SO_4$, seed germination was delayed after incubation with NH_4HCO_3 (Figure A1). This delay in germination may shorten the period for root elongation during germination tests, resulting in reduced root lengths and correspondingly lower GI [26].

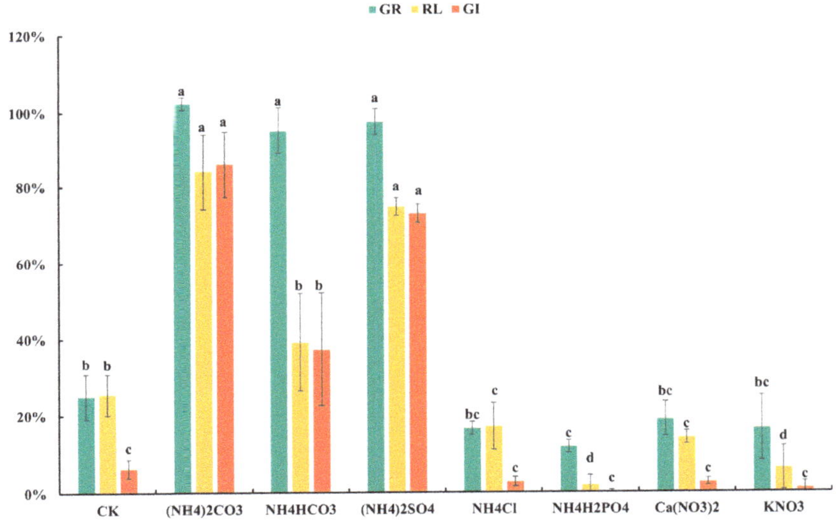

Figure 2. GR, RL, and GI of green wastes before (CK) and after incubation with different ammonium salts and nitrates. Error bars represent standard deviation ($n = 3$). Within each germination index, values that do not share a letter differ significantly according to ANOVA and Duncan's multiple range test ($p < 0.05$).

Nitrogen has a profound impact on plant growth and development by participating in protein synthesis and influencing overall plant growth and development [27,28]. Ammoniacal and nitrate nitrogen are readily soluble in water, serving as quickly available nutrients for plants. To investigate the nutrient effects on seed germination promotion, different forms of nitrogen incubation experiments were conducted. As shown in Figure 2, incubation with $Ca(NO_3)_2$ and KNO_3 did not promote seed germination of green waste seeds, with germination rates (GR) of only 19% and 17% and final seed germination indices (GI) of 2.7% and 1.0%, respectively, indicating that nitrogen in nitrate form does not effectively promote seed germination and growth. Furthermore, even with ammoniacal nitrogen, incubation with NH_4Cl and $NH_4H_2PO_4$ did not increase but rather decreased the seed germination index of green waste, with GI values of 2.9% and 0.2%, respectively. Seed germination was severely inhibited in these incubation treatments, indicating that only appropriate ammonium salts such as $(NH_4)_2CO_3$, NH_4HCO_3, and $(NH_4)_2SO_4$ can effectively reduce or eliminate the phytotoxicity of green waste. Among these, $(NH_4)_2CO_3$ showed the most optimal detoxification effect.

For plants, ammonium ions serve as a rapid nitrogen source, but the previous research found that concentrations exceeding 1 $mg \cdot mL^{-1}$ significantly inhibit seed germination, indicating high levels of phytotoxicity [19]. Therefore, it is crucial to control the dosage of ammonium salts during incubation to ensure effective detoxification of materials without introducing new phytotoxic substances [29]. As shown in Figure 3, this study applied 0.5% to 2.0% $(NH_4)_2CO_3$ for incubating green waste materials. Seed germination in water extracts was normal across all concentrations, with GR reaching 95%, 104%, 106%, and 98%, respectively, and GI significantly increased, achieving 109.3%, 128.4%, 132.8%, and 79.7%. Except for the 2.5% $(NH_4)_2CO_3$ experimental group, all reached the non-phytotoxic standard (GI $\geq$ 80%), with the 1.5% $(NH_4)_2CO_3$ dosage demonstrating optimal detoxification effects, achieving a GI of 132.8%.

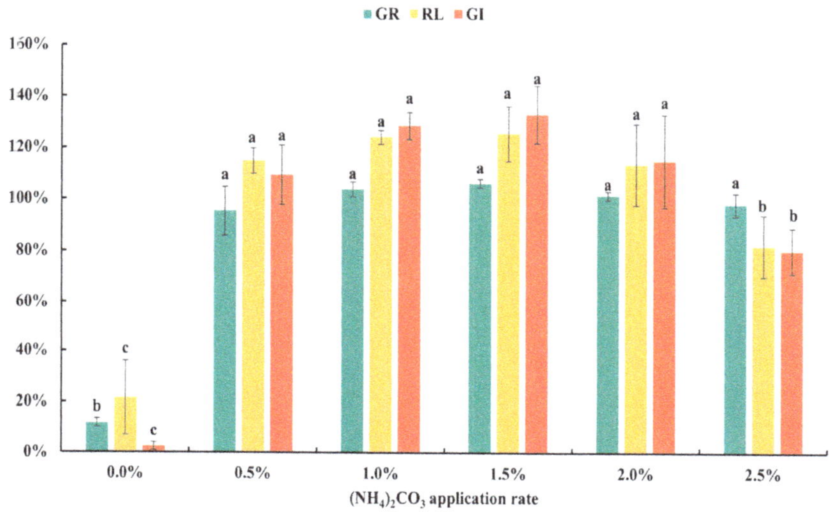

Figure 3. GR, RL, and GI of green waste treated by $(NH_4)_2CO_3$ incubation at different dosages. Error bars represent standard deviation (n = 3). Within each germination index, values that do not share a letter differ significantly according to ANOVA and Duncan's multiple range test ($p < 0.05$).

The variability in detoxification effects due to different ammonium salt dosages can be attributed to the dual role of ammonium carbonate. On the one hand, as a detoxifying agent, $(NH_4)_2CO_3$ may react with major plant phytotoxic substances such as organic acids and phenols, therefore reducing their content; insufficient dosage may lead to incomplete removal of toxic substances. On the other hand, excessive application of $(NH_4)_2CO_3$ can increase ammonia concentrations, which themselves can induce phytotoxicity. Therefore, achieving the optimal balance in the application of detoxifying agents is crucial for effective detoxification without introducing additional phytotoxicity. From a production cost perspective, a 0.5% application rate of the detoxifying agent may be more appropriate. Materials treated at this rate still meet non-phytotoxic standards and can significantly reduce production costs.

3.2. *Effects of Environmental Conditions on Detoxification Efficacy*

As shown in Figure 4, during the incubation of green waste with ammonium salts, there was little change in GI after 1 day, remaining close to 0% similar to the original material. After 1 day of incubation, green waste still significantly inhibited seed germination, with no significant difference in germination rate (GR) compared to the original material. However, by day 3 of incubation, GI rapidly increased to 44.3%, and unlike the 1-day incubation, the material no longer significantly inhibited seed germination, with

GR reaching 80%. By day 5, GR of green waste increased to 96%, relative root length (RL) significantly increased to 97% compared to the 3-day incubation group, and GI reached 92%. From day 5 to 9 of incubation, GI remained stable between 91.3% and 96.4%. However, after 9 days of incubation, there was a decline in GI, reaching 82% by day 11. Although this experimental group did not inhibit seed germination, GR remained at a high level of 97%, whereas RL decreased to 85% compared to the 5–9-day incubation groups. Considering time constraints, a 5-day incubation period appears ideal.

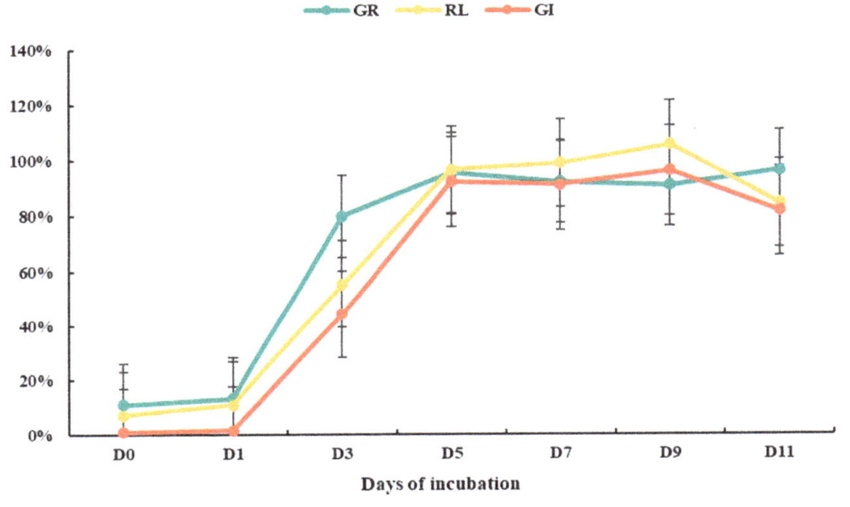

Figure 4. GR, RL, and GI of green wastes treated by ammonium incubation for different days. Error bars represent standard deviation (n = 3).

Previous studies suggested that during ammonium incubation, ammonium ions react chemically with acidic phenols and organic acids in green waste, reducing the content of phytotoxic substances and, therefore, lowering the material's phytotoxicity [19]. However, chemical reactions typically occur quickly. In this study, the significant reduction in phytotoxicity of green waste after 3 days of ammonium incubation suggests that the detoxification process may involve substance transformations possibly mediated by microbial activity rather than solely chemical reactions.

Temperature also significantly affects the detoxification efficiency of ammonium incubation, as shown in Figure 5. GI of green waste after incubation increased with temperature, reaching 2.3% and 23.7% at 20 °C and 25 °C, respectively. At 30 °C, GI peaked at 92%, slightly decreasing to 83.8% at 35 °C. Incubation at 20 °C still inhibited seed germination, with GR only at 20%. At 25 °C, the inhibitory effect on seed germination was significantly reduced (72% GR), but root growth was minimal (RL 33%), resulting in a GI of 23.7%. Incubation at 30 °C and 35 °C did not inhibit seed germination, with GR reaching 97% and 96%, respectively. However, at 35 °C, RL decreased from 95% to 88%. These results indicate that around 30 °C, ammonium incubation achieves optimal detoxification, making it suitable for the harmless treatment of garden waste materials. The optimal temperature range aligns well with the growth conditions of mesophilic microorganisms, whose optimal proliferation temperatures are typically around 30 °C [30,31], further suggesting their critical role in the detoxification process.

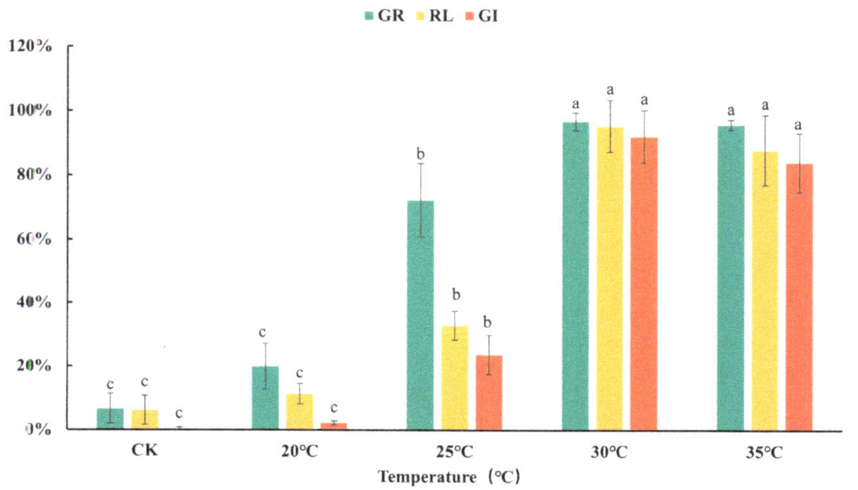

Figure 5. GR, RL, and GI of green waste treated by ammonium incubation at different temperatures. CK represents untreated green waste. Error bars represent standard deviation (n = 3). Within each germination index, values that do not share a letter differ significantly according to ANOVA and Duncan's multiple range test ($p < 0.05$).

Similarly, different ventilation conditions during incubation also resulted in significant differences in detoxification effects, as shown in Figure 6. Both hermetic and once-daily ventilated incubation significantly reduced the phytotoxicity of green waste compared to the original material, with GI reaching 80.5% and 100%, respectively. The difference in GI levels between the two treatments was primarily due to differences in root length growth, with RL values of 84% and 107%, respectively. However, both treatments exceeded the non-toxic standard (GI $\geq$ 80%), indicating that proper ventilation conditions can significantly enhance the detoxification effect. Previous studies indicated that water-soluble substances such as organic acids and amino acids are major contributors to the phytotoxicity of green waste, and reducing their content can significantly decrease the material's phytotoxicity [32]. Composting studies with agricultural waste have shown that water-soluble organic compounds like organic acids and phenolic acids can be utilized as carbon sources by microbes, leading to their decomposition [33–35]. However, microbial decomposition typically requires suitable C/N ratios, pH, and temperature conditions, with green waste composed of plant residues and debris typically having high C/N ratios [36]. During ammonium incubation, the addition of ammonium salts can lower the material's C/N ratio, potentially enhancing microbial activity and promoting the degradation of water-soluble phytotoxic substances such as organic acids and amino acids, therefore reducing phytotoxicity. The detoxification effect of ammonium incubation was evident only after 3 days, similar to the duration required for rapid degradation of soluble substances observed in many composting studies [37,38], further supporting this hypothesis.

Temperature and oxygen content are key factors influencing the activity of aerobic microorganisms. In this study, optimal detoxification effects of ammonium incubation were observed at around 30 °C under appropriate ventilation conditions, suggesting a critical role for microbial processes in ammonium salt-induced detoxification.

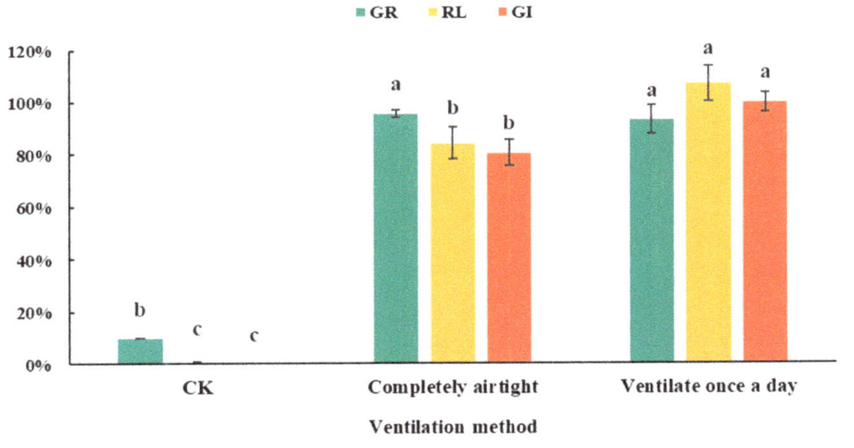

Figure 6. GR, RL, and GI of green waste treated by ammonium incubation at different aeration conditions. CK represents untreated green waste. Error bars represent standard deviation ($n = 3$). Within each germination index, values that do not share a letter differ significantly according to ANOVA and Duncan's multiple range test ($p < 0.05$).

It is noteworthy that while nitrates can also serve as nitrogen sources for microbes and reduce the C/N ratio of materials, this study found that nitrate incubation did not reduce the phytotoxicity of green waste. This may be due to conditions in woody medium environments rich in ammonium salts inducing and promoting nitrification processes [39]. Nitrifying bacteria thrive in neutral to slightly alkaline environments (pH 7.0–8.2) at around 30 °C and require dissolved oxygen levels of 2–5 mg/L for optimal growth and activity. Lack of oxygen or low oxygen content is detrimental to the growth and activity of nitrifying bacteria, conditions which are consistent with the optimal ammonium incubation conditions identified in this study, further indicating the pivotal role of nitrification microbial activity in determining the detoxification effect of ammonium incubation [40–43]. Future research should investigate changes in microbial communities during ammonium incubation to confirm these hypotheses.

3.3. Verification of Detoxification Effects of Ammonium Incubation

A wealth of studies indicates that phenolic compounds are the primary source of phytotoxicity in green waste, making changes in phenolic content a key indicator of phytotoxicity changes [32,34,44,45]. As shown in Figure 7, under the optimized conditions determined in this study (1.5% ammonium carbonate addition, 5 days of incubation at 30 °C with daily ventilation), there was a significant difference in total phenol content of green waste before and after ammonium incubation. The original material had a total phenol content of 5.70 $mg{\cdot}g^{-1}$, whereas, after ammonium incubation, the total phenol content of green waste significantly decreased to 1.49 $mg{\cdot}g^{-1}$, indicating a significant reduction in phytotoxic substances. The change in GI of green waste before and after ammonium incubation, ranging from 0% to 127%, also indicates that after ammonium salt incubation treatment, the phytotoxicity of green waste has been effectively eliminated.

Furthermore, cultivation substrates with phytotoxicity essentially impose stress conditions that significantly affect the plant photosynthesis process. Analysis of chlorophyll fluorescence dynamics provides detailed information about the photosynthetic apparatus, particularly the structure and function of PSII. In this study, we compared the effects of ammonium incubation detoxification treatment (AG) and untreated (OG) horticultural waste substrates on plant chlorophyll fluorescence parameters to assess the mitigating effect of ammonium incubation treatment on phytotoxicity. Table 1 shows that ϕPo (maximum photochemical quantum yield) of AG treatment was 0.83 ± 0.01, while for OG treatment, it was only 0.79 ± 0.02, indicating that ammonium incubation treatment significantly enhanced the photochemical efficiency of PSII, implying that more light energy was effectively used in the photosynthesis process rather than being lost due to damage to the photosystem [46]. In addition, ϕEo (electron transport quantum yield) and ϕRo (total electron transport quantum yield) of AG treatment were also significantly higher than those of OG treatment, indicating that the substrate treated with ammonium incubation not only improved the efficiency of electron transfer from QA- to QB but also enhanced the overall functionality of the electron transport chain, therefore improving the performance of Photosystem II. δRo (electron transfer efficiency from QA- to PQ) was also significantly higher in the AG treatment group than in the OG treatment group (0.16 ± 0.01 vs. 0.13 ± 0.06), further indicating that ammonium incubation treatment effectively reduced the damage to Photosystem II [47].

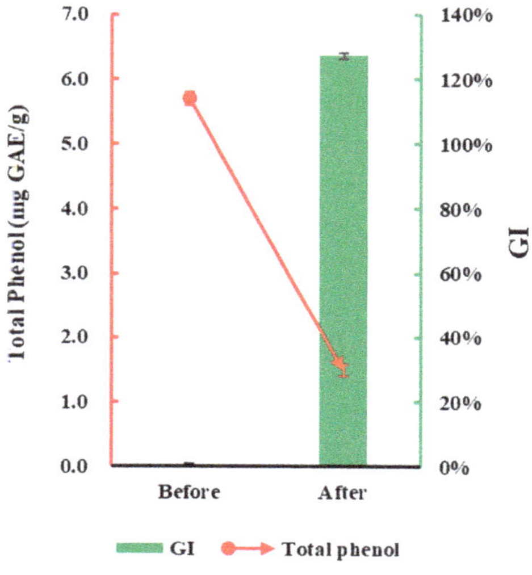

Figure 7. Total phenol content and phytotoxicity of green waste before and after ammonium incubation under optimal conditions. Error bars represent standard deviation (n = 3).

Table 1. Chlorophyll fluorescence parameters of lettuce grown in different growing media.

Growing Media	ϕPo	ϕEo	δRo	ϕRo	PI_{abs}
AG	0.83 ± 0.01	0.50 ± 0.02	0.16 ± 0.01	0.08 ± 0.01	4.80 ± 0.66
OG	0.79 ± 0.02 *	0.44 ± 0.03 **	0.13 ± 0.06 *	0.06 ± 0.03 *	2.24 ± 0.92 **

*, ** represent differences at $p < 0.05$ and $p < 0.01$ levels according to t-test, respectively.

Performance index (PI) is a parameter that comprehensively reflects the absorption, capture, and electron transfer capabilities of Photosystem II. The PI of the AG treatment group was 4.80 ± 0.66, significantly higher than the 2.24 ± 0.92 of the OG treatment group, indicating that ammonium incubation treatment significantly improved the overall performance of Photosystem II. The significant increase in PI implies that plants in the substrate after detoxification treatment can perform photosynthesis more efficiently, therefore alleviating the stress caused by the phytotoxicity of green waste substrate. These results indicate that after ammonium incubation detoxification treatment, the substrate of green waste significantly reduces stress on lettuce, manifested by significantly enhanced photochemical and electron transfer efficiency of Photosystem II, reduced damage to the photosystem, and thus increased photosynthetic capacity. (Refer to Table A1 in Appendix A for an explanation of the chlorophyll fluorescence parameters).

Throughout the cultivation process, lettuce grown in AG (ammonium salt-treated) substrate exhibits full and lush foliage, with leaves displaying the normal purple-green color characteristic of this lettuce variety (Monde, a lettuce field variety from Rijk Zwaan Export Company, De Lier, The Netherlands). In contrast, lettuce grown in OG (untreated) substrate has significantly fewer leaves, and after 28 days of transplanting, the leaves appear yellowish. Additionally, the number and size of lettuce leaves in OG substrate show minimal change between 14 and 28 days after transplanting, indicating a significant slowdown in growth rate after 14 days, reflecting strong phytotoxicity of the OG substrate. On the other hand, lettuce grown in AG substrate, although with lower biomass compared to CG (peat-treated) substrate, exhibits normal growth throughout the growth cycle with good vigor, showing no apparent phytotoxic effects.

As depicted in Figure 8, compared to the untreated green waste substrate (OG), lettuce grown in the detoxified green waste substrate (AG) shows a significant increase in individual plant biomass. Fresh weight per plant increased markedly from 5.60 g to 24.94 g, while dry weight per plant increased from 0.96 g to 2.39 g. This correlates with the earlier findings of chlorophyll fluorescence parameters, indicating that after ammonium incubation detoxification treatment, the phytotoxicity of green waste substrate is significantly reduced, alleviating stress on crops, enhancing photosynthetic efficiency, and ultimately showing improved growth performance.

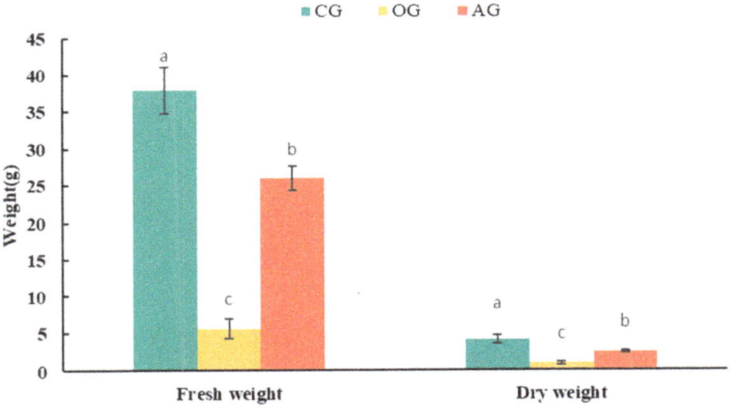

Figure 8. Biomass of lettuce grown in different growing media. Error bars represent standard deviation ($n = 3$). Within each index, values that do not share a letter differ significantly according to ANOVA and Duncan's multiple range test ($p < 0.05$).

Figure 8 also indicates that the biomass of lettuce grown in ammonium salt-treated detoxified green waste substrate remains slightly lower than that in peat control treatment, possibly due to the significant influence of air, water, and nutrient conditions on plant growth in the absence of phytotoxicity. As a soilless cultivation medium, substrate porosity and water retention capacity affect air, water exchange, and nutrient absorption during plant growth [48,49]. As shown in Table 2, using green waste instead of peat as a plant growth substrate shows significant differences in porosity and water retention compared to traditional peat substrates, with higher water retention pore space (close to 70%) and lower aeration pore space (less than 5%), which may lead to inadequate root aeration for lettuce, potentially explaining why the growth performance of lettuce in green waste substrate is inferior to that in peat.

Table 2. Basic physicochemical properties of different growing media.

Growing Media	BD (g/cm^3)	TP	WHC	AFP
CG	0.172 ± 0.006 ^{b}	65.0% ± 4.1% ^{b}	57.0% ± 4.5% ^{b}	8.0% ± 0.5% ^{a}
AG	0.260 ± 0.013 ^{a}	73.7% ± 3.2% ^{a}	69.0% ± 2.4% ^{a}	4.7% ± 1.1% ^{b}
OG	0.259 ± 0.002 ^{a}	70.2% ± 1.4% ^{ab}	68.1% ± 1.5% ^{a}	2.1% ± 0.1% ^{c}

TP represents total porosity; AFP represents air-filled porosity; WHC represents water-holding capacity; BD represents bulk density. Within each index, values that do not share a letter differ significantly according to ANOVA and Duncan's multiple range test ($p < 0.05$).

4. Conclusions

This study systematically investigates the effects of different detoxification agents (types and concentrations) and environmental conditions (including incubation temperature, aeration, and treatment duration) on the detoxification of ammonium salt-treated green waste. Experimental results demonstrate that under appropriate concentrations and incubation conditions, ammonium carbonate significantly reduces the phytotoxicity of green waste. Specifically, treating with 1.5% ammonium carbonate under adequate aeration at 30 °C for five days markedly reduces the total phenol content of the green waste and effectively mitigates its phytotoxic effects on plants.

Furthermore, this study explores the feasibility of using detoxified green waste as a partial substitute for peat in preparing plant-growing media. It was found that untreated original green waste exhibits strong phytotoxicity, leading to significant growth inhibition when used directly as a hydroponic substrate. In contrast, after ammonium incubation, the phytotoxicity of green waste is significantly reduced, alleviating stress on crops and resulting in significantly improved plant growth performance compared to untreated waste.

Ammonium incubation emerges as a promising new technology due to its low energy consumption, lack of pollution, simplicity of operation, and low cost. Compared to traditional composting methods, this technique requires less time and minimally alters the volume of raw materials, which is crucial for the effective utilization of agricultural waste and potential enhancement of production efficiency. Future research could further explore the application of ammonium incubation technology in handling garden waste from different countries and regions, therefore promoting its large-scale application in practical production.

Author Contributions: Conceptualization, W.C. and W.Z.; software, Q.B.; validation, J.L.; formal analysis, W.C. and L.W.; resources, Q.B.; data curation, W.C. and J.L.; writing—original draft preparation, W.C.; writing—review and editing, W.C. and W.Z.; supervision, W.Z.; project administration, Z.Q. and W.Z.; funding acquisition, Z.Q. and W.Z. All authors have read and agreed to the published version of the manuscript.

Funding: This research was funded by Sichuan Science and Technology Program: No. 2021JDRC0025, 2022YFN0028; Chinese Academy of Agricultural Sciences Science and Technology Innovation Project: ASTIP2023–34-IUA-04; Chengdu Agricultural Science and Technology Center Local Finance Special Funds Project: NASC2023ST09, NASC2024TD01.

Data Availability Statement: Data are contained within the article.

Acknowledgments: The authors would like to thank the School of Mechanical Engineering, Chengdu University, and the Institute of Urban Agriculture, Chinese Academy of Agricultural Sciences, Chengdu National Agricultural Science and Technology Center for their support.

Conflicts of Interest: The authors declare no conflicts of interest.

Appendix A

Table A1. Interpretation and calculation of indicators for the determination of chlorophyll fluorescence parameters.

Calculation Formulas	Parameters Interpretation
$FO = F(20\mu s)$	F0 (Minimum Fluorescence): In the state of dark adaptation, when all reaction centers are open, i.e., the primary electron acceptor QA in the reaction center is in an oxidized state, the fluorescence value is observed. This reflects the baseline fluorescence of the reaction center of Photosystem II (PSII), mainly originating from antenna pigments.
$FM = FP = F(t)max$	FM (Maximum Fluorescence): The fluorescence value when all reaction centers are closed in the state of dark adaptation. This reflects the fluorescence of antenna pigments after maximum photochemical quenching.
$FV = FM - FO$	
$ABS/RC = [4(F300\mu s - FO)/(F2ms - FO)](FM/FV) = M0(1/VJ)(1/\phi Po)$	Energy flux absorbed by the active reaction center of PSII per unit.
$MC = (\Delta V/\Delta t)0 = 4(F300\mu s - FO)/(FM - FO) = TR0/RC - ET0/RC$	The approximate initial slope (per millisecond) of transient fluorescence V = f(t). This parameter reflects the closing rate of the active reaction centers (RCs) of PSII.
$TR0/RC = \phi Po(ABS/RC)$	Energy flux captured by the active reaction center of PSII per unit at t = 0
$ET0/RC = \phi Eo(ABS/RC)$	Energy flux transmitted by the active reaction center of PSII per unit at t = 0
$RE0/RC = \phi Ro(ABS/RC) = (1 - VI) \times (M0/VJ)$	At t = 0, the ratio of electron flux reducing the terminal electron acceptor of PSI per active reaction center of PS.II
$DI0/RC = (ABS/RC) - (TR0/RC)$	At t = 0, the energy flux dissipated as heat per active reaction center of PSII.
$\phi Po = TR0/ABS = FV/FM = 1 - FO/FM$	At t = 0, the maximum quantum yield of the primary photochemical reaction, which refers to the quantum yield of photon capture by the PSII reaction center, also represents the maximum photochemical efficiency of the PSII reaction center in the state of dark adaptation.
$\psi Eo = ET0/TR0 = (1 - VJ)$	At t = 0, the ratio of electrons captured by the active reaction center of PSII that are transferred.
$\phi Eo = ET0/ABS = (1 - FO/FM)\ \psi Eo = \phi Po\psi Eo$	At t = 0, the quantum efficiency of electrons transferred from QA- to the electron transport chain.
$\psi Ro = RE0/TR0 = 1 - VI$	At t = 0, the ratio of electrons captured by the active reaction center of PSII that are transferred to PSI.
$\delta Ro = RE0/ET0 = (1 - VI)/(1 - VJ)$	The ratio of electrons reducing the terminal electron acceptor of PSI to the electrons in the electron transport chain.
$\phi Ro = RE0/ABS = \phi Po\psi Ro = \phi Po\psi Eo\delta Ro$	Quantum yield of reducing the terminal electron acceptor on the acceptor side of PSI.
$\phi Do = FO/FM = 1 - \phi Po$	At t = 0, the quantum yield of heat dissipation reflects the proportion of absorbed light energy in Photosystem II (PSII) that is dissipated by the non-photochemical quenching process.
$PIabs = [\gamma o/(1 - \gamma o)][\phi Po/(1 - \phi Po)][\psi Eo/(1 - \psi Eo)]$	Performance index based on absorption: It describes the performance index from the absorption of photons to the reduction of the electron transport chain.

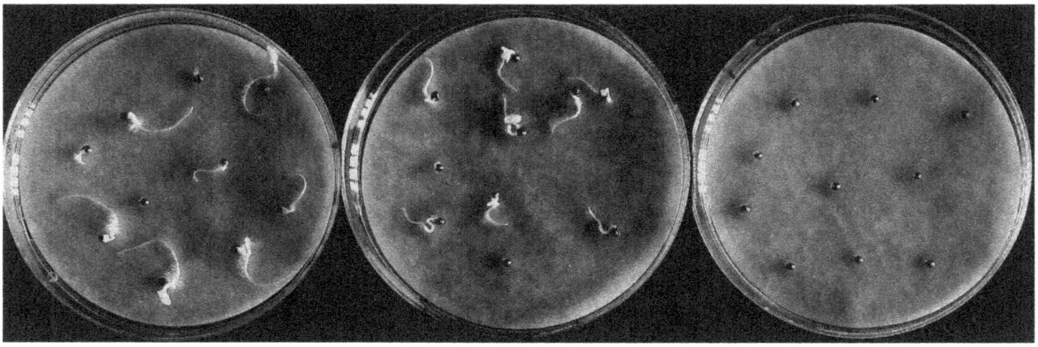

Figure A1. Seed germination at 36 h (from left to right for $(NH_4)_2CO_3$, $(NH_4)_2SO_4$, NH_4HCO_3 treatment groups).

References

1. Caputo, S.; Rumble, H.; Schaefer, M. "I like to Get My Hands Stuck in the Soil": A Pilot Study in the Acceptance of Soil-Less Methods of Cultivation in Community Gardens. *J. Clean. Prod.* **2020**, *258*, 120585. [CrossRef]
2. Amesbury, M.J.; Gallego-Sala, A.; Loisel, J. Peatlands as Prolific Carbon Sinks. *Nat. Geosci.* **2019**, *12*, 880–881. [CrossRef]
3. Gruda, N. Sustainable Peat Alternative Growing Media. *Acta Hortic.* **2012**, *927*, 973–979. [CrossRef]
4. Maher, M.; Prasad, M.; Raviv, M. Organic Soilless Media Components. In *Soilless Culture*; Elsevier: Amsterdam, The Netherlands, 2008; pp. 459–504, ISBN 978-0-444-52975-6.
5. Roehrdanz, M.; Greve, T.; De Jager, M.; Buchwald, R.; Wark, M. Co-Composted Hydrochar Substrates as Growing Media for Horticultural Crops. *Sci. Hortic.* **2019**, *252*, 96–103. [CrossRef]
6. Chemetova, C.; Quilhó, T.; Braga, S.; Fabião, A.; Gominho, J.; Ribeiro, H. Aged Acacia Melanoxylon Bark as an Organic Peat Replacement in Container Media. *J. Clean. Prod.* **2019**, *232*, 1103–1111. [CrossRef]
7. Chemetova, C.; Mota, D.; Fabião, A.; Gominho, J.; Ribeiro, H. Low-Temperature Hydrothermally Treated Eucalyptus Globulus Bark: From by-Product to Horticultural Fiber-Based Growing Media Viability. *J. Clean. Prod.* **2021**, *319*, 128805. [CrossRef]
8. Zhang, D.; Peng, Q.; Yang, R.; Lin, W.; Wang, H.; Zhou, W.; Qi, Z.; Ouyang, L. Slight Carbonization as a New Approach to Obtain Peat Alternative. *Ind. Crops Prod.* **2023**, *202*, 117041. [CrossRef]
9. Altland, J.E.; Owen, J.S.; Jackson, B.E.; Fields, J.S. Physical and Hydraulic Properties of Commercial Pine-Bark Substrate Products Used in Production of Containerized Crops. *HortScience* **2018**, *53*, 1883–1890. [CrossRef]
10. Witcher, A.L.; Blythe, E.K.; Fain, G.B.; Curry, K.J.; Pounders, C.T. Assessing Phytotoxicity in Fresh and Aged Whole Pine Tree Substrates©. *Comb. Proc. Int. Plant Propagators' Soc.* **2011**, *61*, 477–482.
11. Li, Z. Resource Utilization of Agricultural Solid Waste. *J. Integr. Agric.* **2021**, *20*, 1119–1120. [CrossRef]
12. Li, R.; Wang, Q.; Qu, G.; Zhang, Z.; Wang, H. Green Utilization of Organic Waste Resource. *Environ. Sci. Pollut. Res.* **2023**, *30*, 8899–8901. [CrossRef] [PubMed]
13. Wang, J.; Wang, X. A Triumph of Reducing Carbon Emission by Banning Open Straw Burning. *Sci. Bull.* **2023**, *68*, 18–20. [CrossRef] [PubMed]
14. Singh, D.; Dhiman, S.K.; Kumar, V.; Babu, R.; Shree, K.; Priyadarshani, A.; Singh, A.; Shakya, L.; Nautiyal, A.; Saluja, S. Crop Residue Burning and Its Relationship between Health, Agriculture Value Addition, and Regional Finance. *Atmosphere* **2022**, *13*, 1405. [CrossRef]
15. Hu, W.-N.; Wu, S.-D.; Cai, Z.; Lin, Y.-C.; Sun, X.-D. Study on the Recycling Technology of Street Tree Waste Recycling and New Lead Carbon Alloy Processing Technology. *IOP Conf. Ser. Mater. Sci. Eng.* **2020**, *774*, 012136. [CrossRef]
16. Bhuvaneshwari, S.; Hettiarachchi, H.; Meegoda, J. Crop Residue Burning in India: Policy Challenges and Potential Solutions. *Int. J. Environ. Res. Public Health* **2019**, *16*, 832. [CrossRef]
17. Trushna, T.; Diwan, V.; Nandi, S.S.; Aher, S.B.; Tiwari, R.R.; Sabde, Y.D. A Mixed-Methods Community-Based Participatory Research to Explore Stakeholder's Perspectives and to Quantify the Effect of Crop Residue Burning on Air and Human Health in Central India: Study Protocol. *BMC Public Health* **2020**, *20*, 1824. [CrossRef]
18. Cui, W.; Bai, Q.; Liu, J.; Chen, J.; Qi, Z.; Zhou, W. Phytotoxicity Removal Technologies for Agricultural Waste as a Growing Media Component: A Review. *Agronomy* **2023**, *14*, 40. [CrossRef]
19. Zhou, W.; Liao, J.; Zhou, B.; Yang, R.; Lin, W.; Zhang, D.; Wang, H.; Qi, Z. Rapidly Reducing Phytotoxicity of Green Waste for Growing Media by Incubation with Ammonium. *Environ. Technol. Innov.* **2023**, *31*, 103136. [CrossRef]
20. Arshad, M.; Khan, A.H.A.; Hussain, I.; Badar-uz-Zaman; Anees, M.; Iqbal, M.; Soja, G.; Linde, C.; Yousaf, S. The Reduction of Chromium (VI) Phytotoxicity and Phytoavailability to Wheat (*Triticum aestivum* L.) Using Biochar and Bacteria. *Appl. Soil Ecol.* **2017**, *114*, 90–98. [CrossRef]

21. Bohacz, J. Composts and Water Extracts of Lignocellulosic Composts in the Aspect of Fertilization, Humus-Forming, Sanitary, Phytosanitary and Phytotoxicity Value Assessment. *Waste Biomass Valor* **2019**, *10*, 2837–2850. [CrossRef]
22. Luo, Y.; Liang, J.; Zeng, G.; Chen, M.; Mo, D.; Li, G.; Zhang, D. Seed Germination Test for Toxicity Evaluation of Compost: Its Roles, Problems and Prospects. *Waste Manag.* **2018**, *71*, 109–114. [CrossRef]
23. Yakefu, Z.; Huannixi, W.; Ye, C.; Zheng, T.; Chen, S.; Peng, X.; Tian, Z.; Wang, J.; Yang, Y.; Ma, Z.; et al. Inhibitory Effects of Extracts from Cinnamomum Camphora Fallen Leaves on Algae. *Water Sci. Technol.* **2018**, *77*, 2545–2554. [CrossRef]
24. Zhou, L.; Zhou, L.; Wu, H.; Jing, T.; Li, T.; Li, J.; Kong, L.; Zhu, F. Application of Chlorophyll Fluorescence Analysis Technique in Studying the Response of Lettuce (*Lactuca sativa* L.) to Cadmium Stress. *Sensors* **2024**, *24*, 1501. [CrossRef]
25. Shin, Y.K.; Bhandari, S.R.; Jo, J.S.; Song, J.W.; Cho, M.C.; Yang, E.Y.; Lee, J.G. Response to Salt Stress in Lettuce: Changes in Chlorophyll Fluorescence Parameters, Phytochemical Contents, and Antioxidant Activities. *Agronomy* **2020**, *10*, 1627. [CrossRef]
26. Buss, W.; Mašek, O. Mobile Organic Compounds in Biochar—A Potential Source of Contamination—Phytotoxic Effects on Cress Seed (*Lepidium sativum*) Germination. *J. Environ. Manag.* **2014**, *137*, 111–119. [CrossRef] [PubMed]
27. Paschalidis, K.; Tsaniklidis, G.; Wang, B.-Q.; Delis, C.; Trantas, E.; Loulakakis, K.; Makky, M.; Sarris, P.F.; Ververidis, F.; Liu, J.-H. The Interplay among Polyamines and Nitrogen in Plant Stress Responses. *Plants* **2019**, *8*, 315. [CrossRef] [PubMed]
28. Ahmad, N.; Jiang, Z.; Zhang, L.; Hussain, I.; Yang, X. Insights on Phytohormonal Crosstalk in Plant Response to Nitrogen Stress: A Focus on Plant Root Growth and Development. *Int. J. Mol. Sci.* **2023**, *24*, 3631. [CrossRef] [PubMed]
29. Jia, L.; Xie, Y.; Wang, Z.; Luo, L.; Zhang, C.; Pélissier, P.; Parizot, B.; Qi, W.; Zhang, J.; Hu, Z.; et al. Rice Plants Respond to Ammonium Stress by Adopting a Helical Root Growth Pattern. *Plant J.* **2020**, *104*, 1023–1037. [CrossRef] [PubMed]
30. Cáceres, R.; Malińska, K.; Marfà, O. Nitrification within Composting: A Review. *Waste Manag.* **2018**, *72*, 119–137. [CrossRef]
31. Wang, J.; Chadwick, D.R.; Cheng, Y.; Yan, X. Global Analysis of Agricultural Soil Denitrification in Response to Fertilizer Nitrogen. *Sci. Total Environ.* **2018**, *616–617*, 908–917. [CrossRef]
32. Ma, J.; Lu, C.; Bai, L.; Zhang, J.; Shen, Y. Phytotoxic Phenols from the Needles of Cedrus Deodara. *Phytochemistry* **2024**, *219*, 113977. [CrossRef]
33. Ribas-Agustí, A.; Seda, M.; Sarraga, C.; Montero, J.I.; Castellari, M.; Muñoz, P. Municipal Solid Waste Composting: Application as a Tomato Fertilizer and Its Effect on Crop Yield, Fruit Quality and Phenolic Content. *Renew. Agric. Food Syst.* **2017**, *32*, 358–365. [CrossRef]
34. Wang, G.; Yang, Y.; Kong, Y.; Ma, R.; Yuan, J.; Li, G. Key Factors Affecting Seed Germination in Phytotoxicity Tests during Sheep Manure Composting with Carbon Additives. *J. Hazard. Mater.* **2022**, *421*, 126809. [CrossRef] [PubMed]
35. Wang, G.; Kong, Y.; Yang, Y.; Ma, R.; Shen, Y.; Li, G.; Yuan, J. Superphosphate, Biochar, and a Microbial Inoculum Regulate Phytotoxicity and Humification during Chicken Manure Composting. *Sci. Total Environ.* **2022**, *824*, 153958. [CrossRef]
36. Zhang, Y.; Wang, J.; Dai, S.; Zhao, J.; Huang, X.; Sun, Y.; Chen, J.; Cai, Z.; Zhang, J. The Effect of C:N Ratio on Heterotrophic Nitrification in Acidic Soils. *Soil Biol. Biochem.* **2019**, *137*, 107562. [CrossRef]
37. Zhang, L.; Sun, X. Effects of Bean Dregs and Crab Shell Powder Additives on the Composting of Green Waste. *Bioresour. Technol.* **2018**, *260*, 283–293. [CrossRef] [PubMed]
38. Tong, B.; Wang, X.; Wang, S.; Ma, L.; Ma, W. Transformation of Nitrogen and Carbon during Composting of Manure Litter with Different Methods. *Bioresour. Technol.* **2019**, *293*, 122046. [CrossRef]
39. Gibbs, B.M.; Shephard, L.R.; Third, K.A.; Cord-Ruwisch, R. The Presence of Ammonium Facilitates Nitrite Reduction under PHB Driven Simultaneous Nitrification and Denitrification. *Water Sci. Technol.* **2004**, *50*, 181–188. [CrossRef]
40. Prinčič, A.; Mahne, I.; Megušar, F.; Paul, E.A.; Tiedje, J.M. Effects of pH and Oxygen and Ammonium Concentrations on the Community Structure of Nitrifying Bacteria from Wastewater. *Appl. Environ. Microbiol.* **1998**, *64*, 3584–3590. [CrossRef]
41. Ji, B.; Qian, Y.; Zhang, H.; Al-Gabr, H.M.; Xu, M.; Zhang, K.; Zhang, D. Optimizing Heterotrophic Nitrification Process: The Significance of Demand-Driven Aeration and Organic Matter Concentration. *Bioresour. Technol.* **2023**, *376*, 128907. [CrossRef]
42. Phan, D.C.; Vazquez-Munoz, R.; Matta, A.; Kapoor, V. Short-Term Effects of Mn_2O_3 Nanoparticles on Physiological Activities and Gene Expression of Nitrifying Bacteria under Low and High Dissolved Oxygen Conditions. *Chemosphere* **2020**, *261*, 127775. [CrossRef] [PubMed]
43. Courtens, E.N.P.; Boon, N.; De Clippeleir, H.; Berckmoes, K.; Mosquera, M.; Seuntjens, D.; Vlaeminck, S.E. Control of Nitratation in an Oxygen-Limited Autotrophic Nitrification/Denitrification Rotating Biological Contactor through Disc Immersion Level Variation. *Bioresour. Technol.* **2014**, *155*, 182–188. [CrossRef] [PubMed]
44. Mousavi, S.S.; Karami, A.; Haghighi, T.M.; Alizadeh, S.; Maggi, F. Phytotoxic Potential and Phenolic Profile of Extracts from Scrophularia Striata. *Plants* **2021**, *10*, 135. [CrossRef] [PubMed]
45. Pinho, I.A.; Lopes, D.V.; Martins, R.C.; Quina, M.J. Phytotoxicity Assessment of Olive Mill Solid Wastes and the Influence of Phenolic Compounds. *Chemosphere* **2017**, *185*, 258–267. [CrossRef] [PubMed]
46. Baker, N.R. Chlorophyll Fluorescence: A Probe of Photosynthesis In Vivo. *Annu. Rev. Plant Biol.* **2008**, *59*, 89–113. [CrossRef]
47. Kalaji, H.M.; Schansker, G.; Brestic, M.; Bussotti, F.; Calatayud, A.; Ferroni, L.; Goltsev, V.; Guidi, L.; Jajoo, A.; Li, P.; et al. Frequently Asked Questions about Chlorophyll Fluorescence, the Sequel. *Photosynth. Res.* **2017**, *132*, 13–66. [CrossRef]

48. Xu, C.; Zhang, X.; Hussein, Z.; Wang, P.; Chen, R.; Yuan, Q.; Gao, Y.; Song, N.; Gouda, S.G. Influence of the Structure and Properties of Lignocellulose on the Physicochemical Characteristics of Lignocellulose-Based Residues Used as an Environmentally Friendly Substrate. *Sci. Total Environ.* **2021**, *790*, 148089. [CrossRef]
49. Zulfiqar, F.; Allaire, S.E.; Akram, N.A.; Méndez, A.; Younis, A.; Peerzada, A.M.; Shaukat, N.; Wright, S.R. Challenges in Organic Component Selection and Biochar as an Opportunity in Potting Substrates: A Review. *J. Plant Nutr.* **2019**, *42*, 1386–1401. [CrossRef]

plants

MDPI

Review

Kenaf Core as an Alternative Soilless Growing Medium: A Review

Conner C. Austin [1,*], S. Brooks Parrish [1], David G. Clark [1] and Ann C. Wilkie [2,*]

[1] Horticultural Sciences Department, University of Florida-IFAS, Gainesville, FL 32611, USA; brooks.parrish@ufl.edu (S.B.P.); geranium@ufl.edu (D.G.C.)
[2] Department of Soil, Water, and Ecosystem Sciences, University of Florida-IFAS, Gainesville, FL 32611, USA
* Correspondence: conner.austin@ufl.edu (C.C.A.); acwilkie@ufl.edu (A.C.W.)

Abstract

Kenaf (*Hibiscus cannabinus*) core, an abundant renewable byproduct rich in cellulose and hemicellulose, has emerged as a candidate to replace or supplement peat and coco coir in soilless culture. This review synthesizes the physical, chemical, and biological performance of ground kenaf core and benchmarks it against conventional substrates. Kenaf core exhibits low bulk density (0.06 to 0.15 g cm^{-3}), high total porosity (approximately 90%), and substantial plant available water (approximately 42%), supporting root aeration and water supply. Its pH (6.0–7.2) is near optimal for most crops, whereas electrical conductivity (EC) (3.2–4.7 dS m^{-1}) can exceed recommended ranges for salt-sensitive species, which necessitates pre-leaching or blending. Growth studies show comparable shoot and root performance in blends containing 20 to 70% kenaf, with composted kenaf often outperforming raw core. Pure kenaf generally requires more frequent irrigation and may shrink at high proportions. We outline processing variables such as core purity, particle size, composting, and leaching that govern stability and plant response, identify critical data gaps (including standardized EC and pH methods, and long-term shrinkage), and frame a sustainability agenda. Practically, studies to date indicate that pre-leached kenaf core, incorporated at up to about 70% by volume into peat or coir-based blends with structurally stable components such as perlite, can maintain growth and quality for several ornamental and bedding crops under greenhouse and nursery conditions. At the same time, reports of poor performance in some conifers and early suppression in direct-sown vegetables underscore that the suitability of kenaf-based substrates remains crop specific and dependent on material processing and management.

Keywords: soilless culture; container substrates; horticultural sustainability; coco coir; peat

1. Introduction

The science of growing media has evolved significantly since the mid-20th century, with soilless substrates now dominating containerized plant production due to their ability to promote vigorous root growth, enable efficient nutrient uptake, and mitigate risks from soilborne pathogens [1,2]. However, the industry's reliance on two primary components, peat and coco coir, is facing a confluence of critical challenges [3]. Peat is increasingly viewed as ecologically unsustainable due to its slow regeneration rate and the damage that extraction inflicts on vital carbon sinks [4,5]. This has prompted intensifying political initiatives to reduce or eliminate its use, creating regulatory urgency for high-performing alternatives such as wood fiber and compost [6]. At the same time, while coco coir is a renewable byproduct, life cycle assessments confirm that the majority of its environmental footprint stems from the impacts of long-distance sea transport from tropical production zones [7]. Compounding these environmental and political pressures are significant supply chain disruptions, including cost volatility and interconnected barriers of availability, affordability, and performance consistency that affect everyone from producers to consumers [8–11]. Taken together, these factors underscore a pressing need to identify and validate a new class of renewable, locally producible materials that are not only scientifically sound but also economically viable and logistically practical.

Recent work has synthesized the broader landscape of peat alternatives, including wood fiber and other organic byproducts, using formal systematic review methods. For example, Sdao et al. [11] provide a comprehensive assessment of multiple candidate materials in relation to peat reduction targets. In contrast, the present review adopts a deliberately focused scope and concentrates on ground kenaf core as a single, underutilized substrate component. By collating and benchmarking the dispersed literature on kenaf's physical, chemical, biological, and crop performance characteristics, we aim to provide a depth of material-specific analysis that complements these broader evaluations of peat substitutes.

In this context, kenaf (*Hibiscus cannabinus*) is a promising renewable ingredient for soilless media. A fast-growing annual suitable for cultivation in many warm-season production regions, its central stalk comprises an outer bast (35%) and an inner core (65%) [12]. The bast fiber is well suited to applications such as paper pulp and biocomposites due to its high cellulose content and fiber length [13–19]. The woody inner core is often treated as a byproduct, yet its high hemicellulose content and absorbency make it useful for filtration aids, oil absorbents, and as a component of growing media [4,17,19–23]. Early studies show that as a partial replacement in peat-based blends, ground kenaf core delivers comparable performance for many crops, although this varies with species, particle size, and blend ratio.

Existing studies indicate that ground kenaf core can, under certain conditions, support plant growth comparable to conventional peat- and coir-based substrates, particularly when used in blends. However, the available evidence is fragmented across crops and regions, and important aspects of substrate performance such as standardized physical and chemical characterization, long-term structural stability, and quantitative sustainability metrics remain incompletely described or inconsistently reported. In this review, we therefore (i) synthesize published data on the physical, chemical, and biological properties of kenaf core in soilless systems, (ii) benchmark these properties and crop responses against peat and coco coir, and (iii) identify specific knowledge gaps and research priorities needed to assess when and how kenaf can function as a reliable substrate component. By framing what is known, what is uncertain, and what must be measured next, we aim to provide a clear basis for future experimental work and for evidence-based evaluation of kenaf core within modern soilless culture.

2. Review Methodology

This review was conducted as a narrative, topic-focused synthesis rather than a systematic review; nevertheless, we followed a structured approach to identify and select relevant literature. Primary peer-reviewed studies evaluating kenaf (*Hibiscus cannabinus*) core as a component of soilless substrates were identified using Google Scholar, with searches conducted between November 2024 and December 2025. Search queries combined terms such as "kenaf core growing media", "kenaf potting medium", "kenaf container substrate", and "kenaf soilless culture", along with related phrases such as "peat alternative", "coconut coir", and "container substrate physical properties". Reference lists of key articles were also screened (backward citation searching) to capture additional relevant studies.

https://doi.org/10.3390/plants15040666

We focused on studies that (i) reported the use of ground or composted kenaf core or whole-stem kenaf as a component of container media, (ii) provided quantitative data on plant growth, substrate physical or chemical properties, or both. We excluded studies that used kenaf exclusively for non-horticultural applications (such as for biocomposites and paper pulp) unless they provided compositional data directly relevant to substrate performance. Contextual information on peat and coco coir, as well as global trade patterns in growing media constituents, was obtained from peer-reviewed articles, extension publications, and international databases. In particular, trade statistics for peat and coco coir were extracted from the UN Comtrade database (https://comtrade.un.org, accessed on 15 December 2025) using the relevant HS codes.

Because the body of literature specifically addressing kenaf core as a growing medium is relatively limited and spans multiple decades, we did not restrict our search to a fixed publication window; instead, we included all accessible studies up to December 2025 that met the above criteria. Where possible, we prioritized original experimental work over secondary summaries, and we synthesized findings across studies to highlight consistent patterns as well as knowledge gaps in the physical, chemical, biological, and sustainability dimensions of kenaf-based substrates.

3. Desirable Characteristics of Growing Media

To support plant health, growing media must perform four core functions: provide physical support, aerate the root zone, supply water, and deliver nutrients [24,25]. A substrate's ability to perform these functions is determined by its physical, chemical, and biological properties. These characteristics are interdependent and dictate a medium's suitability for a given crop and production system [24,25]. However, since reported values for these properties depend heavily on test protocols, standardized methods and clear reporting are essential for meaningful comparisons. This review therefore uses these three property classes as the framework for comparing kenaf core against peat and coco coir.

3.1. Physical Properties

The physical properties of a growing medium are central to healthy root development and sustained plant growth. A high-quality substrate maintains a balanced relationship between air and water, which enables root respiration, nutrient uptake, and overall plant vigor [26,27]. These properties originate from the size, shape, and packing arrangement of the substrate's particles and are commonly described by metrics including particle size distribution, bulk density, and total porosity. These foundational properties in turn determine functional outcomes such as water-holding capacity and aeration [24–26].

Particle size distribution governs a substrate's pore architecture and therefore its porosity and bulk density. For many container systems, a particle size range of 0.25 to 2.5 mm provides a desirable balance of air-filled porosity and water retention [25,28]. Bulk density, the mass of medium per unit volume, is inversely related to total porosity. Lower bulk density, with typical values for soilless mixes falling near 0.2 to 0.5 g cm^{-3}, indicates a looser structure that promotes aeration and root expansion. Higher bulk density suggests compaction that can restrict air movement and plant growth [25,29–31]. Total porosity, the fraction of volume occupied by pores, usually ranges from 50 to 90% by volume in effective mixes [24,25,30–33].

Water-holding capacity is the fraction of pore space that remains filled with water after gravitational drainage, while aeration is the air-filled fraction in the same state [24]. The two are inversely related; as total porosity increases, aeration generally increases while water-holding capacity decreases. For many crops in small to medium containers, a water-holding capacity of about 50 to 60% and an air-filled porosity of 20 to 30% are common

https://doi.org/10.3390/plants15040666

targets. However, optimal values always vary with container height, irrigation regime, and specific crop requirements [25,30,34].

In summary, a substrate's key physical properties are tightly linked and must be optimized in concert to provide a root environment that supports robust growth. Because reported values can vary with test protocols, container dimensions, and handling, the clear documentation of measurement methods is critical for comparing results across studies.

3.2. Chemical Properties

A substrate's chemical properties govern nutrient availability at the root interface. The key indicators are pH, electrical conductivity (EC), and cation exchange capacity (CEC), which growers can adjust through practices like liming, fertigation, and pre-leaching to optimize plant growth [26,35,36]. Because reported values for these metrics depend on the test method and extraction ratio, clear documentation is essential for meaningful comparisons [24–26].

The pH, which indicates the acidity or alkalinity of the substrate solution, strongly influences nutrient solubility and availability [24]. At high pH, the availability of phosphorus and several micronutrients decline. Conversely, at low pH, aluminum and manganese can become phytotoxic and the availability of key macronutrients may be reduced [25,37,38]. Most container crops perform best within a pH range of 5.2 to 6.5, which ensures broad nutrient availability [25,30,31].

Electrical conductivity measures the concentration of soluble salts in the root-zone solution and serves as a proxy for fertility status [25,39]. While moderate EC indicates adequate nutrients, excessive EC can cause osmotic stress and reduce water uptake, and very low EC suggests nutrient deficiency. Although target ranges are method and crop dependent, many soilless systems aim for an EC between 1.5 and 3.5 dS m^{-1}, with lower targets for salt-sensitive species [25,34].

Cation exchange capacity is the substrate's ability to adsorb and exchange positively charged ions such as K^+, Ca^{2+}, and Mg^{2+}, providing a buffer and reservoir for nutrients [25,40]. For organic soilless media, a CEC between 50 and 200 cmolc kg^{-1} is often desirable, as it promotes nutrient retention without excessive fixation. Substrates with very low CEC are prone to rapid leaching and unstable nutrition, while very high CEC can slow fertilizer adjustments and contribute to imbalances if not managed correctly [25,35,41].

In summary, pH, EC, and CEC jointly regulate nutrient supply and overall plant performance. Optimizing substrates and enabling valid comparisons across studies requires managing these properties within crop-specific targets and consistently reporting the measurement methods used.

3.3. Biological Properties

A substrate's biological properties strongly influence its performance and ability to support healthy plant development [26,42]. The key factors are the presence of harmful organisms, the material's biological stability, and the potential for nutrient immobilization. Each requires active management to maintain a favorable root environment [26]. Pathogens, pests, and weed seeds can severely damage plant health by causing root rots, fungal infections, and direct root damage that impairs water and nutrient uptake [26,43–45]. To minimize these risks, high-quality media must be produced using clean sourcing, screening, and sanitation. Pasteurization or steam treatment is often preferred over complete sterilization because it reduces pathogen loads while preserving beneficial microflora that aid in nutrient cycling and disease suppression. Similarly, preventing weed seed contamination reduces competition for resources and helps ensure uniform crop growth [26,44].

https://doi.org/10.3390/plants15040666

Biological stability is critical for organic substrates, which are subject to microbial decomposition. As microorganisms break down carbon-rich components, the medium can shrink, reducing air-filled porosity and impairing root respiration [46,47]. This decomposition can also alter water relations, sometimes increasing water retention to the point of waterlogging if irrigation is not adjusted [26,48]. Stability can be managed by selecting appropriate particle sizes, blending with structurally stable components, and using pretreatments like composting to reduce the most degradable fractions. High microbial activity can also temporarily immobilize nutrients, particularly nitrogen and phosphorus, as microorganisms consume these ions while metabolizing carbon [26,42]. This process can lead to transient nutrient deficiencies and slow plant growth. To counteract this, growers must monitor substrate nutrient levels and provide supplemental fertilization as needed to maintain adequate nutrition throughout the crop cycle. Managing a substrate's biological profile is integral to its performance. By preventing contamination, maintaining structural stability, and offsetting nutrient immobilization through best practices, growers can sustain a healthy root environment that minimizes risks and supports consistent, high-quality plant growth.

4. Commonly Utilized Growing Media

Peat and coco coir are the dominant substrates in containerized plant production, a status supported by massive international trade flows (Table 1). To illustrate the scale, 2023 peat imports to the United States exceeded 1.1 million metric tons at a value of nearly USD 500 million, which was part of a global trade valued at over USD 1.9 billion [49]. While the market for coco coir is smaller, it remains significant, with global imports valued at over USD 462 million in the same year [49]. These figures underscore the horticultural sector's deep reliance on imported substrates and its vulnerability to supply chain disruptions and price volatility. The primary exporters of peat are Latvia, Estonia, Germany and Canada, while those of coco coir are India and Sri Lanka, resulting in substantial transport distances to supply large markets such as the United States.

Table 1. United States and global imports of peat and coco coir in 2023, quantity (t) and trade values (USD). Source: UN Comtrade, 2023 [49].

Commodity	Net Weight (t)		Approximate Value (Million $)	
	USA	Global	USA	Global
Peat	1,172,104	8,036,966	494	1900
Coco Coir	75,738	1,105,622	52	462

4.1. Peat

Peat remains the dominant substrate in containerized production because it combines high CEC, excellent water retention, and favorable aeration. Together, these properties support robust root growth and provide a strong nutrient buffer [2,25,26,50–52]. Derived from partially decomposed mosses like sphagnum that accumulate in bogs, peat typically exhibits a low bulk density (0.04 to 0.2 g cm^{-3}), high total porosity (83.8 to 96.4%), and a high CEC (90 to 140 cmolc kg^{-1}) [36,52–54]. It also has a low inherent EC of around 0.2 to 0.49 dS m^{-1} [25,52].

However, peat presents significant practical and environmental challenges. Its natural acidity, with a typical pH near 3.9, requires liming for most crops [52]. It also becomes hydrophobic once dry, making it difficult to rewet uniformly and increasing the risk of crop stress [3]. Beyond the greenhouse, peat extraction is increasingly scrutinized for its environmental impact. Peatlands cover only 3% of the global land surface but are vital carbon sinks and hotspots of biodiversity [55–57]. Harvesting peat degrades these functions and habitats, raising questions about its long-term sustainability. These environmental

https://doi.org/10.3390/plants15040666

concerns, combined with the costs of extraction, processing, and long-distance transport, are driving the search for alternatives. While peat's physical and buffering properties make it a valuable substrate, there is a clear and growing need to substitute it with sustainable, locally sourced materials that can maintain high performance.

4.2. Coco Coir

Coco coir (also called coco coir pith or coco peat), a byproduct derived from the fibrous coconut mesocarp, is another widely used substrate component [58]. It offers several advantages, including a near-optimal pH (5.6 to 6.9) that rarely requires adjustment and excellent rewetting properties. Its physical characteristics are also favorable, with a low bulk density (0.04 to 0.08 g cm^{-3}) and high total porosity (85.5 to 89.5%) [52,59]. Coco coir has a higher EC compared to peat, with values ranging from 0.3 to 2.9 dS m^{-1} [59]. However, its CEC (39 to 60 cmolc kg^{-1}) is considerably lower than that of peat [59].

Coco coir's primary challenges are high salt content and inconsistent quality. Material sourced from coastal regions often contains elevated levels of sodium, chloride, and potassium that can impair plant growth unless it is aged and thoroughly washed or buffered [58,60]. The process of creating horticultural-grade coir can take up to two years, leading to significant variability in EC and nutrient profiles between suppliers and even between batches. This inconsistency adds processing costs and management complexity for growers. Although often positioned as a sustainable alternative to peat, these issues of quality control, lengthy processing, and the costs associated with freight from tropical production zones can limit coir's practical utility. Its successful use hinges on sourcing reliable, pre-treated products and using regular substrate tests to guide fertility management.

5. Kenaf as Growing Media

The dried, ground core of the kenaf plant is a promising material for use in growing media. Research has demonstrated its value both as an amendment and as a complete substrate. As an additive in conventional mixes, kenaf core can improve aeration and root development, with performance comparable to peat-based blends at moderate inclusion rates [23,61]. As a complete replacement for traditional media, its performance is more variable and highly dependent on careful management of irrigation and fertility, as well as on pretreatments like composting or leaching [4,23,62].

However, wider adoption of kenaf core is constrained by limited data and a lack of standardized testing protocols. To move forward, research must focus on establishing a more complete and comparable profile of the material. Key priorities include quantifying its long-term structural stability and shrinkage under commercial irrigation, characterizing its nutrient release dynamics, and determining its CEC using standardized methods. Furthermore, given reports of high EC in some batches, best-practice guidelines for pre-leaching and blending are needed. Defining optimal processing variables, such as core purity, particle size, and composting protocols, will ultimately enable growers to realize kenaf's full potential as a sustainable and cost-effective substrate.

5.1. Growth Studies

Growth studies across a range of vegetables and ornamentals support that ground kenaf core is a valuable component in soilless mixes, though performance hinges on inclusion rate, particle size, and crop species (Table 2). For example, when used as an additive for tomato (*Lycopersicon esculentum*), a blend with 20 to 35% finely ground kenaf core in peat yielded the best overall shoot growth [61]. A broader study evaluating a 70% kenaf and 30% peat mix across 17 species showed more nuanced results. While root mass was generally enhanced in the kenaf-amended media, shoot dry weight was higher in the peat control for seven species

https://doi.org/10.3390/plants15040666

and comparable for the remaining ten [63]. Reinforcing this theme of variability, Webber et al. [22] found that when kenaf core was used to substitute pine bark and vermiculite, its effectiveness depended entirely on the grind fineness, blend proportion, and the specific crop being grown. This underscores the need to tailor kenaf blends to a particular application.

Table 2. Summary of published growth studies evaluating ground kenaf core as an additive or replacement in soilless substrates, with species tested, main outcomes, and references.

Species	Main Outcomes	Location	Reference
Lycopersicon esculentum (Tomato)	Finely ground kenaf core at 20 to 35% in peat gave best shoot growth vs. other ratios; supports use as additive at moderate inclusion.	Newark, DE, USA	[61]
Brassaia actinophylla (Umbrella Tree), *Hibiscus rosa-sinensis* (Tropical Hibiscus), *Pittosporum tobira* (Japanese Cheesewood)	100% kenaf supported normal growth but required more irrigation; blends with 70 to 80% kenaf plus peat and vermiculite produced shoot growth comparable to or better than two commercial mixes.	Weslaco, TX, USA	[4]
Nephrolepis exaltata (Boston fern), *Impatiens walleriana* (Impatiens), and *Viola* spp. (Pansies)	Pure kenaf led to slower growth than commercial mixes; 70% kenaf plus 30% peat produced plant growth comparable to commercial media.	Starkville, MS, USA	[23]
Pinus halepensis (Aleppo Pine)	Any mix containing kenaf core performed poorly relative to controls, indicating species-specific limitations for conifers under tested conditions.	Chalkidona, Greece	[62]
Lactuca sativa (Lettuce) and *Capsicum annuum* (Pepper)	Whole-stem kenaf with sand at sowing suppressed early growth even at 10 to 90 kenaf to sand; inhibition partly alleviated by ammonium nitrate soak; normal growth after transplant into kenaf-free media.	Athens, Greece	[64]
Ilex crenata (Japanese Holly) cv. Cherokee	Composted kenaf core compared favorably with pine bark, suggesting composting improves performance.	Poplarville, MS, USA	[65]
Euphorbia pulcherrima (Poinsettia)	In peat-based media, 50 to 70% coarse-ground kenaf by volume produced best growth and quality, comparable to a commercial control (Sunshine #1). Media with more than 70% fine-ground kenaf yielded smaller plants, leaf chlorosis, and container media shrinkage. Kenaf-amended media had lower water-holding capacity than the control and required more frequent irrigation.	College Station, TX, USA	[66]
17 vegetable and ornamental species	70% kenaf core plus 30% peat enhanced root mass; shoot dry weight higher in peat control for 7 species and comparable for 10; final height similar except tomato and *Celosia argentea* favored peat.	Starkville, MS, USA	[63]
Various ornamentals	Kenaf core substituted for pine bark and vermiculite; performance depended on grind fineness, kenaf proportion, and crop, indicating need to tailor particle size and blend ratio.	Lane, OK, USA	[22]

https://doi.org/10.3390/plants15040666

Ground kenaf core has also been tested as a complete replacement for conventional substrates, but results suggest it performs best in blends. For instance, Wang [4] found that while several ornamentals could grow in 100% kenaf core, they required substantially more irrigation. Blends containing 70 to 80% kenaf, however, produced shoot growth comparable to or better than commercial mixes. Williams and Reichert [23] reported a similar pattern: slower growth in pure kenaf but comparable performance in a 70% kenaf and 30% peat mix. Performance is also species-specific, as Tsakaldimi [62] found that kenaf-based media performed poorly for *Pinus halepensis* seedlings, suggesting some species, like conifers, may be less suited to kenaf substrates.

Beyond performance variability, some studies report early growth suppression, particularly when whole-stem kenaf (core plus bark) is used at sowing. In a study on lettuce and pepper, even a low 10% ratio of whole-stem kenaf to sand reduced seedling height, leaf number, and biomass [64]. Although this inhibition was partially alleviated by soaking the kenaf in ammonium nitrate, normal growth resumed only after seedlings were transplanted into kenaf-free media. These findings underscore the importance of material processing and suggest that for sensitive species, kenaf is best introduced at transplant rather than at germination.

Pretreatments like composting can also significantly improve performance. Laiche and Newman [65] reported that *Ilex crenata* cv. Cherokee grown in composted kenaf core compared favorably to plants grown in a standard pine bark medium. Overall, the evidence indicates that ground kenaf core is most effective as a component in blends, typically at up to 70% by volume. Its performance is optimized by selecting an appropriate particle size, pre-leaching to reduce high EC, and using composted material to enhance stability and mitigate potential growth inhibition.

Taken together, these studies span a range of climatic and production contexts, from humid subtropical regions of the southern United States (Texas, Mississippi) to temperate greenhouse environments (Delaware, Oklahoma) and Mediterranean conditions in Greece. These differences in climate and production systems likely influence irrigation demand, substrate drying cycles, and the rate of kenaf core decomposition, all of which can affect crop response. For example, in warmer, more humid regions, higher evapotranspiration and more frequent irrigation may accentuate shrinkage and structural changes in kenaf-based media, whereas in cooler or more controlled environments, the same mixes may remain physically stable for longer crop cycles. Nevertheless, across these diverse locations, a consistent pattern emerges: kenaf core performs best as a component of blended substrates rather than as a sole medium, suggesting that the general conclusions of this review are broadly applicable across production regions, provided that irrigation and fertility are locally optimized.

5.2. *Physical Properties of Kenaf as Growing Media*

Ground kenaf core exhibits favorable physical characteristics that are comparable to both peat and coco coir (Table 3). Its bulk density, at 0.06 to 0.15 g cm^{-3}, is similar to peat and indicates a light, loose structure that resists compaction [22,52,62]. Its total porosity has been reported at 90.7%, a value which aligns with ranges for peat and coir and ensures ample pore space for root aeration [62]. Furthermore, its plant available water content is also high at approximately 41.9%, placing it within the upper range of conventional substrates and suggesting it can supply effective moisture to the root zone [62].

Together, these properties demonstrate that ground kenaf core can provide a suitable physical environment for root growth in soilless systems. It is important to note, however, that reported values for porosity and available water can vary significantly depending on the test protocol, container dimensions, and packing method used. For this reason,

https://doi.org/10.3390/plants15040666

clear documentation of measurement approaches is essential for valid comparisons across different studies and materials.

Table 3. Physical properties of ground kenaf core compared to peat and coco coir.

Growing Media	Bulk Density (g cm^{-3})	Total Porosity (%)	Available Water (%)
Ground Kenaf Core	0.06–0.15 [22,62]	90.7 [62]	41.9 [62]
Peat	0.04–0.2 [52]	83.8–96.4 [52]	6.1–47.8 [52]
Coco Coir	0.04–0.08 [59]	85.5–89.5 [59]	19.9–37.8 [52]

In summary, the physical properties of ground kenaf core, as reported to date, fall broadly within the same general range as those of peat and coco coir for parameters such as bulk density, total porosity, and plant available water. Given the variability in measurement protocols among studies, these comparisons should be interpreted as indicative rather than as demonstrating strict equivalence. Within this context, and with appropriate particle sizing and pretreatment, kenaf core appears capable of providing a suitable root environment and functioning as a substrate component in blends that can reduce reliance on imported materials.

5.3. Chemical Properties of Kenaf as Growing Media

The chemical characteristics of ground kenaf core are promising, though they also highlight areas where further research is needed. Its most significant advantage is a near-optimal pH, which typically ranges from 6.03 to 7.17 (Table 4) [4]. This is comparable to coco coir and falls within the preferred range for most container crops, reducing or eliminating the need for liming that is required for acidic peat-based media [25,30,31].

Table 4. pH and EC of ground kenaf core compared to peat and coco coir.

Growing Media	pH	EC (dS m^{-1})
Ideal	5.2–6.5 [25,30,31]	2.5–3.5 [25,34]
Ground Kenaf Core	6.03–7.17 [4,62]	3.18–4.7 [4]
Peat	3.9 [52]	0.2–0.49 [25,52]
Coco Coir	5.6–6.9 [52,59]	0.3–2.9 [59]

In contrast, kenaf's EC presents a management challenge. Reported EC values of 3.18 to 4.7 dS m^{-1} are considerably higher than those of peat or coir and can exceed the recommended targets for salt-sensitive species [4]. This makes pre-leaching or blending with lower-EC components a necessary practice for many applications. As with other properties, reported EC values are highly dependent on the measurement method and extraction ratio, so consistent reporting is critical for cross-study comparisons.

Ground kenaf core contains measurable amounts of several essential mineral nutrients. For instance, one study reported total calcium and magnesium concentrations (12 and 2.34 mg g^{-1}, respectively) that are considerably higher than typical values for peat and coir (Table 5) [62]. The core's nitrogen content (1.44%) was also found to be comparable to peat and higher than coir. These values suggest that kenaf core may contribute to the baseline supply of several key nutrients, although differences in analytical methods and the limited number of reports mean that these comparisons with peat and coir should be viewed as approximate. However, these figures represent total elemental content, not the plant-available fraction. The release kinetics of these nutrients under typical fertigation regimes remain uncharacterized. Therefore, while kenaf has the potential to supplement crop nutrition and perhaps reduce initial fertilizer inputs, existing evidence does not justify reducing fertigation inputs without careful monitoring. In practice, it is prudent to base

https://doi.org/10.3390/plants15040666

fertility management on regular substrate testing rather than to assume a substantial nutrient contribution from the kenaf itself.

Table 5. Inorganic nutrient content of phosphorus (P), potassium (K), nitrogen (N), calcium (Ca), and magnesium (Mg) of ground kenaf core compared to peat and coco coir.

Nutrient	Ground Kenaf Core	Peat	Coco Coir
P (mg g^{-1})	0.14 [62]	0.25 [67]	0.27 [58]
K (mg g^{-1})	0.5 [62]	0.13 [67]	0.33 [58]
N (%)	1.44 [62]	0.0–5.0 [68]	0.44 [69]
Ca (mg g^{-1})	12.0 [62]	5.86 [67]	0.058 [70]
Mg (mg g^{-1})	2.34 [62]	0.97 [67]	0.055 [70]

To more reliably characterize the chemical performance of ground kenaf core, a targeted research agenda on its chemical behavior is needed that uses standardized analytical methods. This agenda should prioritize three key areas. First, its nutrient dynamics must be characterized using standardized measurement protocols for CEC and EC, along with a thorough analysis of nutrient release kinetics over time, so that future comparisons with peat and coir are based on directly comparable data. Second, the long-term stability of its pH and EC must be assessed under commercial growing conditions. To support this, all studies must consistently report measurement methods, such as saturated paste or pour-through, to ensure results are comparable. Finally, the effects of pretreatments like composting and leaching should be systematically evaluated to quantify their impact on salt levels and potential phytotoxic compounds. By generating this body of method-rich, comparable data, kenaf core can be optimized for widespread adoption as a sustainable and efficient growing medium.

5.4. Biological Properties of Kenaf as Growing Media

The biological properties of kenaf core are primarily defined by its chemical composition: approximately 32.1% cellulose, 41% hemicellulose, and 25.2% lignin (Table 6). Its lignin content is significantly lower than that of coco coir and is at the lower end of the range reported for peat. Because lignin provides structural rigidity and resists microbial breakdown, kenaf's lower lignin content makes it more susceptible to decomposition than more stable substrates [25].

Table 6. Cellulose, hemicellulose, and lignin content of ground kenaf core compared to peat and coco coir.

Growing Media	Cellulose (%)	Hemicellulose (%)	Lignin (%)
Ground Kenaf Core	32.1 [18]	41 [18]	25.2 [18]
Peat	5–20 [71]	10–30 [71]	5–40 [71]
Coco Coir	32–43 [18]	0.15–0.25 [18]	40–50 [18]

This higher rate of decomposition has two major consequences for growers. The first is volumetric shrinkage, which reduces total porosity and air-filled pore space, potentially impairing root respiration. This effect is most pronounced at high inclusion rates, with one study reporting significant shrinkage when kenaf exceeded 70% of the mix by volume [4]. The second consequence is the potential for nutrient release as microorganisms break down organic matter. While this can contribute to plant nutrition, it may also lead to imbalances if not monitored.

Managing these biological properties is key to the successful use of kenaf. Shrinkage can be mitigated by blending kenaf with more structurally stable components like perlite, managing particle size, or using composted core, which has already undergone

https://doi.org/10.3390/plants15040666

initial decomposition. To manage nutrient dynamics, monitoring substrate pH and EC and adjusting fertigation accordingly appear important, particularly in kenaf-rich mixes where decomposition and salt levels may change over time. Further research is needed to quantify shrinkage rates under commercial irrigation and to define optimal particle size and pretreatment protocols to maximize long-term stability.

Taken together, the studies reviewed here suggest a process–property–performance framework for kenaf-based substrates (Figure 1). Processing decisions such as core purity (removal of bark and extraneous stem tissue), particle size distribution, and the use of composting and pre-leaching directly influence key physical and chemical properties, including bulk density, porosity and available water, shrinkage behavior, pH, and EC. These property shifts, in turn, help explain the observed variation in plant responses, such as differences in root mass, shoot growth, irrigation demand, and sensitivity to salinity across species and production systems. Finally, management choices at the grower level, including blend ratio with peat or coir, the inclusion of structurally stable components like perlite, and irrigation and fertigation regimes, can be used to moderate these effects and align kenaf-based media with crop-specific requirements. This framework provides a basis for interpreting the diverse results summarized in Tables 2–6 as parts of a coherent substrate system rather than as isolated findings.

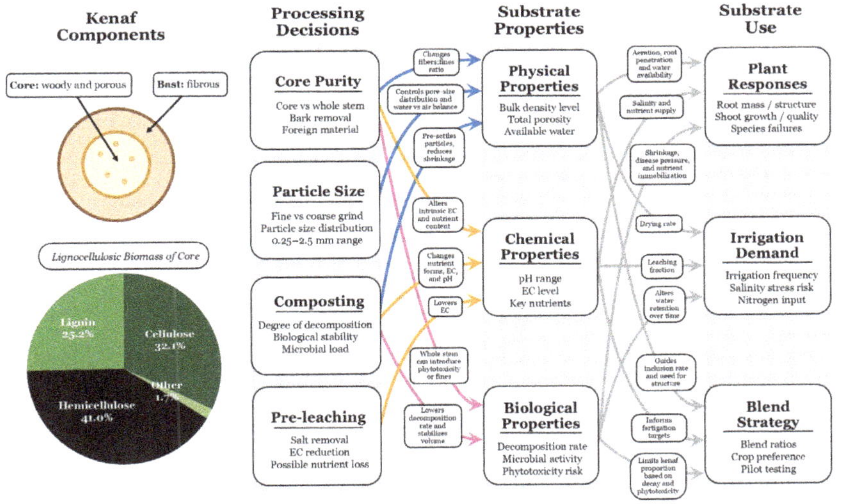

Figure 1. ***Conceptual process–property–performance framework for ground kenaf core in soilless substrates.*** The figure highlights that kenaf core performance emerges from the interaction of structure, processing, and management rather than from any single factor. The leftmost panel illustrates key structural and compositional features of kenaf core (high hemicellulose and cellulose, lower lignin, high porosity and low bulk density) [18] that underlie its behavior as a growing medium component. Processing decisions (core purity, particle size, composting, pre-leaching) modify these features through specific mechanisms (for example, altering pore-size distribution and water–air balance, stabilizing or accelerating decomposition and shrinkage, or removing soluble salts and reducing EC). These changes in physical, chemical, and biological properties in turn influence plant responses (root development, shoot growth, species-specific failures), irrigation and fertigation demand, and blend strategy. Colored arrows indicate processing-driven mechanisms affecting substrate properties: blue arrows highlight changes that primarily alter physical properties, orange arrows highlight mechanisms that primarily affect chemical properties, and pink arrows highlight mechanisms affecting biological properties. Gray arrows represent downstream impacts of these properties for plant responses, irrigation demand, and blend strategy.

https://doi.org/10.3390/plants15040666

Notwithstanding these promising indications, the current evidence base on kenaf core as a substrate component has important limitations. Most growth studies have been conducted on a relatively narrow set of ornamental and bedding crops under greenhouse or nursery conditions in a limited number of geographic regions, and documented failures in species such as *Pinus halepensis* and in direct-sown lettuce and pepper show that responses can be strongly crop- and management-specific. Structurally, the relatively low lignin content and high hemicellulose fraction raise concerns about decomposition and volumetric shrinkage at high inclusion rates, particularly under frequent irrigation. Chemically, high native EC in some batches constrains use for salt-sensitive species unless effective pre-leaching is implemented, and long-term trajectories of pH and EC under commercial fertigation regimes remain poorly quantified. Economically and environmentally, potential advantages related to local cultivation and dual-use of bast and core are still hypothetical in the absence of kenaf-specific life-cycle and cost analyses. Addressing these limitations through standardized characterization, long-term performance trials across diverse environments, and quantitative sustainability assessments will be essential to define where kenaf-based substrates are truly fit for purpose and where alternative materials may be more appropriate.

6. Discussion

The search for sustainable growing media has become urgent as producers confront the environmental, economic, and logistical drawbacks of relying on peat and coco coir [9,25]. In this context, ground kenaf core emerges as a credible, locally sourced candidate substrate component. The studies synthesized here indicate that, when properly managed and used in appropriate blends, kenaf core can supplement conventional substrates without compromising crop performance for several tested ornamental and bedding species [4,22,23,61–63,65,66]. However, these findings also indicate that the performance of kenaf core is constrained by crop- and system-specific responses, potential shrinkage at high inclusion rates, and uncertainties about its long-term chemical behavior and sustainability.

The evidence presents a mixed but interpretable profile of benefits and constraints. Physically, reported values for kenaf core's bulk density and total porosity fall broadly within the same order of magnitude as those reported for peat [52] and coco coir [59], and its near-neutral pH [4,62] reduces or eliminates the need for liming [25,30,31]. At the same time, some lots exhibit relatively high native EC that often requires pre-leaching [4], and the lower lignin content [18] may increase susceptibility to microbial decomposition and volumetric shrinkage, particularly at high inclusion rates [4,46–48]. Growth studies in ornamental and bedding crops suggest that blends containing roughly 20 to 70% kenaf core by volume can produce shoot and root growth similar to commercial mixes under greenhouse and nursery conditions [4,22,23,61,63,65,66], when particle size, irrigation, and fertility are adjusted appropriately. In contrast, species such as *Pinus halepensis* and direct-sown lettuce and pepper have performed poorly in kenaf-containing media under the reported conditions [62,64], which underscores that kenaf is not a universally suitable substrate component and that its use must be evaluated on a crop- and system-specific basis. More generally, because the reported physical and chemical properties of kenaf, peat, and coco coir derive from heterogeneous measurement protocols, the numerical comparisons presented here should be interpreted as indicative of broad ranges rather than as standardized benchmarks that demonstrate strict equivalence among substrates.

From a sustainability perspective, kenaf core has several potential advantages that warrant further quantitative evaluation. As a renewable annual crop, kenaf can be cultivated in many warm-season production regions, which may reduce the transport distances and associated emissions compared with imported substrates such as peat and

https://doi.org/10.3390/plants15040666

coco coir [4,5,7,8,50–52]. In addition, the dual-use model in which the bast fiber serves higher-value industrial applications (paper pulp, biocomposites), leaving the core as a byproduct for horticultural use, suggests a possible economic and resource-efficiency benefit. However, robust kenaf-specific life cycle or carbon footprint data for growing media applications are not yet available, so these putative advantages remain to be demonstrated formally. Realizing and documenting any such benefits at scale will require not only optimized processing infrastructure and quality assurance but also dedicated life-cycle assessments that compare kenaf-based substrates with conventional mixes under consistent functional units.

For practice, a cautious implication of the current evidence is that pre-leached and, where feasible, composted kenaf core can function as an amendment in blends with peat or coir for certain crops, particularly ornamentals and bedding plants, under greenhouse and nursery conditions similar to those in the published studies [4,22,23,61–63,65,66]. Performance depends on tailoring the particle size and blend ratio to the crop and production system, and on managing irrigation to account for the water-holding characteristics and EC of kenaf-based mixes. Regular monitoring of substrate pH and EC remains essential, and it would be advisable to evaluate new formulations at small scale before broader adoption.

Despite these potential benefits, there is a notable lack of quantitative life-cycle assessment data specific to kenaf core used as a growing medium. Existing assessments of growing media focus mainly on peat, coco coir, or other organic constituents and do not consider kenaf core within a comparable functional unit, such as per cubic meter of substrate or per crop cycle. As a result, robust indicators such as global warming potential, cumulative energy demand, and water use cannot yet be directly compared between kenaf-based substrates and conventional peat- or coir-based mixes. This absence of kenaf-specific life-cycle data represents an important research gap and currently limits the ability to rigorously quantify the environmental benefits that are suggested by its agronomic performance and supply chain characteristics.

To transition kenaf from a niche alternative to a mainstream component, a targeted research roadmap is required. This roadmap must prioritize three areas: (i) Standardization, including universal methods for measuring and reporting CEC, EC, and pH to ensure cross-study comparability; (ii) Performance optimization, through studies that quantify long-term shrinkage and determine optimal particle sizes and pretreatment protocols for representative crops; and (iii) Economic validation, via life-cycle assessments that clarify the cost and carbon savings of kenaf relative to conventional substrates.

In conclusion, ground kenaf core shares many of the essential attributes of a high-quality growing medium and appears to be a promising candidate substrate component. Current evidence suggests that, when pre-treated and used in appropriate blends, kenaf core can help reduce reliance on non-renewable peat and globally shipped coco coir for certain crops and production systems. At the same time, further work is needed to standardize methods for physical and chemical characterization, quantify long-term structural stability and shrinkage, and generate kenaf-specific life-cycle and economic assessments. With these research needs addressed, kenaf-based substrates could more reliably support high-quality crop production while contributing to a more sustainable and resilient future for soilless culture.

Author Contributions: Conceptualization, C.C.A. and A.C.W.; methodology, C.C.A. and A.C.W.; validation, A.C.W. and S.B.P.; investigation, C.C.A. and A.C.W.; resources, A.C.W. and D.G.C.; data curation, C.C.A. and A.C.W.; writing—original draft preparation, C.C.A.; writing—review and editing, A.C.W., S.B.P. and D.G.C.; visualization, C.C.A. and A.C.W.; supervision, A.C.W. and D.G.C. All authors have read and agreed to the published version of the manuscript.

https://doi.org/10.3390/plants15040666

Funding: This research received no external funding.

Data Availability Statement: The original contributions presented in the study are included in the article, further inquiries can be directed to the corresponding authors.

Conflicts of Interest: The authors declare no conflicts of interest.

Abbreviations

The following abbreviations are used in this manuscript:

CEC	Cation exchange capacity
EC	Electrical conductivity

References

1. Alexander, P.D.; Bragg, N.C.; Meade, R.; Padelopoulos, G.; Watts, O. Peat in horticulture and conservation: The UK response to a changing world. *Mires Peat* **2008**, *3*, 8. [CrossRef]
2. Kitir, N.; Yildirim, E.; Şahin, Ü.; Turan, M.; Ekinci, M.; Ors, S.; Kul, R.; Ünlü, H.; Ünlü, H. Peat use in horticulture. In *Peat*; IntechOpen: London, UK, 2018; pp. 75–90. [CrossRef]
3. Robbins, J.A. *Growing Media for Container Production in a Greenhouse or Nursery: Part I—Components and Mixes*; Cooperative Extension Service, University of Arkansas: Little Rock, AR, USA, 2001. Available online: https://www.uaex.uada.edu/publications/pdf/FSA-6097.pdf (accessed on 14 December 2024).
4. Wang, Y.T. Using ground kenaf stem core as a major component of container media. *J. Am. Soc. Hortic. Sci.* **1994**, *119*, 931–935. [CrossRef]
5. Neumaier, D.; Meinken, E. Peat substitutes in growing media—Options and limitations. In *II International Symposium on Horticulture in Europe*; Acta Horticultura: Leuven, Belgium, 2012; Volume 1099, pp. 159–166. [CrossRef]
6. Gruda, N.S.; Hirschler, O.; Stuart, J. Peat reduction in horticulture: An overview of Europe. In *IX South-Eastern Europe Symposium on Vegetables and Potatoes*; Acta Horticulturae: Leuven, Belgium, 2023; Volume 1391, pp. 545–560. [CrossRef]
7. Toboso-Chavero, S.; Madrid-López, C.; Villalba, G.; Gabarrell Durany, X.; Hückstädt, A.B.; Finkbeiner, M.; Lehmann, A. Environmental and social life cycle assessment of growing media for urban rooftop farming. *Int. J. Life Cycle Assess.* **2021**, *26*, 2085–2102. [CrossRef]
8. Hwang, S.J.; Jeong, B.R. Growth of *Phalaenopsis* plants in five different potting media. *J. Jpn. Soc. Hortic. Sci.* **2007**, *76*, 319–326. [CrossRef]
9. Hullet, J. Peat Supply Update: Why Communication is Key. Greenhouse Product News. 2023. Available online: https://www.greenhousegrower.com/production/peat-supply-update-why-communication-is-key/ (accessed on 14 December 2024).
10. Koseoglu, M.N.; Roberts, M. Supply chain dynamics of moving from peat-based to peat-free horticulture. *Sustainability* **2025**, *17*, 6159. [CrossRef]
11. Sdao, A.E.; Gruda, N.S.; De Lucia, B. Beyond peat: Wood fiber and two novel organic byproducts as growing media—A systematic review. *Plants* **2025**, *14*, 1945. [CrossRef]
12. Webber, C.L. Crude protein and yield components of six kenaf cultivars as affected by crop maturity. *Ind. Crops Prod.* **1993**, *2*, 27–31. [CrossRef]
13. Kaldor, A.F.; Karlgren, C.; Verwest, H. Kenaf—A fast growing fiber source for papermaking. *TAPPI J.* **1990**, *73*, 205–208.
14. Nishino, T.; Hirao, K.; Kotera, M.; Nakamae, K.; Inagaki, H. Kenaf reinforced biodegradable composite. *Compos. Sci. Technol.* **2003**, *63*, 1281–1286. [CrossRef]
15. Wambua, P.; Ivens, J.; Verpoest, I. Natural fibers: Can they replace glass in fiber reinforced plastics? *Compos. Sci. Technol.* **2003**, *63*, 1259–1264. [CrossRef]
16. Rowell, R.M.; Sanadi, A.; Jacobson, R.; Caulfield, D. Properties of kenaf/polypropylene composites. In *Kenaf: Properties, Processing and Products*; Ag & Bio Engineering, Mississippi State University: Mississippi State, MS, USA, 1999; Volume 1, pp. 381–392.
17. Webber, C.L.; Bledsoe, V.K. Kenaf yield components and plant composition. In *Trends in New Crops and New Uses*; Janick, J., Whipkey, A., Eds.; ASHS Press: Alexandria, VA, USA, 2002; pp. 348–357.
18. Akil, H.M.; Omar, M.F.; Mazuki, A.A.M.; Safiee, S.; Ishak, Z.A.M.; Abu Bakar, A. Kenaf fiber reinforced composites: A review. *Mater. Des.* **2011**, *32*, 4107–4121. [CrossRef]
19. Austin, C.C.; Mondell, C.N.; Clark, D.G.; Wilkie, A.C. Kenaf: Opportunities for an ancient fiber crop. *Agronomy* **2024**, *14*, 1542. [CrossRef]
20. Lee, S.A.; Eiteman, M.A. Ground kenaf core as a filtration aid. *Ind. Crops Prod.* **2001**, *13*, 155–161. [CrossRef]

21. Goforth, C.E. The evaluation of kenaf as an oil sorbent. In *A Summary of Kenaf Production and Product Development Research 1939–1993*; Goforth, C.E., Fuller, M.J., Eds.; Mississippi State University Bulletin: Mississippi State, MS, USA, 1994; Volume 1011, p. 25.
22. Webber, C.L.; Whitworth, J.; Dole, J. Kenaf (*Hibiscus cannabinus* L.) core as a containerized growth medium component. *Ind. Crops Prod.* **1999**, *10*, 97–105. [CrossRef]
23. Williams, R.D.; Reichert, N.A. Growth and economic assessments of kenaf core as the major component in greenhouse potting media. *HortScience* **1995**, *30*, 426c. [CrossRef]
24. Landis, T.D.; Nisley, R.G. *The Container Tree Nursery Manual*; No. 674; U.S. Department of Agriculture, Forest Service: Washington, DC, USA, 1990.
25. Agarwal, P.; Saha, S.; Hariprasad, P. Agro-industrial-residues as potting media: Physicochemical and biological characters and their influence on plant growth. *Biomass Convers. Biorefin.* **2021**, *13*, 9601–9624. [CrossRef]
26. Barrett, G.E.; Alexander, P.D.; Robinson, J.S.; Bragg, N.C. Achieving environmentally sustainable growing media for soilless plant cultivation systems—A review. *Sci. Hortic.* **2016**, *212*, 220–234. [CrossRef]
27. Owen, J.S., Jr.; Warren, S.L.; Bilderback, T.E.; Cassel, D.K.; Albano, J.P. Physical properties of pine bark substrate amended with industrial mineral aggregate. *Acta Hortic.* **2008**, *779*, 131–138. [CrossRef]
28. Benito, M.; Masaguer, A.; Moliner, A.; De Antonio, R. Chemical and physical properties of pruning waste compost and their seasonal variability. *Bioresour. Technol.* **2006**, *97*, 2071–2076. [CrossRef]
29. Bunt, A.C. Some physical and chemical characteristics of loamless pot-plant substrates and their relation to plant growth. *Acta Hortic.* **1974**, *37*, 131–140. [CrossRef]
30. Abad, M.; Noguera, P.; Burés, S. National inventory of organic wastes for use as growing media for ornamental potted plant production: Case study in Spain. *Bioresour. Technol.* **2001**, *77*, 197–200. [CrossRef]
31. Noguera, P.; Abad, M.; Puchades, R.; Maquieira, A.; Noguera, V. Influence of particle size on physical and chemical properties of coconut coir dust as container medium. *Commun. Soil Sci. Plant Anal.* **2003**, *34*, 593–605. [CrossRef]
32. Beardsell, D.V.; Nichols, D.G.; Jones, D.L. Physical properties of nursery potting mixtures. *Sci. Hortic.* **1979**, *11*, 1–8. [CrossRef]
33. Wisdom, B.; Nyembezi, M.; Agathar, K. Effect of different vermiculite and pine bark media substrates mixtures on physical properties and spiral rooting of radish (*Raphanus sativus* L.) in float tray system. *Rhizosphere* **2017**, *3*, 67–74. [CrossRef]
34. Nguyen, V.T.; Wang, C.H. Use of organic materials as growing media for honeydew melon seedlings in organic agriculture. *Commun. Soil Sci. Plant Anal.* **2017**, *48*, 2137–2147. [CrossRef]
35. Handreck, K.; Black, N. *Growing Media for Ornamental Plants and Turf*, 4th ed.; NewSouth Publishing: Sydney, Australia, 2010; pp. 1–560.
36. Silber, A.; Bar-Tal, A. Nutrition of substrate-grown plants. In *Soilless Culture: Theory and Practice*, 2nd ed.; Raviv, M., Lieth, J.H., Bar-Tal, A., Eds.; Academic Press: Cambridge, MA, USA, 2019; pp. 197–257. [CrossRef]
37. Ellsworth, D.S.; Xuan, L. Photosynthesis and canopy nutrition of four sugar maple forests on acid soils in northern Vermont. *Can. J. For. Res.* **1994**, *24*, 2118–2127. [CrossRef]
38. St. Clair, S.B.; Lynch, J.P. Differences in the success of sugar maple and red maple seedlings on acid soils are influenced by nutrient dynamics and light environment. *Plant Cell Environ.* **2005**, *28*, 874–885. [CrossRef]
39. Visconti, F.; de Paz, J.M. Electrical conductivity measurements in agriculture: The assessment of soil salinity. In *New Trends and Developments in Metrology*; Cocco, L., Ed.; IntechOpen: London, UK, 2016; pp. 99–126. [CrossRef]
40. Ugochukwu, U.C. Characteristics of clay minerals relevant to bioremediation of environmental contaminated systems. In *Modified Clay and Zeolite Nanocomposite Materials*; Mercurio, M., Sarkar, B., Langella, A., Eds.; Elsevier: Amsterdam, The Netherlands, 2019; pp. 219–242. [CrossRef]
41. Lemaire, F. Physical, chemical and biological properties of growing medium. *Acta Hortic.* **1995**, *396*, 273–284. [CrossRef]
42. Carlile, W.R.; Wilson, D.P. Microbial activity in growing media—A brief review. In *Proceedings of the II Symposium on Horticultural Substrates and Their Analysis, XXIII International Horticultural Congress, Florence, Italy, 27 August–1 September 1990*; Acta Horticulturae: Leuven, Belgium, 1991; Volume 294, pp. 197–206. [CrossRef]
43. Noble, R.; Elphinstone, J.G.; Sansford, C.E.; Budge, G.E.; Henry, C.M. Management of plant health risks associated with processing of plant-based wastes: A review. *Bioresour. Technol.* **2009**, *100*, 3431–3446. [CrossRef]
44. Carlile, B.; Schmilewski, G. Life in growing media—The good, the bad and the ugly. In Proceedings of the International Peat Symposium, Peat in Horticulture—Life in Growing Media, Amsterdam, The Netherlands, 11 October 2010; pp. 7–14.
45. Kühne, S.; Heller, K. Sciarid fly larvae in growing media—Biology, occurrence, substrate and environmental effects and biological control measures. In Proceedings of the International Peat Symposium, Peat in Horticulture—Life in Growing Media, Amsterdam, The Netherlands, 11 October 2010; pp. 95–102.
46. Cattivello, C.; Della Donna, E.; Pantanali, R. Behavior of peat substrates during cyclamen and poinsettia cultivation. *Acta Hortic.* **1996**, *450*, 439–448. [CrossRef]

https://doi.org/10.3390/plants15040666

47. Aendekerk, T.G.L. Decomposition of peat substrates in relation to physical properties and growth of *Chamaecyparis*. *Acta Hortic.* **1996**, *450*, 191–198. [CrossRef]
48. Nash, M.A.; Pokorny, F.A. Shrinkage of selected two-component container media. *HortScience* **1990**, *25*, 930–931. [CrossRef]
49. United Nations, Department of Economic and Social Affairs, Statistics Division. UN Comtrade Database. Available online: https://comtrade.un.org (accessed on 2 December 2025).
50. Schmilewski, G. Growing medium constituents used in the EU. *Acta Hortic.* **2009**, *819*, 33–45. [CrossRef]
51. Caron, J.; Rochefort, L. Use of peat in growing media: State of the art on industrial and scientific efforts envisioning sustainability. *Acta Hortic.* **2011**, *982*, 15–22. [CrossRef]
52. Carlile, W.R.; Cattivello, C.; Zaccheo, P. Organic growing media: Constituents and properties. *Vadose Zone J.* **2015**, *14*, vzj2014-09. [CrossRef]
53. Basiliko, N.; Knowles, R.; Moore, T.R. Roles of moss species and habitat in methane consumption potential in a northern peatland. *Wetlands* **2004**, *24*, 178–185. [CrossRef]
54. Puustjärvi, V. Physical and chemical properties. In *Peat in Horticulture*; Robinson, D.W., Lamb, J.G.D., Eds.; Academic Press: London, UK, 1975; pp. 23–38.
55. Xu, J.; Morris, P.J.; Liu, J.; Holden, J. PEATMAP: Refining estimates of global peatland distribution based on a meta-analysis. *Catena* **2018**, *160*, 134–140. [CrossRef]
56. Waddington, J.M.; Warner, K. Atmospheric CO_2 sequestration in restored mined peatlands. *Ecoscience* **2001**, *8*, 359–368. [CrossRef]
57. Juutinen, A.; Saarimaa, M.; Ojanen, P.; Sarkkola, S.; Haara, A.; Karhu, J.; Nieminen, M.; Minkkinen, K.; Penttilä, T.; Laatikainen, M.; et al. Trade-offs between economic returns, biodiversity, and ecosystem services in the selection of energy peat production sites. *Ecosyst. Serv.* **2019**, *40*, 101027. [CrossRef]
58. Mariotti, B.; Martini, S.; Raddi, S.; Tani, A.; Jacobs, D.F.; Oliet, J.A.; Maltoni, A. Coconut coir as a sustainable nursery growing media for seedling production of the ecologically diverse *Quercus* species. *Forests* **2020**, *11*, 522. [CrossRef]
59. Evans, M.R.; Konduru, S.; Stamps, R.H. Source variation in physical and chemical properties of coconut coir dust. *HortScience* **1996**, *31*, 965–967. [CrossRef]
60. Noguera, P.; Abad, M.; Puchades, R.; Noguera, V.; Maquieira, A.; Martínez, J. Physical and chemical properties of coir waste and their relation to plant growth. *Acta Hortic.* **1996**, *450*, 365–374. [CrossRef]
61. Pill, W.G.; Tilmon, H.D.; Taylor, R.W. Nitrogen-enriched ground kenaf (*Hibiscus cannabinus* L.) stem core as a component of soilless growth media. *J. Hortic. Sci.* **1995**, *70*, 673–681. [CrossRef]
62. Tsakaldimi, M. Kenaf (*Hibiscus cannabinus* L.) core and rice hulls as components of container media for growing *Pinus halepensis* M. seedlings. *Bioresour. Technol.* **2006**, *97*, 1631–1639. [CrossRef]
63. Reichert, N.A.; Baldwin, B.S. Growth of bedding plants in a kenaf-based potting medium. *Proc. Int. Kenaf Assoc.* **1996**, *7*, 25–30.
64. Tsakonas, A.; Stergiou, V.; Polissiou, M.; Akoumianakis, K.; Passam, H.C. Kenaf (*Hibiscus cannabinus* L.) based substrates for the production of compact plants. *Ind. Crops Prod.* **2005**, *21*, 223–227. [CrossRef]
65. Laiche, A.J.; Newman, S.E. Kenaf fiber core as a component of growth media for containerized woody landscape plants. *Proc. South. Nurs. Assoc. Res. Conf.* **1994**, *39*, 77–80.
66. Lang, H.J. Growth of poinsettia in potting media amended with ground kenaf stem core. *HortScience* **1997**, *32*, 483F–484F. [CrossRef]
67. Amha, Y.; Bohne, H.; Schmilewski, G.; Picken, P.; Reinikainen, O. Physical, chemical and botanical characteristics of peats used in the horticultural industry. *Eur. J. Hortic. Sci.* **2010**, *75*, 177–183. [CrossRef]
68. Huat, B.B.K.; Kazemian, S.; Prasad, A.; Barghchi, M. State of an art review of peat: General perspective. *Int. J. Phys. Sci.* **2011**, *6*, 1988–1996. [CrossRef]
69. Prabhu, S.R.; Thomas, G.V. Biological conversion of coir pith into a value-added organic resource and its application in agri-horticulture: Current status, prospects and perspective. *J. Plant. Crops* **2002**, *30*, 1–17.
70. Xiong, J.; Tian, Y.; Wang, J.; Liu, W.; Chen, Q. Comparison of coconut coir, rockwool, and peat cultivations for tomato production: Nutrient balance, plant growth and fruit quality. *Front. Plant Sci.* **2017**, *8*, 1327. [CrossRef]
71. Delicato, D. Physical-Chemical Properties and Sorption Characteristics of Peat. Ph.D. Thesis, Dublin City University, Dublin, Ireland, 1995.

https://doi.org/10.3390/plants15040666

plants

Review

Beyond Peat: Wood Fiber and Two Novel Organic Byproducts as Growing Media—A Systematic Review

Anna Elisa Sdao [1,*], Nazim S. Gruda [2,*] and Barbara De Lucia [1]

1 Department of Soil, Plant and Food Sciences, University of Bari Aldo Moro, Via Amendola 165/A, 70126 Bari, Italy
2 Institute of Crop Science and Resource Conservation (INRES), Horticultural Sciences, University of Bonn, Auf dem Hügel 6, 53121 Bonn, Germany
* Correspondence: anna.sdao@uniba.it (A.E.S.); ngruda@uni-bonn.de (N.S.G.)

Abstract: Environmental concerns drive the search for sustainable organic alternatives in horticultural substrates. This review critically examines three agro-industry renewable byproducts—wood fiber, coffee silverskin, and brewer's spent grain—as partial peat substitutes. We aimed to comprehensively analyze their origin, processing methods, current applications, and key physical, hydrological, and chemical properties relevant to horticultural use. In soilless culture, wood fiber can be used as a stand-alone substrate. When incorporated at 30–50% (v/v) in peat mixtures, it supports plant growth comparable to peat; however, higher proportions may restrict water and nutrient availability. Coffee silverskin demonstrates high water retention and nutrient content, but its inherent phytotoxicity requires pre-treatment (e.g., co-composting); at concentrations up to 20%, it shows promise for potted ornamental crops. Brewer's spent grain is nutrient-rich but demands careful management due to its rapid decomposition and potential salinity issues; inclusion rates around 10% have shown beneficial effects. In conclusion, when used appropriately in blends, these bio-based byproducts represent viable alternatives to reduce peat dependence in vegetable and ornamental cultivation, contributing to more sustainable horticultural practices. Future research should optimize pre-treatment methods for coffee silverskin and brewer's spent grain, investigate long-term stability in diverse cropping systems, and explore novel combinations with other organic waste streams to develop circular horticultural substrates.

Keywords: brewer's spent grain; bioresources; circular economy; coffee silverskin; containerized cultivation peat alternative; renewable raw material; sawdust

1. Introduction

Soilless culture systems refer to all plant cultivation methods without using natural soil. These systems are generally classified into two main categories: liquid-based and solid media systems. Among liquid-based approaches, hydroponics is the most common, where plant roots grow directly in a nutrient solution that supplies water and essential minerals. In contrast, solid and aggregate media systems use substrates that provide physical support and a medium for delivering water and nutrients [1]. Here, we will use the terms "substrate" and "growing media" interchangeably.

Solid media systems typically arrange substrates in containers, foil-lined beds, rows, or shallow layers. One of the most significant applications of this method is in the production of ornamental plants. The containerized cultivation of trees, shrubs, and herbaceous species has become a central practice in the decorative sector, offering flexibility, efficient space utilization, and better environmental control. Containerized ornamental plants are used in urban and rural environments, improving aesthetic appeal, supporting biodiversity, and contributing to human well-being [2].

A growing medium placed in a container provides anchorage for roots while maintaining suitable levels of water and air to support healthy plant growth. However, container systems inherently limit root expansion, gas exchange, and water buffering capacity. To overcome these constraints, substrates typically comprise blended organic and inorganic components that balance the physical and chemical characteristics needed for optimal plant performance [3–5].

Inorganic materials commonly used in substrate mixtures include both natural and processed materials. Natural materials, such as sand, pumice, and tuff, provide good drainage and aeration. On the other hand, processed materials like perlite, vermiculite, rock wool, expanded clay, and zeolite are valued for their structural consistency and long-term stability [4,6,7]. Although they often comprise a smaller portion of substrate blends [8], these materials remain essential for ensuring reliable performance, especially in intensive horticultural systems [9].

Increasingly, there is a shift toward sourcing substrate components locally and sustainably. Renewable and regionally available materials are encouraged to reduce environmental impact and production costs while promoting circular practices in horticulture [7,8,10]. Consequently, the careful selection and formulation of substrates are crucial for the success of soilless cultivation systems, particularly in high-value sectors such as ornamental plant production.

Peat is the most widely used material in horticultural substrates [8,11]. From an agronomic perspective, it is considered an ideal substrate for propagation and cultivation due to its ability to retain water and nutrients and its porosity, which provides oxygen to the roots and facilitates drainage. Adding specific amendments can easily correct its low pH and nutrient content [11].

However, the use of peat in horticulture has been strongly criticized because of environmental and climate change concerns [7] and because it is a non-renewable organic matrix [12]. The areas where peat accumulates over time, called peatlands, cover an estimated 4 million km^2, which accounts for 2.8% of the Earth's surface. It is a unique ecosystem that develops in oxygen-deficient environments with groundwater or heavy precipitation, forming carbon- and nitrogen-rich organic soils. Their carbon storage capacity is significant, estimated at approximately 644 Gt, accounting for 21% of global organic carbon [13]. Therefore, peatlands are crucial to climate stability [14]. Also, the "2030 Agenda for Sustainable Development" in target 15.1 refers to "Protect, restore and promote the sustainable use of the earth's ecosystem", among which the protection of peatlands as environments rich in biodiversity and useful for CO_2 sequestration [15].

In *Pinus engelmannii* Carr. (Apache Pine) seedling production [16], the substrate is a critical factor, as the properties of its constituent materials significantly influence growth and future plant development. In general, the root system of the seedlings produced in sawdust-based growing media presented good physiognomy based on the points of growth observed in the root, possibly due to the ratio of the dry root biomass and the growth potential.

Moreover, the organic and inorganic components also directly impact production costs. González-Orozco et al. [16] pointed out the difference in cost of USD 2.34 between 50% composted pine bark (USD 3.49) and 70% pine sawdust (USD 1.15) peat-based growing media. Therefore, the substrate's cost has led to research on replacing peat with local materials.

The physical properties of growing media depend on the morphological characteristics of the particles composing them [17]. A growing medium must strike a balance between water retention, air supply, and drainage to prevent waterlogging and disease, while ensuring sufficient porosity for root aeration. In addition, it must guarantee a stable structure over time, resistant to shrinkage or expansion [18]. Chemically, the substrate should have a neutral to slightly acidic pH [19], low salinity, and a slow decomposition rate, and contain essential nutrients or allow for adequate fertilization [18]. Other desirable qualities include the absence of weed seeds, nematodes, pathogens, and phytotoxins [20]. From a practical perspective, the material used for the substrate should be readily available, low-cost, and easy to mix with different matrices [5,7,8].

Agriculture-based industries generate approximately 1.3 billion tons of waste annually [21,22]. Most of this waste is disposed of through incineration or landfilling, which presents considerable environmental, economic, and social challenges. Capanoglu et al. [23] emphasize that the carbon footprint associated with agro-industrial waste is significant, contributing approximately 3.3 billion tons of CO_2 emissions to the atmosphere each year. This excessive generation of agro-industrial waste disproportionately impacts developing countries, where viable alternatives for recovering and reusing these valuable resources remain limited [24], underscoring a critical gap in sustainable waste management.

Forestry byproducts are residual logging and wood processing materials, including bark, branches, sawdust, and industrial wood residues. Agro-industry byproducts are materials left over after processing agricultural goods and can be valuable rather than wasteful. Often considered waste, they hold potential for bioenergy, bioproducts, and sustainable materials, supporting circular economy approaches while reducing the environmental impact of forest-based industries [25]. Utilizing food, drink, and agricultural waste presents both a challenge and a necessity in daily life. This aligns with the core concepts of the circular economy, which are based on the three rules: Reduce, Reuse, and Recycle. Furthermore, target 12.5 is a key component of Sustainable Development Goal 12, which focuses on achieving sustainable consumption and production patterns. It emphasizes moving from a "take, make, and dispose" linear economy towards a more circular one [4,26].

2. Organic Byproducts as Growing Media

Organic byproduct sources derived from forestry and agro-industrial processes—such as wood fiber, coffee silverskin, and brewer's spent grain—are gaining attention for their potential use in horticultural growing media (Figure 1).

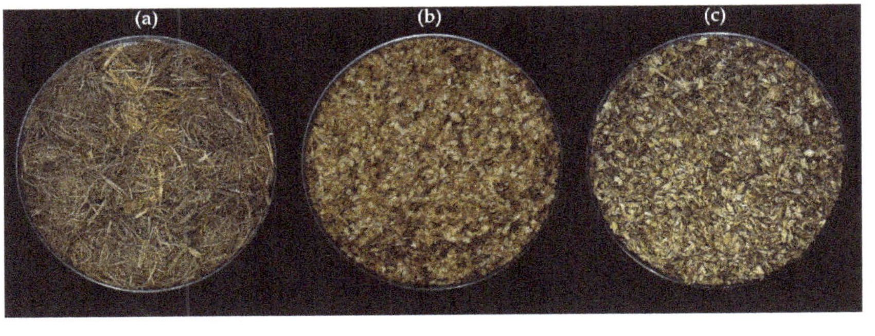

Figure 1. Wood fiber (**a**), coffee silverskin (**b**), and brewer's spent grain (**c**).

Several standard growing media contain up to 30 to 40% by volume of wood fiber (WF), and its potential for co-use in growing media has not yet been fully exploited. Coffee silverskin (CS), a thin tegument directly covering the coffee bean, is a byproduct of the coffee roasting process. Considering the large quantities produced globally and their potential environmental impact if not effectively utilized, developing sustainable applications for CS becomes an apparent necessity. Similarly, brewer's spent grain (BSG), the solid residue remaining after brewing, stands out as an attractive and cost-effective raw material. Its year-round availability and rich composition of carbon and nitrogen sources make it particularly appealing for diverse agricultural and horticultural applications.

Despite the recognized potential of these byproducts, a comprehensive synthesis of their diverse properties and efficacy in horticultural applications remains a critical research gap. While wood fiber has established its presence as a growing media component, its full potential for co-utilization with other novel organic byproducts, especially in diverse horticultural contexts, is still underexplored.

This systematic review addresses this gap by examining three organic byproducts, two of which are unconventional—coffee silverskin and brewer's spent grain—alongside wood fiber, all sourced from the forestry and agro-industrial sectors, for their potential use in growing media. The objective is to provide a comprehensive overview of their origin, processing, current applications, and key chemical and hydrological properties, focusing on their individual and synergistic effectiveness in supporting the cultivation of a wide range of horticultural crops.

3. Methodology

3.1. *Formulation of the Problem*

The use of peat in horticultural substrates raises environmental and climate change-related concerns. Therefore, it is urgent to identify organic replacement byproducts that can be used in substrate composition and validated in cultivation.

3.2. *Keyword Identification*

Thomson Reuters' Web of Science, Elsevier's Scopus, and Google Scholar were queried from 1972 to April 2025 for worldwide published scientific publications, written in English, with subject areas limited to Agricultural and biological sciences, and Environmental science and they were identified using the following keywords: (i) potted growing media (Figure 2a); (ii) agro-environmental sustainability (Figure 2b); (iii) forestry and agro-industry byproducts (Figure 2c); (iv) unconventional organic matrices (Figure 2d). Subsequently, the process of identification of bibliographical references took place by title/keywords/abstract and discarding those irrelevant.

About 67% of the references are categorized as organic and inorganic, peat-free, and sustainable growing media (Figure 2a). Regarding agro-environmental sustainability (Figure 2b), 30% of the references are placed in themes related to peatland C-rich ecosystems, and 30% in sustainable development. In Figure 2c (forestry and agro-industry byproducts), the agro-industry byproducts account for 66% of the references. Figure 2d (unconventional organic matrices) shows that wood fiber (WF) references consist of 28%, those of coffee silverskin (CS) of 38%, and those of brewer's spent grain (BSG) of 34%.

Figure 3 shows the references (%) of agricultural use: in WF (a), 91% represents ornamentals and vegetables; in CS (b), soil amendments and compost represent 49% of uses, while peat replacement reaches 17%. The BSG (c) agricultural use consists of 44% as fertilizer.

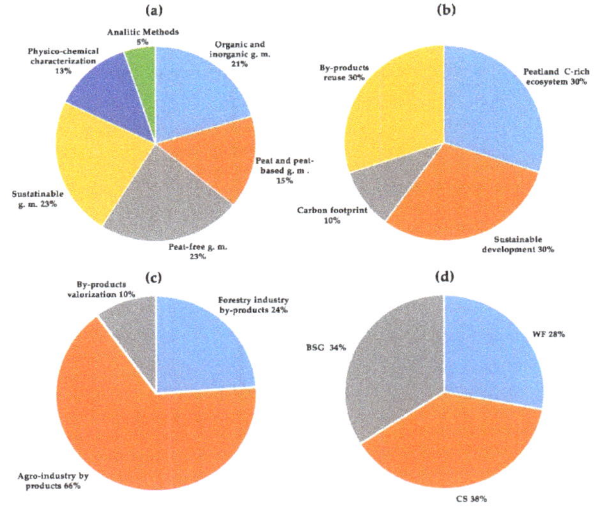

Figure 2. Keywords references (%) divided into categories of interest: (**a**) potted growing media; (**b**) agro-environmental sustainability; (**c**) forestry and agro-industry byproducts; (**d**) conventional and unconventional organic matrices as wood fiber (WF), coffee silverskin (CS), and brewer's spent grain (BSG).

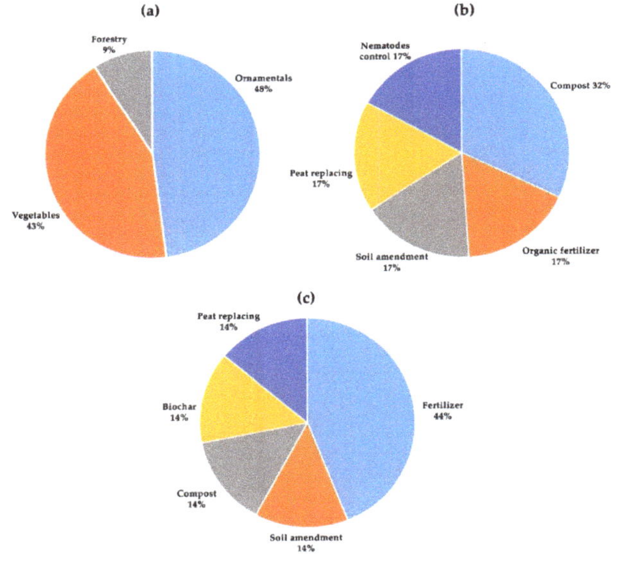

Figure 3. Overview of agricultural applications for the three organic materials. (**a**) Wood fiber (WF) is used predominantly in horticultural practices, with 91% applied to ornamental plants and vegetable production; (**b**) presents data on coffee silverskin (CS), where soil amendments and composting account for 49% of its agricultural uses, and peat replacement represents 17%; (**c**) shows that brewer's spent grain (BSG) is primarily utilized as a fertilizer, making up 44% of its reported agricultural application.

4. Wood Fiber

Wood-based growing media, including bark, chips, biochar, sawdust, and wood fibers, have been used in Europe for a long time. Their market is expected to reach 30 Mm^3 $year^{-1}$ in 2050 [27]. Wood fiber (WF) is generally made from lumber, and as a substrate, it contains little or no bark [28,29]. It is considered a renewable resource that can be obtained from sustainably managed forests, thus, as an organic material resulting from industrial wood processing. Production typically involves subjecting wood to high pressure and temperature. Wood fiber is often processed through grinding, where the size of the wood pieces is reduced and the fibers separate from each other. It appears as filaments. It is predominantly sourced from coniferous species such as Norway spruce (*Picea abies* (L.) H. Karst.), pine (*Pinus* spp.), lodgepole pine (*Pinus contorta* Douglas ex Loudon), or broadleaf trees such as chestnut (*Castanea sativa* Mill.) [30]. In the late 1980s, the first to grind pine wood for use as a soilless culture substrate were Laiche Jr. and Nash [31]. Following their promising results, interest in commercial WF grew in the early 1990s. In 1994, Benoit and Ceustermans [32] described WF in cultivation as the phenomenon of degradation by microorganisms, which leads to a worsening of the physical properties and nitrogen immobilization [33–36]. In Europe, WF growing media were particularly popular in Germany in the early 2000s, as reported by Gumy [37] and Bobo and Jackson [38]. In the same period, they also attracted the attention of the USA [8].

4.1. The Production Process

Wood fiber (WF)-based growing media production involves turning raw wood, often a byproduct of sawmilling or wood processing, into a stable, safe, and horticulturally suitable substrate. The process varies depending on the desired fiber quality and end use. Wood is mechanically defibered using disc refiners, hammer mills, or high-speed grinders [39], and then treated with steam at temperatures between 80 °C and 120 °C (stabilized WF), ensuring the elimination of plant pathogens, reducing the resin content (which could be phytotoxic), and partially stabilizing the material biologically.

WF can be produced in both coarse and fine grades, and the characteristics of the source material and fiber structure are significant in plant cultivation. During this process, nitrogen (N) is often used to enrich it, particularly when derived from wood chips or bark, and tends to have a relatively low N content [28]. The most commercially available WF products have included various trade names over the years, although some are no longer actively produced or widely distributed. Notable examples include Toresa, Hortifibre, Cultifibre, Piatel, Torbella, SunFiber Nuova Flesan, Ecofibrex, and Brill Substrate Fiber.

4.2. Agricultural and Non-Agricultural Uses

In farming systems, wood fiber is used to cultivate vegetables [40–43], ornamentals [12,44–46], forest seedlings [47], and trees [48]. WF use is consolidated above all in the production of insulating panels for construction and laminates [25], as a mechanical buffer, and electrolyte reservoir for sodium–ion batteries [49].

4.3. Physical, Hydrological, and Chemical Characterizations

4.3.1. Physical and Hydrological Characterizations

Gruda and Schnitzler [50,51] found a low bulk density, good drainability, and rewettability of wood fiber (WF). Domeño et al. [52] describe a bulk density of 0.07 g cm^{-3} (recommended range (R.R.) < 0.4) and a particle density of 1.56 g cm^{-3} (R.R. = 1.4–2.6). Commercial products are typically mixtures of particle size fractions in varying percentages.

Wood fiber (WF) moisture content is between 10.3% [53] and 41% [41]. Gruda and Schnitzler [50] reported that pure or 50–70% wood fiber: white peat (*v:v*) showed a similar total pore space to each other and in line with the R.R. >85% found by De Boodt and Verdonck [54]. These WF-based growing media showed lower readily available water (E.A.W. calculated as the amount of free water of the tested materials when the water tension increased from pF = 1.0 to 1.7) than peat, according to Domeño et al. [52], who report an E.A.W. of 13.8 vol. %. In mixtures of peat and WF (*v:v*, 50:50), the readily available water reached 25.1%, which is lower than that of peat (33.5%) but is still within the ideal range of 20–30%, as noted by De Boodt and Verdonck [54] (1972), with rapid drainage of surplus water.

Reineke et al. [45] reported that in petunia, the thermo-hydrolytically and thermally treated WF, at 30–60% v/v, decreased shoot biomass and flower number with increasing WF content due to lower water-holding capacity. Therefore, the WF proportion and irrigation regime are key to achieving a peat substitution without compromising quality.

Finally, wettability characterization, defined by Michel [55] as the degree of hydrophilicity, is particularly crucial in containerized crops, in which fluctuations can quickly lead to oxygen stress and root asphyxiation in cases of excessive irrigation, or conversely, to water stress and potential hydrophobicity when water inputs are insufficient. Barrett et al. [4] argued that WF exhibits a high degree of wettability: in fact, it is used, in blended growing media, to optimize physical properties, reduce bulk density, increasing air space and, thus, improving re-wetting capacity; for the same reason, it is not used as a stand-alone component for the constitution of a substrate [56].

Table 1 reports the main physical and hydrological properties determined for wood fiber, as described in previous studies.

Table 1. Wood fiber physical and hydrological characterizations.

Parameters	Unit of Measure	Value	References
B.D.	g cm^{-3}	0.07	[52]
P.D.	g cm^{-3}	1.56	[52]
M.	%	10.3–41.0	[41,53]
T.P.S.	%	>85	[54]
E.A.W.	vol. %	13.8	[52]

B.D. = Bulk Density; P.D. = Particle Density; M = Moisture; T.P.S. = Total Pore Space; E.A.W. = Readily Available Water.

4.3.2. Chemical Characterization

Regarding the chemical properties, wood fiber has a subacid pH, between 4.1 [53] and 6.0 [30], and a variable but still very low EC: from 4.0–6.0 [30] to 18.6 [57] to 28 mS cm^{-1} [53]. This wide variability is due to the heterogeneity of the raw wood material. The ash content was estimated at 0.15% (w/w) [53]. Domeño et al. [52] stated a cation exchange capacity of 0.22 meq g^{-1}; Zaccheo et al. [30] reported a total organic matter content of 95–98%, based on the findings of Zawadzińska et al. [53], along with an organic carbon content of 64–67%. Zaccheo et al. [30] estimated the total nitrogen content to be 0.20–0.40%, consistent with Zawadzińska et al. [53], who found 0.16%. Čepulienė et al. [41] identified 0.22%, while Sdao et al. [57] measured 0.45%. The C/N ratio noted by Zaccheo et al. [30] of 168–336 differs little from that observed by Čepulienė et al. [41] of 435 but is in line with the value documented by Domeño et al. [52], equal to 456. Wood fiber is primarily made of cellulose and lignin, rich in C and poor in nitrogen. Čepulienė et al. [41] recorded a P content of 57 mg kg^{-1} and a K of 472 mg kg^{-1}. Finally, Zawadzińska et al. [53] indicate Ca = 1574 mg kg^{-1}, Mg = 346 mg kg^{-1}, and Fe = 46.5 mg kg^{-1}. Maher et al. [28] determined a lignin content of 360 g kg^{-1} (dry weight), while Domeño et al. [52] quantified a cellulose content of 180 g kg^{-1} (dry weight) and a hemicellulose content of 489 g kg^{-1}

(dry weight). Table 2 summarizes the key chemical characteristics of wood fiber, including C/N ratio and pH, based on the relevant literature.

Table 2. Wood fiber chemical characterization.

Parameters	Unit of Measure	Value	References
pH		4.1–6.0	[30,53]
E.C.	mS cm^{-1}	4.0–28	[30,53,57]
C.E.C.	meq g^{-1}	0.22	[53]
T.O.M.	%	95–98	[30,53]
O.C.	%	64–67	[30]
T.N.	%	0.20–0.45	[30,41,53,57]
C/N		168–456	[30,41,52]
P	mg kg^{-1}	57	[41]
K	mg kg^{-1}	472	[41]
Ca	mg kg^{-1}	1574	[53]
Mg	mg kg^{-1}	346	[53]
Fe	mg kg^{-1}	46.5	[53]
Lignin	g kg^{-1} d.w.	360	[28]
Cellulose	g kg^{-1} d.w.	180	[52]
Hemicellulose	mg kg^{-1} d.w.	489	[52]

E.C. stands for Electrical Conductivity; C.E.C. stands for Cation Exchange Capacity; T.O.M. stands for Total Organic Matter; O.C. stands for Organic Carbon, T.N. stands for Total Nitrogen; and C/N stands for Carbon–Nitrogen ratio.

4.3.3. Practical Applications

Wood fiber's low bulk density, good trainability, and high wettability make it an excellent component for lightening growing media and enhancing aeration in containerized crops, particularly when blended with peat. While its lower readily available water capacity and high C/N ratio necessitate careful irrigation management and potential nitrogen supplementation, WF-based mixtures optimize physical properties and improve re-wetting capacity, positioning wood fiber as a valuable partial peat substitute for diverse horticultural applications.

4.4. *Crop's Performance*

In recent years, wood fiber (WF) has gained popularity and become a renewable raw material in peat-based substrates [58–60]. Studies on vegetables and ornamentals have shown differences in performance and fertilizer requirements, so the substrate composition needs to be optimized.

4.4.1. Wood Fiber Peat Replacement up to 50%

In *Leucanthemum vulgare* [12], wood fiber (WF), at moderate levels (30–40% v/v) and combined with coir and compost, supported plant growth and nutrient uptake comparable to, or better than, peat-based control. Beretta and Ripamonti [61] report that a peat substitution rate of between 30% and 50% showed promising results in the cultivation of *Juniperus communis* 'Repanda', *J. communis* 'Prince of Wales', and *Ribes uva-crispa* 'Hinnonmaki Rot'. In *Petunia* × *hybrida*, peat-reduced mixes up 35% WF yielded comparable plant vitality and flowering to peat control. In basil, the buffering capacity of WF may help maintain more stable pH levels [44]. Laun et al. [62] also stated that wood fiber in vegetable seedlings could replace up to 30 to 50% of peat volume.

4.4.2. Wood Fiber Peat-Replacement Beyond 50%

Substrates containing 70% wood fiber (WF) led to significant biomass reduction and impaired performance, highlighting that excessive doses may compromise water retention and nitrogen availability [61]. Woznicki et al. [63] (2024) showed that substrates with high WF content were less acidic and rapidly increased pH in petunia cultivation, resulting in iron malabsorption. In Petunia and basil, WF $\geq$ 75% led to chlorosis, mainly growth reduction, unless pH and nutrition supply were taken care of. In the hydroponic strawberry system [42], the 100% spruce fiber (*Picea abies*) significantly advanced the ripening and commercial marketable yield, compared to the control peat and perlite. Similar results were achieved in studies by Gruda and Schnitzler [29] with lettuce transplants. Many authors agree that WF in the hydroponic cultivation of strawberries has given good results as a stand-alone substrate [42,64–66].

4.4.3. Wood Fiber and Fertilization

Incorporating wood fiber components into peat-based soilless substrates can increase the amount of fertilizer N needed during production as a result of greater N immobilization [5,67]. Čepulienė et al. [41] conducted a study to explore the effect of WF as an alternative to peat in growing media for cucumber (*Cucumis sativus* L.) cultivation, focusing on its impact on nutrient content and uptake. This experiment compared different growing media combinations of peat and WF (0, 25, 50, 100 WF) with additional nitrogen fertilizer (13, 23, 30 g plant^{-1}). Results showed that in the substrates with 50 and 100% WF content and with the increased additional fertilization, there was a significant reduction in dry biomass of the aerial part compared to the peat-based substrate without fertilization.

5. Coffee Silverskin

Coffee is among the most widely consumed beverages globally [68]. Coffee has become a beloved staple in people's lives worldwide; over 2.25 billion cups of coffee are consumed worldwide each day [69]. Cultivation mainly occurs within the "coffee belt" between the Tropics of Cancer and Capricorn, at 25° North and 25° South latitudes. In the coffee industry, a "bag" refers to a standard unit of measurement for coffee, typically 60 kg of raw coffee beans, from Central and South America, accounting for 96.93 million bags, followed by Asia, with 41.91 million bags. Additional contributions came from Africa: 15.0 million bags. The global green coffee market was valued at approximately USD 39.97 billion in 2024 [70].

There are two main coffee species: *Coffea arabica* cultivated in Latin America and Ethiopia, accounting for 57,4% of production (102.2 million bags), and *C. canephora* (syn. *C. robusta*) cultivated in Southeast Asia and Africa, which constitutes 42,6% (75.8 million bags) [71]. *C. arabica* grows at higher altitudes (usually 600–2000 m) and requires cooler temperatures. Arabica plants are more delicate and require more agronomic maintenance; *C. robusta* grows at lower altitudes, is more resistant to heat and drought, and is easier to cultivate [72]. Green coffee, along with its byproducts and residues, represents a valuable source of macromolecules (e.g., carbohydrates, proteins, pectin) and bioactive compounds. These can be cost-effectively extracted and repurposed for novel applications in the food, nutraceutical, agricultural, cosmetic, and pharmaceutical sectors, aligning with FAO's bioeconomy and circular economy principles [73].

5.1. The Production Process

The *testa* surrounds the coffee bean, a thin, fibrous, and protective layer, often referred to as the silverskin, due to its silvery appearance. Once the green beans are picked and processed, the silverskin is typically eliminated during roasting, leaving the roasted beans used to make coffee [74]. Coffee silverskin (CS) represents the only byproduct released during the coffee roasting process, accounting for approximately 4.2% (w/w) of seed total weight. About 400,000 tons of this waste is generated each year [75,76]. For coffee producers, managing this waste represents a significant cost [77], and its storage also poses a risk, as CS is highly flammable [78].

5.2. Agricultural and Non-Agricultural Uses

Although coffee silverskin (CS) is primarily considered industrial waste, it is a biomass with a valuable raw input into a wide range of potential applications [79].

The favorable composition of coffee processing waste, its high availability, and low cost make it an excellent source of various bioactive compounds with nutritional value [80].

In recent times, the Farm-to-Fork strategy [81], Salbitani et al. [82], and Carnier et al. [83] have shown the use of CS for organic fertilization purposes due to its high content of macro and micronutrients. Other uses in agriculture have also been evaluated: Picca et al. [84] and González-Moreno et al. [85] evaluated their use in compost and vermicompost. Finally, Sdao et al. [57] evaluated its use as a substitute matrix for peat in containerized ornamental plants.

Regarding the non-agricultural uses, CS has been studied in the food sector [86] to enrich preparations such as cookies, chocolate cake [87], and candies, as a fat replacer for chicken hamburgers, and as a source of antioxidants [88]. Industrially, it can be used as a material for polymers [89] and packaging [90].

In cosmetics utilization, dos Santos et al. (2021) [91] showed that CS in sunscreen is promising, as is in other cosmetic applications [92]. CS has also been studied as a feed for another critical sector, such as insect breeding [93].

5.3. Physical, Hydrological, and Chemical Characterization

5.3.1. Physical and Hydrological Characterization

Prandi et al. [94] describe the structural characterization of coffee silverskin (CS), supplied by Illycaffè S.p.A. (Trieste, Italy), as consisting of flakes ranging in size from 2 to 5 mm, with a bulk density of 0.18 g cm^{-3} and a particle density of 0.71 g cm^{-3}. Moisture content ranged between 7.3 and 7.1, in agreement with Pourfarzad et al. [95] and Arya et al. [96]. Ballesteros et al. [97] report a water holding capacity of 5.11 g H_2O g^{-1} dry sample, probably due to the high fiber content [96]. Sdao et al. [57] report a water retention of 91.4%. Table 3 presents the physical and hydrological parameters of coffee silverskin.

Table 3. Coffee silverkin physical and hydrological characterization.

Parameters	Unit of Measure	Value	References
F.S.	mm	2–5	[94]
B.D.	g cm^{-3}	0.18	[94]
P.D.	g cm^{-3}	0.71	[94]
M.	%	7.1–7.3	[94–96]
W.H.C.	g H_2O g^{-1} dry sample	5.11	[97]
W.R.	%	91.4	[97]

F.S. = Flake Size; B.D. = Bulk Density; P.D. = Particle Density; M. = Moisture content; W.H.C. = Water-Holding Capacity; W.R. = Water Retention.

5.3.2. Chemical Characterization

In coffee silverskin (CS), the pH, determined by Sdao et al. [98], was found to be 5.7; this value is consistent with that reported by Thligene et al. [99] of 5.6 and Del Pozo et al. [78] of 5.3. Electrical conductivity (EC), as reported by Sdao et al. [57], was measured at 1.8 dS m^{-1}, which is in agreement with González-Moreno et al. [85]. However, this contrasts with the findings of Carnier et al. [83], who reported a significantly higher EC of 4.9 dS m^{-1}. This discrepancy underscores the necessity of monitoring soil conditions during the application of CS to prevent salinization. The ash content, evaluated by Sdao et al. [98] with 6.6, aligns with the range reported by Pourfarzad et al. [95], Ballesteros et al. [97], and Arya et al. [96] of 5–7 g 100 g^{-1} dry weight.

The total organic matter content, as reported by Sdao et al. [98], was found to be 93.4 g 100 g^{-1}, consistent with the 91.3 g 100 g^{-1} reported by González-Moreno et al. [85]. Carnier et al. [83] reported 43.3 g 100 g^{-1}. The C/N ratio also varies: González-Moreno et al. [85] reported a minimum value of 11, while Ballesteros et al. [97] reported 14.4, and Carnier et al. [83] reported 19.1. The total Kjeldahl nitrogen is reported by Carnier et al. [82] as 0.14 g 100 g^{-1}; in contrast, González-Moreno et al. [85] reported 5.1 g 100 g^{-1}, and Sdao et al. [57] 2.39 g 100 g^{-1}.

Sdao et al. [57] reported a total phosphorus content of 0.76 g 100 g^{-1}. The presence of a considerable concentration of potassium, equal to 7,6 g kg^{-1}, in this waste makes it particularly interesting for tropical soils, which are often poor in K, and many crops. Carbohydrates, including cellulose, hemicellulose, and lignin, were also analyzed. Arya et al. [96] assessed the hemicellulose and cellulose contents of 62.1 and 23.7 g 100 g^{-1}; they also reported a lignin content of 28.6 g 100 g^{-1}, which agrees with the findings of Mussatto et al. [100], who documented 30.2 g 100 g^{-1}. Finally, protein content was determined by Arya et al. [96] to be 18.7 g 100 g^{-1}, aligning with 16.2 g 100 g^{-1} reported by Mussatto et al. [100].

Coffee silverskin is rich in inhibitory substances, such as caffeine, tannins, and polyphenols, which are responsible for insufficient nitrogen mineralization. Consequently, while these residues are rich in organic matter and exhibit promising soil conditioning properties, their use as primary nitrogen sources is not recommended. Table 4 outlines the main chemical properties of coffee silverskin, including organic matter content and nutrient profile.

Table 4. Coffee silverskin chemical characterization.

Parameters	Unit of Measure	Value	References
pH		5.3–5.6	[78,98,99]
E.C.	dS m^{-1}	1.8–4.9	[95–97]
Ash	g 100 g^{-1}	5–7	[95–98]
T.O.M.	g 100 g^{-1}	43.3–93.4	[83,85,98]
T.N.	g 100 g^{-1}	0.14–5.1	[57,83,85]
C/N		11–19.1	[83,85,97]
P	g 100 g^{-1}	0.76	[57]
K	g kg^{-1}	7.6	[57]
Lignin	g 100 g^{-1}	28.6–30.2	[96,100]
Cellulose	g 100 g^{-1}	23.7	[96]

Table 4. *Cont.*

Parameters	Unit of Measure	Value	References
Hemicellulose	g 100 g^{-1}	62.1	[96]
Protein	g 100 g^{-1}	16.2–18.7	[96,100]
T.P.	mg kg^{-1}	140	[101]

E.C. = Electrical conductivity; T.O.M. = Total Organic Matter; T.N. = Total Nitrogen; C/N = Carbon–Nitrogen ratio; T.P.= Total Polyphenol.

5.3.3. Phytotoxicity

Another fundamental aspect of coffee silverskin (CS) is phytotoxicity. Picca et al. [84] showed that it is due to high polyphenols and chlorogenic acid content, which can limit its direct application as a soil amendment, despite its beneficial agronomic properties, such as high nitrogen and potassium content, and excellent water retention capacity. The co-composting CS with carbon-rich materials eliminates phytotoxicity while maintaining its agronomic benefits. The antioxidant properties of CS derive from its polyphenolic content: Ziemah et al. [101] report a total polyphenol content (GAE) of 140 mg g^{-1}. Al-Charchafchi et al. [102] results indicate that chlorogenic acid possesses phytotoxic properties, as it negatively affects seed germination.

5.3.4. Practical Applications

Owing to its high water holding capacity and significant organic matter content, coffee silverskin is a promising amendment for improving soil structure and water retention in growing media. However, its inherent phytotoxicity and potential for variable electrical conductivity (EC) and nitrogen immobilization primarily necessitate its use after co-composting carbon-rich materials. When properly treated, its potassium richness makes it particularly beneficial for K-deficient soils or specific crop requirements.

5.4. *Crop's Performance*

Coffee silverskin (CS) was used as organic fertilizer, compost, vermicompost, and peat-based growing medium to cultivate herbaceous and ornamental crops.

5.4.1. Coffee Silverskin as Organic Fertilizer

Salbitani et al. [82] show that at a 2% dose, CS improves soil properties and promotes barley growth and physiology. In comparison, at a 10% dose, it negatively impacts leaf proteins. It reduces Chl b and Chl a/Chl b ratios, probably due to potential disruption of the photosynthetic apparatus, decreasing chlorophyll synthase activity, or nitrogen availability. Carnier et al. [83] investigated CS. They experimented with coffee powder for maize (*Zea mays* L.) at the dose of 450 mg of Kjeldahl nitrogen per pot, compared to a control fertilized with ammonium nitrate. Maize growth in soil treated with coffee residues did not significantly differ from the control but was lower than in ammonium nitrate-treated plants, suggesting limited nitrogen availability.

5.4.2. Coffee Silverskin as Co-Composting Matrix

Composting organic residues for agricultural soil amendment offers significant benefits. Through enhanced water retention, composting reduces inorganic fertilizer demand and irrigation requirements. This positions composting as a cost-effective and industrially scalable technology for organic waste recycling, yielding a valuable soil conditioner with fertilizing properties [103].

Picca et al. [84] demonstrated that coffee silverskin (CS) co-composting with gardening pruning waste and biochar produced nutrient-rich, non-phytotoxic compost with phytostimulant properties. González-Moreno et al. [85] examined the vermicomposting process of CS and spent coffee grounds (SCG), mixed with horse manure (HM). The vermicomposting was conducted at varying ratios of CS/SCG/HM (25, 50, 75, and 100%) and lasted 60 days. The biological results showed that the growth and reproduction of the worms were significantly affected by the different waste combinations. The total biomass growth was highest in CS = 50, while treatment with 100% CS showed the lowest growth. The worm growth rate followed a similar trend. Regarding *Triticum aestivum* seed germination, used as a compost maturity and non-phytotoxicity indicator, no significant differences were found between the various mixtures. However, the germination, at a dose of CS 50%, showed the best results.

5.4.3. Coffee Silverskin as Peat-Based Growing Medium

Sdao et al. [57] evaluated the coffee silverskin (CS) at 0, 10, 20, and 40% doses in seedlings and sale pot plants, as a partial peat substitute in ornamental marigold (Tagetes patula) and nasturtium (Tropaeolum *minus*) cultivation. In marigold seedlings, the peat-based substrate achieved the highest chlorophyll content; the values decreased as the CS dose increased. In nasturtium, however, the chlorophyll content at the CS 10% dose was comparable to that of the control, exhibiting the highest values. As sale plants, the dry weight in tagetes did not show significant differences between treatments; in nasturtium, 20% CS reached results comparable to peat.

6. Brewer's Spent Grain

Beer is one of the oldest and most popular alcoholic beverages in the world. According to the 2024 European Beer Trends Statistics Report [104], beer production reached 346 million hL in 2023, marking a decline of more than 3% compared to 2022. Italy ranked ninth in beer production, with a total output of 17,430 thousand hL.

6.1. The Production Process

Brewer's spent grain (BSG) is the primary byproduct of the brewing industry, accounting for 85% of total brewing residues [105], with a global annual production of 12 million t. In producing the most common beer type, light lager, malt consumption is estimated at approximately 20 kg hL^{-1} [106,107]. BSG is made through a fermentation process in which the sugars contained in malt are transformed into alcohol and carbon dioxide through the action of yeasts. During the malting phase, barley grains are germinated and dried to develop the necessary enzymes for starch conversion. In mashing, the malt is mixed with hot water, allowing the enzymes to break down starches into fermentable sugars; this stage generates BSG [108]. The brewing industry significantly impacts the environment by generating substantial waste. Per 1000 tons of beer, approximately 137–173 tons of solid waste are produced, with spent grain accounting for roughly 20 kg per 1 hL of beer [109].

6.2. Agricultural and Non-Agricultural Uses

Traditionally considered waste, brewer's spent grain (BSG) is now the focus of a few research studies aimed at developing sustainable strategies for its valorization. Several studies highlight its potential applications across various sectors [110]. BSG has also been studied and applied in agricultural uses, such as producing biochar [111], organic soil amendment [112], compost [113], and peat replacement [57].

Regarding non-agricultural uses, brewers' spent grain (BSG) for animal feed is already consolidated and favored by breweries because it is environmentally sustainable and economically favorable [114]. BSG is used in the food industry for bakery products [115]; in plant-based meat [116], pasta [117], and yogurt [118], due to its high content of fiber, protein, and bioactive compounds. Moreover, it is utilized in the production of bioplastics and biocomposites [119], bioenergy [120], and reducing sugars for biofuels [121]. In recent years, the use of BSG in the medical-pharmaceutical field has been interesting and increasingly studied by researchers [122]. In animal nutrition, BSG supplementation has improved growth performance, immune responses, and antioxidant capacity, without detrimental effects on metabolic profiles [123,124].

6.3. Physical, Hydrological, and Chemical Characterization

Brewer's spent grain (BSG) is a complex lignocellulosic material composed of the seed coat, pericarp, and husk layers of barley grain [125].

6.3.1. Physical and Hydrological Characterization

Naibaho et al. [126] reported a bulk density of 0.129–0.159 g cm^{-3}. Meneses et al. [127], Castro et al. [128], Cacace et al. [112], and Belardi et al. [129] agree on a moisture ranging between 78% and 84%. This high moisture content and fermentable sugars, mainly maltose and maltotriose [130], make BSG highly susceptible to microbial growth, leading to rapid spoilage within a short storage period of 7 to 10 days. Refrigeration or drying is advised to maintain physico-chemical properties and prevent deterioration [131]. Sdao et al. [57] report a water retention of 207%. Table 5 displays brewers' spent grain's principal physical and hydrological features, focusing on structure, moisture retention, and air capacity.

Table 5. Brewer's spent grain physical and hydrological characterization.

Parameters	Unit of Measure	Value	References
B.D.	g cm^{-3}	0.129–0.159	[126]
M.	%	78–84	[112,127–129]
W.R.	%	207	[57]

B.D. = Bulk Density; M. = Moisture Content; W.R. = Water Retention.

6.3.2. Chemical Characterization

Brewer's spent grain (BSG) pH is sub-acidic, with attested values of 5.5 [128], 4.5 [112], and 6.0 [98]. Sdao et al. [57] evaluated the electrical conductivity (EC) as 3480 µS cm^{-1}. The ash content is variable, and it was accounted for in 2021 by Castro et al. (2.3%) [128], in 2023 by Belardi et al. (3.5%) [129], in 2024 by Mainali et al. (8.5%) [132], and in 2025 by Garcia et al. (1.9%) [133]. Sdao et al. [57] evaluated the Organic Matter content as 93.8 g 100 g^{-1}. Organic Carbon content ranged between 47.3 [132] and 53% [134], and C/N ratio ranged from 10.8 [132] to 13.9 [134]. Nitrogen content ranged between 3.8 [134] and 5.1% [132].

According to Meneses et al. [127], the most abundant elements in BSG are phosphorus (P = 6000 mg kg^{-1}), calcium (Ca = 3600 mg kg^{-1}), magnesium (Mg = 1900 mg kg^{-1}), and sulfur (S = 2900 mg kg^{-1}). Meneses et al. [127] pointed out a sodium (Na) content of 137 mg kg^{-1}, according to Ikram et al. [105]. The negative impacts of high Na levels on plant metabolism include cellular dehydration, nutrient deficiency, growth inhibition, stomatal rupture, and plant death [135].

Chemically, 100 g d.w. BSG contains about 15–25 g protein, 50–70 g fiber (hemicellulose, cellulose, and lignin), and 5–10 g fat [136].

Bachmann et al. [131] reported a cellulose content between 16.8 and 26.0% and 19.2 and 41.9% of hemicellulose. Lignin content was reported in a range of 15.4 [137]–28 g 100 g^{-1} [138]. Bachmann et al. [131] argue that a high lignin content makes BSG interesting for obtaining biochar. The high protein content is mainly characterized by the abundant presence of amino acids (2571.7 mg kg^{-1}) [117,139]. The presence of lysine, histidine, and arginine is linked to anti-inflammatory potential and antimicrobial properties [140]. Belardi et al. [129] investigated the total polyphenols content, which reported

7.41 mg GAE (gallic acid equivalent) g^{-1} d.m. Niabaho et al. [126] reported that the type of grain, production process, sun exposure, soil type, and climatic conditions caused differences in the amounts of phenolic compounds. Table 6 shows the chemical composition of brewer's spent grain, including nitrogen content, carbon content, and pH, as reported by several authors.

Table 6. Brewer's spent grain chemical characterization.

Parameters	Unit of Measurement	Value	References
pH		4.5–6.0	[98,113,128]
E.C.	dS m^{-1}	3480	[57]
Ash	%	1.9–8.5	[128,129,132,133]
T.O.M.	g 100 g^{-1}	93.8	[57]
O.C.	%	47.3–53.0	[132–134]
T.N.	%	3.8–5.1	[132–134]
C/N		10.8–13.9	[132–134]
P	mg kg^{-1}	6000	[127]
Ca	mg kg^{-1}	3600	[127]
Mg	mg kg^{-1}	1900	[127]
S	mg kg^{-1}	2900	[127]
Na	mg kg^{-1}	137	[105,127]
Lignin	g 100 g^{-1}	15.4–28.0	[137,138]
Cellulose	%	16.8–26.0	[131]
Hemicellulose	%	19.2–41.9	[131]
Protein	g 100 g^{-1} d.w.	15–25	[136]
Amino acids	mg kg^{-1}	2571.7	[117,139]
Fiber	g 100 g^{-1} d.w.	50–70	[136]
Fat	g 100 g^{-1} d.w.	5–10	[136]
T.P.	mg GAE g^{-1} d.m.	7.41	[129]

E.C. = Electrical conductivity; T.O.M. = Total Organic Matter; O.C. = Organic Carbon content; T.N. = Total Nitrogen; C/N= Carbon–Nitrogen ratio; T.P.= Total Polyphenol.

6.3.3. Practical Applications

Brewer's spent grain, with its high-water retention capacity, significant organic matter content, and richness in essential macronutrients (such as P, Ca, Mg, and S), presents as a promising organic component for growing substrates. However, its high moisture content and rapid degradability necessitate careful management to prevent spoilage, suggesting that preliminary treatments like drying or composting are required before incorporation into substrates. Furthermore, its lignocellulosic composition and high lignin content make it of interest for biochar production, which could further enhance its properties as a long-term amendment.

6.4. Crop's Performance

According to Assandri et al. [141], the use of brewer's spent grain (BSG) for agricultural purposes began to be studied in the early 90s, following the research of Mbagwu and Ekwealor [142] who used it as a fertilizer when combined with mineral fertilizers on maize.

Subsequently, Yoo et al. [111] studied BSG-based biochar as an organic soil amendment for lettuce in response to soil acidification. The 5% application rate yielded the best lettuce fresh shoot weight increase. Cacace et al. [112] also examined its agronomic potential as an organic soil amendment in the escarole (*Cichorium endivia* var. Cuartana) cultivation. Results showed that BSG amendments increased leaf number and fresh/dry weight, soil organic carbon, total nitrogen, and water retention. Ebido et al. [113] evaluated BSG composted with poultry manure for amaranth (*Amaranthus viridis* L.) grown, which significantly increased total fresh biomass (143%) and dry matter (58%) compared to the control, correlated with increased exchangeable Ca and total N. Finally, Sdao et al. [57] evaluated BSG in container-grown ornamental marigold and nasturtium, at 10, 20, and 40% peat replacement. In marigold seedlings, 10% BSG improved chlorophyll content and photosynthesis, while 40% BSG led to growth depression. In marketable plants, 10% BSG produced root length and dry matter similar to peat control. Meanwhile, 20% and 40% doses reduced plant height and flower production due to increased electrical conductivity and phosphorus deficiency. The researchers concluded that BSG is a potential partial P substitute at low replacement rates. Table 7 compares peat, wood fiber, coffee silverskin, and brewer's spent grain. It summarizes their origin, main advantages, and limitations as growing media components, along with relevant bibliographic references.

Table 7. Comparative Analysis of *Sphagnum* Peat, Coffee Silverskin, Brewer's Spent Grain, and Wood Fiber as Alternative Organic Materials for Growing Media.

Material	Origin	Advantages	Disadvantages	References
Sphagnum peat	Surface of the rewetted peat soils	Good T.P.S., air content, and W.H.C. The acidic pH value can easily be adjusted.	Contain herbal or grass seeds, poorly re-wettable; investment costs are still high.	[8]
Wood fiber	Byproducts from the wood industry (sawdust, chips, bark).	Good drainability, re-wetting. Low B.D. is used to optimize the physical properties of the blend. It reduces B.D., increases airspace, and improves rewettability.	May cause nitrogen immobilization (due to degradation by microorganisms). High C/N. Degradation by microorganisms leads to deterioration of physical properties. Relatively low N content. It can impair W.R. and nitrogen availability if used in excessive proportions. High proportions (≥75%) can lead to chlorosis and reduced growth unless pH and nutrient supply are managed.	[4,5,8,12,25,27–62]
Coffee silverskin	Byproduct of coffee roasting.	Rich in nutrients (N, P, K, etc.). High W.R. capacity, high O.M., acidic pH. Antioxidant properties. Compostable to eliminate phytotoxicity.	Phytotoxicity (polyphenols, chlorogenic acid) inhibits germination. Low C/N. Compositional variability. Low nitrogen mineralization.	[57,68–102]
Brewer's spent grain	Byproduct of brewing (mashing).	Rich in fiber, protein, and bioactives (amino acids). Biochar, soil conditioner, compost, and peat substitute (low doses) are used in agriculture. Good W.R. Sub-acid pH.	High M. and sugar content (rapid deterioration: 7–10 days); needs refrigeration/drying. High Na levels (negative impact on plants). At high doses, it reduces growth and flowering due to high EC and P deficiency.	[57,98,104–143]

T.P.S. = Total Pore Space; W.H.C. = Water-Holding Capacity; B.D. = Bulk Density, W.R. = Water Retention; O.M. = Organic Matter content; M. = Moisture content; E.C. = Electrical conductivity; P = Phosphorous.

7. Conclusions

This systematic review highlights the significant potential of wood fiber, coffee silverskin, and brewer's spent grain as viable components for horticultural growing media, offering sustainable alternatives to peat. While many of the findings presented in this review are interesting for practical applications, several research gaps and key technological challenges remain for their widespread and optimized adoption. For wood fiber, a well-established component, optimizing its inclusion rates for different species and overcoming potential water and nutrient availability limitations at higher concentrations are key research avenues. Long-term degradation patterns of wood fiber in growing media and strategies for improved nitrogen management and pH stabilization also warrant further investigation. Although coffee silverskin shows promising water retention and high nutrient content, the technological challenge of eliminating its intrinsic phytotoxicity at a large scale is key for broader application. Future research must fully elucidate specific inhibitory compounds and develop robust, industrial-scale pre-treatment methods. Similarly, for brewer's grains, overcoming the obstacles of microbial content that leads to rapid spoilage and high salinity is critical to ensuring substrate performance and quality plants. Studies will also assess their integration into advanced cultivation systems such as hydroponics, aeroponics, and vertical farming, emphasizing enhancing sustainability and efficiency.

A central goal remains ensuring the environmental sustainability of wood fiber production relative to peat. Future research needs to go beyond the morphophysiological quality to truly confirm the use of these innovative organic matrices as viable peat alternatives. It also needs to assess their agronomic, environmental, and economic sustainability comprehensively. This means conducting Life Cycle Assessments (LCA) for the entire production process of plants grown in these new organic matrices, and thoroughly analyzing the production costs of the plant products. This will provide a complete and holistic picture of their sustainability compared to traditional peat-based cultivation.

Author Contributions: Conceptualization, A.E.S., N.S.G. and B.D.L.; methodology, A.E.S., N.S.G. and B.D.L.; investigation, A.E.S., N.S.G. and B.D.L.; writing—original draft preparation, A.E.S., N.S.G. and B.D.L.; writing—review and editing, A.E.S., N.S.G. and B.D.L. All authors have read and agreed to the published version of the manuscript.

Funding: This research received no external funding.

Conflicts of Interest: The authors declare no conflicts of interest.

Abbreviations

WF	Wood fiber
CS	Coffee silverskin
BSG	Brewer's spent grain
R.R	Recommended range
G.M.	Growing Media

References

1. Gruda, N.S. Soilless culture systems and growing media in horticulture: An overview. In *Advances in Horticultural Soilless Culture*; Gruda, N., Ed.; Burleigh Dodds Science Publishing Limited: Cambridge, UK, 2021. [CrossRef]
2. Francini, A.; Romano, D.; Toscano, S.; Ferrante, A. The Contribution of Ornamental Plants to Urban Ecosystem Services. *Earth* **2022**, *3*, 1258–1274. [CrossRef]
3. Baumgarten, A. Analytical methods for growing media–challenges and perspectives. *Acta Hortic.* **2008**, *779*, 97–104. [CrossRef]
4. Barrett, G.E.; Alexander, P.D.; Robinson, J.S.; Bragg, N.C. Achieving environmentally sustainable growing media for soilless plant cultivation systems—A review. *Sci. Hortic.* **2016**, *212*, 220–234. [CrossRef]

5. Gruda, N.S. Increasing sustainability of growing media constituents and stand-alone substrates in soilless culture systems. *Agronomy* **2019**, *9*, 298. [CrossRef]
6. Kaudal, B.B.; Chen, D.; Madhavan, D.B.; Downie, A.; Weatherley, A. An Examination of Physical and Chemical Properties of Urban Biochar for Use as Growing Media Substrate. *Biomass Bioenergy* **2016**, *84*, 49–58. [CrossRef]
7. Gruda, N.S. Advances in Soilless Culture and Growing Media in Today's Horticulture—An Editorial. *Agronomy* **2022**, *12*, 2773. [CrossRef]
8. Gruda, N.S.; Hirschler, O.; Stuart, J. Peat reduction in horticulture—An overview of Europe. *Acta Hortic.* **2024**, *1391*, 545–560. [CrossRef]
9. Bar-Tal, A.; Saha, U.K.; Raviv, M.; Tuller, M. Inorganic and synthetic organic components of soilless culture and potting mixtures. In *Soilless Culture*, 2nd ed.; Raviv, M., Lieth, J.H., Eds.; Elsevier: Amsterdam, The Netherlands, 2019; pp. 259–301. [CrossRef]
10. Jozinović, A.; Šubarić, D.; Ačkar, Đ.; Babić, J.; Orkić, V.; Guberac, S.; Miličević, B. Food Industry By-Products as Raw Materials in the Production of Value-Added Corn Snack Products. *Foods* **2021**, *10*, 946. [CrossRef]
11. Kitir, N.; Yildirim, E.; Şahin, Ü.; Turan, M.; Ekinci, M.; Ors, S.; Ünlü, H. Peat use in horticulture. In *Peat Use in Horticulture*; IntechOpen: London, UK, 2018; pp. 75–90. [CrossRef]
12. Di Lonardo, S.; Cacini, S.; Becucci, L.; Lenzi, A.; Orsenigo, S.; Zubani, L.; Rossi, G.; Zaccheo, P.; Massa, D. Testing new peat-free substrate mixtures for the cultivation of perennial herbaceous species: A case study on *Leucanthemum vulgare* Lam. *Sci. Hortic.* **2021**, *289*, 110472. [CrossRef]
13. Dargie, G.C.; Lewis, S.L.; Lawson, I.T.; Mitchard, E.T.; Page, S.E.; Bocko, Y.E.; Ifo, S.A. Age, extent and carbon storage of the central Congo Basin peatland complex. *Nature* **2017**, *542*, 86–90. [CrossRef]
14. Harenda, K.M.; Lamentowicz, M.; Samson, M.; Chojnicki, B.H. The role of peatlands and their carbon storage function in the context of climate change. In *Interdisciplinary Approaches for Sustainable Development Goals*; Zielinski, T., Sagan, I., Surosz, W., Eds.; Springer: Cham, Switzerland, 2018; pp. 169–187. [CrossRef]
15. Leifeld, J.; Menichetti, L. The underappreciated potential of peatlands in global climate change mitigation strategies. *Nat. Commun.* **2018**, *9*, 1071. [CrossRef] [PubMed]
16. González-Orozco, M.M.; Prieto-Ruíz, J.Á.; Aldrete, A.; Hernández-Díaz, J.C.; Chávez-Simental, J.A.; Rodríguez-Laguna, R. Nursery Production of Pinus engelmannii Carr. with Substrates Based on Fresh Sawdust. *Forests* **2018**, *9*, 678. [CrossRef]
17. Durand, S.; Jackson, B.E.; Fonteno, W.C.; Michel, J.-C. Quantitative Description and Classification of Growing Media Particle Morphology through Dynamic Image Analysis. *Agriculture* **2023**, *13*, 396. [CrossRef]
18. Abad, M.; Noguera, V.; Martínez-Corta, J.; Martínez-Herrero, M.D.; Fornes, F. Physical and Chemical Properties of Sedge Peat-Based Media and Their Relation to Plant Growth. *Acta Hortic.* **1989**, *238*, 45–56. [CrossRef]
19. Jackson, B.E.; Wright, R.D.; Gruda, N. Container medium pH in a pine tree substrate amended with peat moss and dolomitic limestone affects plant growth. *HortScience* **2009**, *44*, 1983–1987. [CrossRef]
20. Gruda, N.; Rau, B.; Wright, R.D. Laboratory bioassay and greenhouse evaluation of a pine tree substrate used as a container substrate. *Eur. J. Hortic. Sci.* **2009**, *74*, 73–78. [CrossRef]
21. Sadh, P.K.; Duhan, S.; Duhan, J.S. Agro-Industrial Wastes and Their Utilization Using Solid-State Fermentation: A Review. *Bioresour. Bioprocess.* **2018**, *5*, 1. [CrossRef]
22. Kour, R.; Singh, S.; Sharma, H.B.; Naik, T.S.K.; Shehata, N.; Ali, W.; Kapoor, D.; Dhanjal, D.S.; Singh, J.; Khan, A.H.; et al. Persistence and Remote Sensing of Agri-Food Wastes in the Environment: Current State and Perspectives. *Chemosphere* **2023**, *317*, 137822. [CrossRef]
23. Capanoglu, E.; Nemli, E.; Tomas-Barberan, F. Novel Approaches in the Valorization of Agricultural Wastes and Their Applications. *J. Agric. Food Chem.* **2022**, *70*, 6787–6804. [CrossRef]
24. Freitas, L.C.; Barbosa, J.R.; da Costa, A.L.C.; Bezerra, F.W.F.; Pinto, R.H.H.; de Carvalho Junior, R.N. From Waste to Sustainable Industry: How Can Agro-Industrial Wastes Help in the Development of New Products? *Resour. Conserv. Recycl.* **2021**, *169*, 105466. [CrossRef]
25. Asdrubali, F.; Evangelisti, L.; Guattari, C.; Roncone, M.; Milone, D. Experimental analysis of the thermal performance of wood fiber insulating panels. *Sustainability* **2023**, *15*, 1963. [CrossRef]
26. Hirschler, O.; Osterburg, B. Peat extraction, trade and use in Europe: A material flow analysis. *Mires Peat* **2022**, *28*, 24–27. [CrossRef]
27. Blok, C.; Eveleens, B.; van Winkel, A. Growing Media for Food and Quality of Life in the Period 2020–2050. *Acta Hortic.* **2021**, *1305*, 341–356. [CrossRef]
28. Maher, M.; Prasad, M.; Raviv, M. Organic soilless media components. In *Soilless Culture*; Elsevier: Amsterdam, The Netherlands, 2008; pp. 459–504. [CrossRef]
29. Gruda, N.; Schnitzler, W.H. Wood fiber substrates as a peat alternative for vegetable production. *Eur. J. Wood Wood Prod.* **2006**, *64*, 347–350. [CrossRef]

30. Zaccheo, P.; Cattivello, C.; Longo, C.; Crippa, L.; Notaristefano, P.; Orfeo, D. A Comparative Study on Some Locally Available Organic Materials for Their Potential Utilization in Sustainable Growing Media Blends. *Acta Hortic.* **2024**, *1389*, 123–130. [CrossRef]
31. Laiche, A.J., Jr.; Nash, V.E. Evaluation of Pine Bark, Pine Bark with Wood, and Pine Tree Chips as Components of a Container Plant Growing Media. *J. Environ. Hortic.* **1986**, *4*, 22–25. [CrossRef]
32. Benoit, F.; Ceustermans, N. Growing Cucumber on Ecologically Sound Substrates. *Hydroponics Transpl. Prod.* **1994**, *396*, 55–66. [CrossRef]
33. Gruda, N.; Schnitzler, W.H. The influence of organic substrates on growth and physiological parameters of vegetable seedlings. *Acta Hortic.* **1997**, *450*, 487–494. [CrossRef]
34. Gruda, N.; Schnitzler, W.H. Influence of wood fiber substrates and application rates on the growth of tomato transplants. *Adv. Hortic. Sci.* **1999**, *13*, 20–24.
35. Wu, L.; Li, R.; Liu, J.; Cui, W.; Qi, Z.; Zhou, W. Nitrogen Immobilization by Wood Fiber Substrates Strongly Affects the Photosynthetic Performance of Lettuce. *Plants* **2025**, *14*, 1518. [CrossRef]
36. Gruda, N.; Tucher, S.v.; Schnitzler, W.H. N-Immobilisierung in Holzfasersubstraten bei der Anzucht von Tomatenjungpflanzen (*Lycopersicon lycopersicum* (L.) Karst. ex Farw.). *J. Appl. Bot.* **2000**, *74*, 32–37. [CrossRef]
37. Gumy, N. Toresa and other wood-fibre products: Advantages and drawbacks when used in growing media. In Proceedings of the International Peat Symposium Peat in Horticulture: Peat and Its Alternatives in Growing Media, Amsterdam, The Netherlands, 30 October 2001; pp. 39–44.
38. Bobo, J.G.; Jackson, B.E. North American and European Conifer Species Evaluated for Use as Wood Components in Growing Media: A Mini Review. *Acta Hortic.* **2024**, *1389*, 131–138. [CrossRef]
39. Bunt, B.R. *Media and Mixes for Container-Grown Plants: A Manual on the Preparation and Use of Growing Media for Pot Plants*; Springer Science & Business Media: Berlin, Germany, 2012.
40. Paquet, G.; Nguyen, T.T.A.; Brégard, A.; Barrada, A.; Poirier, G.; Dorais, M. The agronomic potential of a new vertical growing farming system using biofertilizers and peat-based wood fibre growing medium: Lettuce as a case study. *Acta Hortic.* **2022**, *1355*, 455–462. [CrossRef]
41. Čepulienė, R.; Butkevičienė, L.M.; Steponavičienė, V. Nutrient Use Efficiency and Cucumber Productivity as a Function of the Nitrogen Fertilization Rate and the Wood Fiber Content in Growing Media. *Plants* **2024**, *13*, 2911. [CrossRef]
42. Woznicki, T.; Jackson, B.E.; Sønsteby, A.; Kusnierek, K. Wood Fiber from Norway Spruce—A Stand-Alone Growing Medium for Hydroponic Strawberry Production. *Horticulturae* **2023**, *9*, 815. [CrossRef]
43. Robert, F.S.; Thériault, L.; Allaire, M.; Charland, M. Characterization of selected growing media for greenhouse-grown strawberries. *Acta Hortic.* **2024**, *1389*, 91–98. [CrossRef]
44. Haraldsen, T.K.; Aurdal, S.M.; Woznicki, T.L. Peat-free and peat-reduced growing media for greenhouse petunia (*Petunia* × *hybrida* Vilm.) production. *Acta Hortic.* **2024**, *1389*, 41–50. [CrossRef]
45. Reineke, T.; Muhammed, H.H.A.; Anlauf, R.; Daum, D. Impact of thermo-hydrolytically treated wood fibers as a substrate component on the growth of petunias. *Acta Hortic.* **2024**, *1389*, 105–112. [CrossRef]
46. Di Lonardo, S.; Sodini, M.; Massa, D.; Nesi, B.; Orsenigo, S.; Zubani, L.; Cacini, S. The effect of different peat-free growing media and fertilization levels on the plant nutrition of *Leucanthemum vulgare* (Lam.) and *Dianthus barbatus* (L.). *J. Plant Nutr.* **2025**, *48*, 114–129. [CrossRef]
47. Mariotti, B.; Oliet, J.A.; Andivia, E.; Tsakaldimi, M.; Villar-Salvador, P.; Ivetić, V.; Montagnoli, A.; Janković, I.K.; Bilir, N.; Bohlenius, H.; et al. A Global Review on Innovative, Sustainable, and Effective Materials Composing Growing Media for Forest Seedling Production. *Curr. For. Rep.* **2023**, *9*, 413–428. [CrossRef]
48. Heiskanen, J.; Ruhanen, H.; Himanen, K.; Kivimäenpää, M.; Silvan, N. Growth of Nordic container forest tree seedlings in some peatless and peat-reduced growing media. *New For.* **2024**, *55*, 1499–1517. [CrossRef]
49. Zhu, H.; Jia, Z.; Chen, Y.; Weadock, N.; Wan, J.; Vaaland, O.; Han, X.; Li, T.; Hu, L. Tin Anode for Sodium-Ion Batteries Using Natural Wood Fiber as a Mechanical Buffer and Electrolyte Reservoir. *Nano Lett.* **2013**, *13*, 3093–3100. [CrossRef] [PubMed]
50. Gruda, N.; Schnitzler, W.H. Suitability of wood fiber substrate for production of vegetable transplants: I. Physical properties of wood fiber substrates. *Sci. Hortic.* **2004**, *100*, 309–322. [CrossRef]
51. Gruda, N.; Schnitzler, W.H. Suitability of wood fiber substrates for production of vegetable transplants II. The effect of wood fiber substrates and their volume weights on the growth of tomato transplants. *Sci. Hortic.* **2004**, *100*, 333–340. [CrossRef]
52. Domeño, I.; Irigoyen, I.; Muro, J. New Wood Fibre Substrates Characterization and Evaluation in Hydroponic Tomato Culture. *Eur. J. Hortic. Sci.* **2010**, *75*, 89–94. [CrossRef]
53. Zawadzińska, A.; Salachna, P.; Nowak, J.S.; Kowalczyk, W. Response of Interspecific Geraniums to Waste Wood Fiber Substrates and Additional Fertilization. *Agriculture* **2021**, *11*, 119. [CrossRef]
54. De Boodt, M.; Verdonck, O. The physical properties of the substrates in horticulture. *Acta Hortic.* **1972**, *26*, 37–44. [CrossRef]
55. Michel, J.C. Wettability of organic growing media used in horticulture: A review. *Vadose Zone J.* **2015**, *14*, vzj2014-09. [CrossRef]

56. Fain, G.B.; Gilliam, C.H.; Sibley, J.L.; Boyer, C.R. WholeTree substrates derived from three species of pine in production of annual vinca. *HortTechnology* **2008**, *18*, 13–17. [CrossRef]
57. Sdao, A.E.; Mondelli, D.; Piscitelli, L.; Scaltrito, E.; Leoni, B.; Cristiano, G.; De Lucia, B. Morpho-Physiological Response of Bedding Plants Quality to Unconventional Agro-Industrial Organic Matrices to Peat Replacing: Preliminary Outcomes. *Acta Hortic.* **2025**, *1417*, 125–132. [CrossRef]
58. Theurl, M.C.; Hörtenhuber, S.J.; Lindenthal, T.; Palme, W. Unheated Soil-Grown Winter Vegetables in Austria: Greenhouse Gas Emissions and Socio-Economic Factors of Diffusion Potential. *J. Clean. Prod.* **2017**, *151*, 134–144. [CrossRef]
59. Gruda, N.; Bisbis, M.; Tanny, J. Impacts of protected vegetable cultivation on climate change and adaptation strategies for cleaner production—A review. *J. Clean. Prod.* **2019**, *225*, 324–339. [CrossRef]
60. Hirschler, O.; Osterburg, B.; Weimar, H.; Glasenapp, S.; Ohmes, M.F. *Peat Replacement in Horticultural Growing Media: Availability of Bio-Based Alternative Materials*; Thünen Working Paper No. 190; Thünen Institute: Braunschweig, Germany, 2022. [CrossRef]
61. Beretta, D.; Ripamonti, M. Evaluation of Wood Fiber as Component of Substrates for Container Nursery Crops. *Acta Hortic.* **2021**, *1305*, 77–82. [CrossRef]
62. Laun, N.; Weinheimer, S.; Lutz, F.; Emmel, M.; Gruda, N. Peat reduced substrates for vegetable seedlings. *Acta Hortic.* **2021**, *1321*, 23–30. [CrossRef]
63. Woznicki, T.L.; Sønsteby, A.; Aurdal, S.M.; Kusnierek, K.; Haraldsen, T.K. Optimizing Peat and Wood Fiber Blends: Impacts of Liming and Fertilization on Growth of Petunia (*Petunia x hybrida* Vilm.) and Basil (*Ocimum basilicum* L.). *Horticulturae* **2024**, *10*, 895. [CrossRef]
64. Woznicki, T.; Kusnierek, K.; Roos, U.M.; Andersen, S.; Zimmer, K.; Sønsteby, A. Exploration of Alternative Growing Media in Strawberry Production with Focus on Wood Fiber from Norway Spruce. *Acta Hortic.* **2021**, *1305*, 15–22. [CrossRef]
65. Aurdal, S.M.; Woznicki, T.L.; Haraldsen, T.K.; Kusnierek, K.; Sønsteby, A.; Remberg, S.F. Wood fiber-based growing media for strawberry cultivation: Effects of incorporation of peat and compost. *Horticulturae* **2022**, *9*, 36. [CrossRef]
66. Woznicki, T.; Kusnierek, K.; Vandecasteele, B.; Sønsteby, A. Reuse of Coir, Peat, and Wood Fiber in Strawberry Production. *Front. Plant Sci.* **2024**, *14*, 1307240. [CrossRef]
67. Harris, C.N.; Dickson, R.W.; Fisher, P.R.; Jackson, B.E.; Poleatewich, A.M. Evaluating Peat Substrates Amended with Pine Wood Fiber for Nitrogen Immobilization and Effects on Plant Performance with Container-grown Petunia. *HortTechnology* **2020**, *30*, 107–116. [CrossRef]
68. Lachenmeier, D.W.; Teipel, J.; Scharinger, A.; Kuballa, T.; Walch, S.G.; Grosch, F.; Bunzel, M.; Okaru, A.O.; Schwarz, S. Fully automated identification of coffee species and simultaneous quantification of furfuryl alcohol using NMR spectroscopy. *J. AOAC Int.* **2020**, *103*, 306–314. [CrossRef]
69. Coffee Statistics: Consumption, Preferences, & Spending. Available online: https://www.driveresearch.com/market-research-company-blog/coffee-survey (accessed on 16 May 2025).
70. Coffee Production 2023/2024. Available online: https://www.fas.usda.gov/data/production/commodity/0711100 (accessed on 18 May 2025).
71. Coffee Report and Outlook December 2023—International Coffee Organization. Available online: https://icocoffee.org/documents/cy2023-24/Coffee_Report_and_Outlook_December_2023_ICO.pdf (accessed on 16 May 2025).
72. Haro, N.; Meza-Mori, G.; Lopez, J.L.Z.; Rascón, J.; Pariente, E.; Condori-Apfata, J.A.; Granda-Santos, M.; Inga, B.M.F.; Oliva-Cuz, M.; Lopez, R.Y.R.; et al. Influence of agroforestry systems on *Coffea arabica* L. yield and quality at different altitudes in Amazonas, Peru. *J. Agric. Food Res.* **2025**, *19*, 101574. [CrossRef]
73. Bojórquez-Quintal, E.; Xotlanihua-Flores, D.; Bacchetta, L.; Diretto, G.; Maccioni, O.; Frusciante, S.; Rojas-Abarca, L.M.; Sánchez-Rodríguez, E. Bioactive Compounds and Valorization of Coffee By-Products from the Origin: A Circular Economy Model from Local Practices in Zongolica, Mexico. *Plants* **2024**, *13*, 2741. [CrossRef] [PubMed]
74. Brzezińska, R.; Wirkowska-Wojdyła, M.; Piasecka, I.; Górska, A. Application of Response Surface Methodology to Optimize the Extraction Process of Bioactive Compounds Obtained from Coffee Silverskin. *Appl. Sci.* **2023**, *13*, 5388. [CrossRef]
75. Klingel, T.; Kremer, J.I.; Gottstein, V.; de Rezende, T.R.; Schwarz, S.; Lachenmeier, D.W. A review of coffee by-products including leaf, flower, cherry, husk, silver skin, and spent grounds as novel foods within the European Union. *Foods* **2020**, *9*, 665. [CrossRef]
76. Martuscelli, M.; Esposito, L.; Di Mattia, C.D.; Ricci, A.; Mastrocola, D. Characterization of coffee silver skin as potential food-safe ingredient. *Foods* **2021**, *10*, 1367. [CrossRef] [PubMed]
77. Ansanelli, G.; Fiorentino, G.; Chifari, R.; Meisterl, K.; Leccisi, E.; Zucaro, A. Sustainability assessment of coffee silverskin waste management in the metropolitan city of Naples (Italy): A life cycle perspective. *Sustainability* **2023**, *15*, 16281. [CrossRef]
78. del Pozo, C.; Rego, F.; Yang, Y.; Puy, N.; Bartrolí, J.; Fàbregas, E.; Bridgwater, A.V. Converting coffee silverskin to value-added products by a slow pyrolysis-based biorefinery process. *Fuel Process. Technol.* **2021**, *214*, 106708. [CrossRef]
79. Pongsiriyakul, K.; Wongsurakul, P.; Kiatkittipong, W.; Premashthira, A.; Kuldilok, K.; Najdanovic-Visak, V.; Adhikari, S.; Cognet, P.; Kida, T.; Assabumrungrat, S. Upcycling coffee waste: Key industrial activities for advancing circular economy and overcoming commercialization challenges. *Processes* **2024**, *12*, 2851. [CrossRef]

80. Pyrzynska, K. Useful Extracts from Coffee By-Products: A Brief Review. *Separations* **2024**, *11*, 334. [CrossRef]
81. Farm to Fork Strategy: The Role of Nutrients. Available online: https://www.fertilizerseurope.com/farm-to-fork-strategy-the-role-of-nutrients/ (accessed on 12 May 2025).
82. Salbitani, G.; Chianese, M.R.; Bossa, R.; Bencivenga, T.; Carraturo, F.; Nappo, A.; Guida, M.; Loreto, F.; Carfagna, S. Cultivation of barley seedlings in a coffee silverskin-enriched soil: Effects in plants and in soil. *Plant Soil* **2024**, *498*, 199–211. [CrossRef]
83. Carnier, R.; Berton, R.S.; Coscione, A.R.; Pires, A.M.M.; Corbo, J.Z. Coffee Silverskin and Expired Coffee Powder Used as Organic Fertilizers. *Coffee Sci.* **2019**, *14*, 24–32. [CrossRef]
84. Picca, G.; Plaza, C.; Madejón, E.; Panettieri, M. Composting of coffee silverskin with carbon rich materials leads to high quality soil amendments. *Waste Biomass Valor.* **2024**, *14*, 297–307. [CrossRef]
85. González-Moreno, M.A.; García Gracianteparaluceta, B.; Marcelino Sádaba, S.; Zaratiegui Urdin, J.; Robles Domínguez, E.; Pérez Ezcurdia, M.A.; Seco Meneses, A. Feasibility of vermicomposting of spent coffee grounds and silverskin from coffee industries: A laboratory study. *Agronomy* **2020**, *10*, 1125. [CrossRef]
86. Lorbeer, L.; Schwarz, S.; Franke, H.; Lachenmeier, D.W. Toxicological Assessment of Roasted Coffee Silver Skin (Testa of Coffea sp.) as Novel Food Ingredient. *Molecules* **2022**, *27*, 6839. [CrossRef]
87. Franca, A.S.; Basílio, E.P.; Resende, L.M.; Fante, C.A.; Oliveira, L.S. Coffee Silverskin as a Potential Ingredient for Functional Foods: Recent Advances and a Case Study with Chocolate Cake. *Foods* **2024**, *13*, 3935. [CrossRef] [PubMed]
88. Lestari, W.; Hasballah, K.; Listiawan, M.Y.; Sofia, S. Coffee by-products as the source of antioxidants: A systematic review. *F1000Research* **2022**, *11*, 220. [CrossRef]
89. Ramos, F.J.H.T.V.; Marques, M.D.F.V.; de Oliveira Aguiar, V.; Gondim, F.F.; dos Santos Gomes, L.; de Oliveira Gomes, P.H. Geopolymer composites reinforced with silverskin fibers from the coffee industry waste. *J. Mater. Res. Technol.* **2024**, *31*, 3287–3300. [CrossRef]
90. Garcia, C.V.; Kim, Y.T. Spent coffee grounds and coffee silverskin as potential materials for packaging: A review. *J. Polym. Environ.* **2021**, *29*, 2372–2384. [CrossRef]
91. Dos Santos, É.M.; de Macedo, L.M.; Tundisi, L.L.; Ataide, J.A.; Camargo, G.A.; Alves, R.C.; Oliveira, M.B.P.P.; Mazzola, P.G. Coffee by-products in topical formulations: A review. *Trends Food Sci. Technol.* **2021**, *111*, 280–291. [CrossRef]
92. Grigolon, G.; Nowak, K.; Poigny, S.; Hubert, J.; Kotland, A.; Waldschütz, L.; Wandrey, F. From coffee waste to active ingredient for cosmetic applications. *Int. J. Mol. Sci.* **2023**, *24*, 8516. [CrossRef]
93. Ruschioni, S.; Duca, D.; Tulli, F.; Zarantoniello, M.; Cardinaletti, G.; Corsi, L.; Olivotto, I.; Basili, D.; Naspetti, S.; Truzzi, C.; et al. Evaluation of Growth Performance and Environmental Impact of *Hermetia illucens* Larvae Reared on Coffee Silverskins Enriched with *Schizochytrium limacinum* or *Isochrysis galbana* Microalgae. *Animals* **2024**, *14*, 609. [CrossRef] [PubMed]
94. Prandi, B.; Ferri, M.; Monari, S.; Zurlini, C.; Cigognini, I.; Verstringe, S.; Schaller, D.; Walter, M.; Navarini, L.; Tassoni, A.; et al. Extraction and chemical characterization of functional phenols and proteins from coffee (*Coffea arabica*) by-products. *Biomolecules* **2021**, *11*, 1571. [CrossRef]
95. Pourfarzad, A.; Mahdavian-Mehr, H.; Sedaghat, N. Coffee silverskin as a source of dietary fiber in bread-making: Optimization of chemical treatment using response surface methodology. *LWT-Food Sci. Technol.* **2013**, *50*, 599–606. [CrossRef]
96. Arya, S.S.; Venkatram, R.; More, P.R.; Vijayan, P. The wastes of coffee bean processing for utilization in food: A review. *J. Food Sci. Technol.* **2022**, *59*, 429–444. [CrossRef]
97. Ballesteros, L.F.; Teixeira, J.A.; Mussatto, S.I. Chemical, functional, and structural properties of spent coffee grounds and coffee silverskin. *Food Bioproc. Technol.* **2014**, *7*, 3493–3503. [CrossRef]
98. Sdao, A.E.; Cristiano, G.; De Lucia, B. Department of Soil, Plant and Food Sciences. University of Bari Aldo Moro: Bari, Italy, 2025; *Unpublished work*.
99. Thligene, N.; Mezzapesa, G.N.; Mondelli, D.; Trani, A.; Veronico, P.; Melillo, M.T.; Dumontet, S.; Miano, T.; Sasanelli, N. Effect of Coffee Silver Skin and Brewers' Spent Grain in the Control of Root-Knot Nematodes. *Helminthologia* **2019**, *56*, 30–41. [CrossRef]
100. Mussatto, S.I.; Machado, E.M.S.; Martins, S.; Teixeira, J.A. Production, Composition, and Application of Coffee and Its Industrial Residues. *Food Bioprocess Technol.* **2011**, *4*, 661–672. [CrossRef]
101. Ziemah, J.; Ullrich, M.S.; Kuhnert, N. Antibacterial Activity Potential of Industrial Food Production Waste Extracts Against Pathogenic Bacteria: Comparative Analysis and Characterization. *Foods* **2024**, *13*, 1902. [CrossRef]
102. Al-Charchafchi, F.; Al-Quadan, F. Effect of chlorogenic acid on germination and seedling growth, and on the enzymes activity extracted from *Artemisia herba alba* ASSO. Part I: Germination and seedling growth. *Dirasat Pure Sci.* **2006**, *33*, 168–175.
103. Serna-Jiménez, J.A.; Siles, J.A.; de los Ángeles Martín, M.; Chica, A.F. A Review on the Applications of Coffee Waste Derived from Primary Processing: Strategies for Revalorization. *Processes* **2022**, *10*, 2436. [CrossRef]
104. European Beer Trends—2024 Edition. Available online: https://brewersofeurope.eu/european-beer-trends/ (accessed on 11 May 2025).
105. Ikram, S.; Huang, L.; Zhang, H.; Wang, J.; Yin, M. Composition and nutrient value proposition of brewers spent grain. *J. Food Sci.* **2017**, *82*, 2232–2242. [CrossRef] [PubMed]

106. Wen, C.; Zhang, J.; Duan, Y.; Zhang, H.; Ma, H. A Mini-Review on Brewer's Spent Grain Protein: Isolation, Physicochemical Properties, Application of Protein, and Functional Properties of Hydrolysates. *J. Food Sci.* **2019**, *84*, 3330–3340. [CrossRef] [PubMed]
107. Hejna, A.; Szulc, J.; Błaszak, B. Brewers' spent grain—Simply waste or potential ingredient of functional food? *Zywnosc Nauka Technol. Jakosc/Food Sci. Technol. Qual.* **2023**, *30*, 5–23. [CrossRef]
108. Bamforth, C.W. Progress in brewing science and beer production. *Annu. Rev. Chem. Biomol. Eng.* **2017**, *8*, 161–176. [CrossRef]
109. Amoriello, T.; Ciccoritti, R. Sustainability: Recovery and Reuse of Brewing-Derived By-Products. *Sustainability* **2021**, *13*, 2355. [CrossRef]
110. Pasquet, P.-L.; Villain-Gambier, M.; Trébouet, D. By-Product Valorization as a Means for the Brewing Industry to Move toward a Circular Bioeconomy. *Sustainability* **2024**, *16*, 3472. [CrossRef]
111. Yoo, J.H.; Luyima, D.; Lee, J.H.; Park, S.Y.; Yang, J.W.; An, J.Y.; Yun, Y.U.; Oh, T.K. Effects of Brewer's Spent Grain Biochar on the Growth and Quality of Leaf Lettuce (*Lactuca sativa* L. var. crispa.). *Appl. Biol. Chem.* **2021**, *64*, 10. [CrossRef]
112. Cacace, C.; Cocozza, C.; Traversa, A.; Coda, R.; Rizzello, C.G.; Pontonio, E.; De Mastro, F.; Brunetti, G.; Verni, M. Potential of Native and Bioprocessed Brewers' Spent Grains as Organic Soil Amendments. *Front. Sustain. Food Syst.* **2022**, *6*, 1010890. [CrossRef]
113. Ebido, N.E.; Nnadi, A.L.; Adeoluwa, O.O.; Ndubuaku, U.M.; Obalum, S.E.; Ugwuoju, C.L.; Ajoagu, G.M.; Baiyeri, K.P. Influence of brewery waste and animal manure-based compost on the growth of green amaranth in sandy tropical soils. *Org. Farming* **2024**, *10*, 69–79. [CrossRef]
114. Eliopoulos, C.; Arapoglou, D.; Chorianopoulos, N.; Markou, G.; Haroutounian, S.A. Conversion of brewers' spent grain into proteinaceous animal feed using solid state fermentation. *Environ. Sci. Pollut. Res.* **2022**, *29*, 29562–29569. [CrossRef]
115. Paiva, C.F.; Almeida, T.D.S.F.D.; Arelhano, G.E.; Alvarado, A.V.R.; De Menezes, M.B.; Argandoña, E.J.S.; De Alencar Gomes, I.L.; Moya, A.M.T.M.; Loubet Filho, P.S.; Dos Santos, E.F. Potential of brewer's spent grain as a nutritional ingredient in bakery products. *Plant Food Hum. Nutr.* **2025**, *80*, 12. [CrossRef]
116. Da Silva, A.M.M.; Lima, M.A.; Koksel, F.; Sato, A.C.K. Incorporation of brewer's spent grain into plant-based meat analogues: Benefits to physical and nutritional quality. *Int. J. Food Sci. Technol.* **2024**, *59*, 3870–3882. [CrossRef]
117. Naibaho, J.; Setiawan, R.D.; Korzeniowska, M. Biological properties of bioactive compounds from brewers' spent grain: Current trends, challenges, and perspectives. *Curr. Opin. Food Sci.* **2024**, *62*, 101268. [CrossRef]
118. Naibaho, J.; Butula, N.; Jonuzi, E.; Korzeniowska, M.; Laaksonen, O.; Föste, M.; Kütt, M.L.; Yang, B. Potential of brewers' spent grain in yogurt fermentation and evaluation of its impact in rheological behaviour, consistency, microstructural properties and acidity profile during the refrigerated storage. *Food Hydrocoll.* **2022**, *125*, 107412. [CrossRef]
119. Qazanfarzadeh, Z.; Masek, A.; Chakraborty, S.; Kumaravel, V. Development of brewer's spent grain-derived bio nanocomposites through a multiproduct biorefinery approach for food packaging. *Ind. Crop Prod.* **2024**, *220*, 119226. [CrossRef]
120. Panić, S.; Đurišić-Mladenović, N.; Petronijević, M.; Stijepović, I.; Milanović, M.; Kozma, G.; Kukovecz, Á. Valorization of waste biomass towards biochar production—Characterization and perspectives for sustainable applications in Serbia. *Environ. Technol. Innov.* **2025**, *37*, 104043. [CrossRef]
121. Belardi, I.; De Francesco, G.; Alfeo, V.; Bravi, E.; Sileoni, V.; Marconi, O.; Marrocchi, A. Advances in the valorization of brewing by-products. *Food Chem.* **2025**, *465*, 141882. [CrossRef]
122. Birk, L.; Dos Santos, B.P.; Ossanes, D.S.; de Souza Schwarz, P.; Bachmann, S.A.L.; Sebben, V.C.; Eller, S.; de Oliveira, T.F. Brewer's Spent Grain as a Potential Sorbent for Toxicology Methods: Application to Antidepressant Analysis in Urine. *J. Pharm. Biomed. Anal.* **2025**, *254*, 116564. [CrossRef]
123. Estevão-Rodrigues, T.; Fernandes, H.; Moutinho, S.; Ferreira, M.; Castro, C.; Belo, I.; Salgado, J.M.; Oliva-Teles, A.; Peres, H. Effect of Solid-Fermented Brewer's Spent Grain on Growth, Metabolism, and Oxidative Status of European Seabass (*Dicentrarchus labrax*). *Fishes* **2025**, *10*, 49. [CrossRef]
124. Ferri, I.; Dametti, M.R.; Frazzini, S.; Dell'Anno, M.; Rossi, L. Valorization of Carob and Brewer's Spent Grain as Growth-Substrate Supplements in Tenebrio molitor Rearing. *Animals* **2025**, *15*, 1697. [CrossRef]
125. Karlović, A.; Jurić, A.; Ćorić, N.; Habschied, K.; Krstanović, V.; Mastanjević, K. By-products in the malting and brewing industries—Re-usage possibilities. *Fermentation* **2020**, *6*, 82. [CrossRef]
126. Naibaho, J.; Korzeniowska, M. The Variability of Physico-Chemical Properties of Brewery Spent Grain from 8 Different Breweries. *Heliyon* **2021**, *7*, e06583. [CrossRef] [PubMed]
127. Meneses, N.G.; Martins, S.; Teixeira, J.A.; Mussatto, S.I. Influence of extraction solvents on the recovery of antioxidant phenolic compounds from brewer's spent grains. *Sep. Purif. Technol.* **2013**, *108*, 152–158. [CrossRef]
128. Castro, L.E.N.; Colpini, L.M.S. All-Around Characterization of Brewers' Spent Grain. *Eur. Food Res. Technol.* **2021**, *247*, 3013–3021. [CrossRef]

129. Belardi, I.; Marrocchi, A.; Alfeo, V.; Sileoni, V.; De Francesco, G.; Paolantoni, M.; Marconi, O. Sequential Extraction and Attenuated Total Reflection–Fourier Transform Infrared Spectroscopy Monitoring in the Biorefining of Brewer's Spent Grain. *Molecules* **2023**, *28*, 7992. [CrossRef]
130. Mussatto, S.I. Biotechnological potential of brewing industry by-products. In *Biotechnology for Agro-Industrial Residues Utilisation: Utilisation of Agro-Residues*; Springer: Dordrecht, The Netherlands, 2009; pp. 313–326.
131. Bachmann, S.A.L.; Calvete, T.; Féris, L.A. Potential applications of brewery spent grain: Critical an overview. *J. Environ. Chem. Eng.* **2022**, *10*, 106951. [CrossRef]
132. Mainali, K.; Yadav, M.P.; Sharma, B.K.; Sarker, M.I.; Ngo, H.; Hotchkiss, A.; Simon, S. Isolation and Characterization of the Physiochemical Properties of Brewer's Spent Grain. *Agriculture* **2024**, *15*, 47. [CrossRef]
133. García, D.C.; Villalba, I.; Savino, N.; Nazareno, M.A. Nutritional and Functional Characterization of Different Types of Brewer's Spent Grain Flours. *Food Biosci.* **2025**, *64*, 105890. [CrossRef]
134. Kebede, Y. Analysis of the physico-chemical characteristics of brewers spent grain (BSG). *J. Nat. Sci. Res.* **2020**, *11*, 9–12. [CrossRef]
135. Abbas, A.; Siddiq, Z.; Hayyat, M.U.; Zhang, Y.J.; Ghaffar, R.; Gatasheh, M.K. Na^+ and K^+ compartmentalization in *Spinacea oleracea* and their effects on growth, water relations, endogenous melatonin, and non-structural carbohydrates. *Sci. Hortic.* **2024**, *323*, 112467. [CrossRef]
136. González-Díaz, R.L.; Saldarriaga-Hernandez, S.; Ramírez-Aguirre, M.D.; Guerrero-Higareda, S.; García-Cayuela, T.; Garcia-Amezquita, L.E.; Carrillo-Nieves, D. Valorization of brewer's spent grains through *Ganoderma lucidum* cultivation: Functional food ingredients and bread prototypes. *LWT* **2024**, *214*, 117131. [CrossRef]
137. Parchami, M.; Agnihotri, S.; Taherzadeh, M.J. Aqueous ethanol organosolv process for the valorization of brewer's spent grain (BSG). *Bioresour. Technol.* **2022**, *362*, 127764. [CrossRef] [PubMed]
138. Mussatto, S.I.; Roberto, I.C. Chemical Characterization and Liberation of Pentose Sugars from Brewer's Spent Grain. *J. Chem. Technol. Biotechnol.* **2006**, *81*, 268–274. [CrossRef]
139. Jin, Z.; Lan, Y.; Ohm, J.B.; Gillespie, J.; Schwarz, P.; Chen, B. Physicochemical composition, fermentable sugars, free amino acids, phenolics, and minerals in brewers' spent grains obtained from craft brewing operations. *J. Cereal Sci.* **2022**, *104*, 103413. [CrossRef]
140. Aguilar-Toalá, J.E.; Cruz-Narváez, Y.; Quintanar-Guerrero, D.; Liceaga, A.M.; Zambrano-Zaragoza, M.L. In silico bioactivity analysis of peptide fractions derived from brewer's spent grain hydrolysates. *Int. J. Food Sci. Technol.* **2024**, *59*, 2804–2815. [CrossRef]
141. Assandri, D.; Pampuro, N.; Zara, G.; Cavallo, E.; Budroni, M. Suitability of Composting Process for the Disposal and Valorization of Brewer's Spent Grain. *Agriculture* **2021**, *11*, 2. [CrossRef]
142. Mbagwu, J.S.C.; Ekwealor, G.C. Agronomic potential of brewers' spent grains. *Biol. Wastes* **1990**, *34*, 335–347. [CrossRef]
143. Tumbure, A.; Pulver, C.; Black, L.; Walsh, L.; Prasad, M.; Leahy, J.J.; Corbett, E.; Gaffney, M.T. Bio-Resource Availability in Ireland: A Practical Review of Potential Replacement Materials for Use in Horticultural Growth Media. *Horticulturae* **2025**, *11*, 378. [CrossRef]

plants

MDPI

Article

Assessment of Vegetable Species for Microgreen Production in Unheated Greenhouses: Yield, Nutritional Composition, and Sensory Perception

Pabla Rebolledo [1], Gilda Carrasco [1,*], Claudia Moggia [1], Pedro Gajardo [1], Gabriela Rodrigues Sant'Ana [2], Fernando Fuentes-Peñailillo [3], Miguel Urrestarazu [4] and Eduardo Pradi Vendruscolo [2,*]

[1] Departamento de Horticultura, Facultad de Ciencias Agrarias, Universidad de Talca, Talca 3460000, Chile; prebolle@utalca.cl (P.R.); cmoggia@utalca.cl (C.M.); pgajardo19@alumnos.utalca.cl (P.G.)

[2] Departamento de Agronomia, Universidade Estadual de Mato Grosso do Sul, Cassilândia 79540-000, Brazil; gabrielasant_ana@icloud.com

[3] Instituto de Investigación Interdisciplinaria (I3), Vicerrectoría Académica (VRA), Universidad de Talca, Talca 3460000, Chile; ffuentesp@utalca.cl

[4] Departamento Agronomía, Universidad de Almería, 04120 Almería, Spain; mgavilan@ual.es

* Correspondence: gcarrasc@utalca.cl (G.C.); eduardo.vendruscolo@uems.br (E.P.V.)

Abstract: Cultivating microgreens in central-southern Chile in unheated greenhouses offers a viable and productive alternative to growers. In 2023, two experiments were conducted in autumn and spring. These experiments involved the production of microgreens of eleven vegetable species. The tray system with the substrate was employed. Subsequently, agronomic, nutritional, and sensory perception variables were assessed. Despite notable fluctuations in external temperatures between these seasons, a diverse array of microgreens can be successfully cultivated, meeting local consumer preferences. Research indicates that microgreens grown under these conditions exhibit high nutritional quality, serving as a rich source of essential nutrients and bioactive compounds beneficial to human health. This nutritional value remains consistent across autumn and spring, establishing microgreens as a reliable and valuable food option. The observed acceptance and purchasing intentions among the surveyed population suggest a promising market opportunity for introducing these products regionally. Consumers appreciate microgreens' quality and nutritional advantages, underscoring their potential.

Keywords: microgreens; vegetables; food composition; sensory analysis; functional foods

1. Introduction

Microgreens, also known as microplants, are young vegetables or herbs characterized by fully expanded cotyledon leaves and the emergence or partial expansion of the first true leaf [1,2]. They originated in California, United States, in the 1980s and were initially grown commercially. The surge in marketing efforts was due to their remarkable taste, colors, appearances, and textures [3]. They can be used in salads, soups, sandwiches, or as visually appealing, edible garnishes, enhancing various dishes [4]. Additionally, microgreens are rich in diverse nutrients, driving increased consumer interest due to their nutritional value [1,5].

Microgreens are rich in minerals and phytochemicals that promote human health, containing higher nutrient levels than adult plants and earning them the label of superfoods or functional foods. Numerous studies have underscored their vital impact on human health [1,6–8]. In Chile, the cultivation of microgreens has markedly increased in recent years, especially in urban areas with a rising demand for fresh, nutritious products [3]. Recent reviews, such as [9], have highlighted the nutritional value of various microgreen species, emphasizing that their nutrient composition and content can significantly vary based on the growth stage and are influenced by factors such as cultivars and growing conditions [10,11]. Additionally, microgreens possess superior phytochemical profiles compared to baby vegetables, making them an even more potent source of health-promoting compounds [12].

Microgreen production is advantageous for productivity as it optimizes physical space and inputs, resulting in minimal resource losses [13]. This demonstrates the increased sustainability of microgreen production compared to the traditional cultivation of mature plants [14,15]. Moreover, microgreen production allows for a diverse range of species to be cultivated and quickly commercialized due to their short shelf-life. Consequently, the cultivation environment must be close to consumer centers [16]. As various methodologies and technologies continue to gain prominence, there is a growing need to enhance the approaches used in producing microgreens. Essentially, evaluating the cultivation of different species under accessible conditions encourages local investments, bolsters the food supply, and elevates quality. Furthermore, conducting assessments on mineral content and consumer preferences leads to better reception in the local market, enhancing commercialization and helping producers to direct their activities.

This study aimed to evaluate the performance and nutritional composition of various microgreens produced locally and assess consumer perception regarding their organoleptic characteristics.

2. Results

2.1. Earliness and Yield

During the study, only chard had a low germination rate (20%), while the other species had germination rates above 80%. Additionally, the sowing density ranged from 90 to 640 mg m^{-2}, and the average germination time varied between three and nine days due to the distinct characteristics of each species (Table 1).

Table 1. Characteristics of germination rate, sowing density, and germination time of the species used to grow microgreens.

	Commercial Name	Sowing Density ($g\ m^{-2}$)	Germination Index (%)	Germination Time (days)
Experiment 1	Chard	640	20	9
	Carrot	130	85	9
	Radish	320	85	3
	Lettuce	90	85	8
	Kale	190	85	3
	Arugula	128	95	3
Experiment 2	Mizuna	255	100	3
	Fennel	155	80	8
	Spinach	255	100	3
	Broccoli	320	95	3
	Anise	180	75	8

2.2. Biometric and Productive Analysis

In the autumn experiment, radishes had significantly greater height than other species, while carrots and lettuce had the lowest average for plant height (Table 2). In the spring experiment, mizuna, fennel, and broccoli showed no significant differences in plant height, and all three species were taller than spinach. However, anise exhibited growth issues, which were attributable to the elevated temperatures recorded within the greenhouse during the spring period. Additionally, regarding green mass harvested per tray and productivity, radish and lettuce exhibited superiority in the autumn experiment, whereas mizuna and broccoli excelled in the spring experiment (Table 2).

Table 2. Average height values harvested fresh mass and productivity of different species used to grow microgreens.

Commercial Name	Height (cm) *	Fresh Mass (g tray^{-1}) *	Yield (g m^{-2}) *
	Autumn experiment		
Chard	3.6 d	116 bc	756 bc
Carrot	2.9 e	96 c	624 c
Radish	6.6 a	341 a	2216 a
Lettuce	2.9 e	295 a	1994 a
Kale	4.4 c	88 c	673 c
Arugula	6.2 b	136 b	1171 b
p-value	0.000	0.000	0.000
	Spring experiment		
Mizuna	7.9 a	1367 a	8877 a
Fennel	8.4 a	808 b	5247 b
Spinach	6.7 b	995 b	6461 b
Broccoli	8.5 a	1346 a	8805 a
p-value	0.004	0.003	0.000

* The means followed by the same letter in the column do not differ from each other using the Tukey test at a 5% probability.

2.3. Nutritional and Mineralogical Analysis

It was found that the total proteins and raw fiber contents of the species used in the autumn experiment varied between them, ranging from 3.7% to 4.8% and 0.4% to 0.7%, respectively. For these same characteristics, the variation presented in the spring experiment was 2.42% to 3.50% and 0.5% to 0.6%, respectively, with higher averages of proteins and fiber being observed for fennel (Table 3).

Table 3. Average values of total protein, total phenols, nitrates, raw fiber, and beta-carotene of different species used to grow microgreens.

Commercial Name	Total Proteins (%)	Total Phenols (mg kg^{-1} of GAE)	Nitrates (g L^{-1})	Raw Fiber (%)	β-Carotene (mg equiv 100 g^{-1})
		Autumn experiment [1]			
Chard	4.4 a	3335.9 a	3281.9 a	0.5 a	39,966 a
Carrot	4.6 a	3083.6 a	2508.5 a	0.5 a	15,379 d
Radish	3.7 a	2001.2 b	1099.5 b	0.6 a	17,438 c
Lettuce	3.8 a	2060.5 b	4169.3 a	0.7 a	12,550 e
Kale	4.6 a	3269.6 a	2489.1 a	0.6 a	19,369 b
Arugula	4.8 a	3304.6 a	2442.2 a	0.4 a	12,612 e
p-value	n.s.	0.007	0.000	n.s.	0.000
		Spring experiment [2]			
Mizuna	2.94	5663.8	4707.8	0.5	1451
Fennel	3.50	6164.5	5600.0	0.6	3474
Spinach	2.42	11,025.0	3866.7	0.5	3629
Broccoli	2.67	4538.9	4266.0	0.5	2498

GAE: Gallic acid equivalent; n.s.: not significant. [1] For the autumn experiment, means followed by the same letter in the column do not differ using the Tukey test at a 5% probability. [2] For the spring experiment, no statistical analyses were performed.

For total phenol contents, two groups were observed: one formed by chard, carrot, kale, and arugula, with values above 3000 mg kg^{-1} GAE, which was higher than the second group, formed by radish and lettuce, for which values close to 2000 mg kg^{-1} GAE were observed. In the spring experiment, the highest value of total phenol content was observed for spinach (11,000 mg kg^{-1}), followed by fennel, mizuna, and broccoli (4500 to 6000 mg kg^{-1}), respectively.

Only radish, with the lowest average, showed a significant difference from the other species for the nitrate characteristic in the autumn. Chard stood out for beta-carotene levels, with values over twice the concentrations verified for nitrate.

In the autumn experiment, only radish showed significantly lower average beta-carotene levels than other species. On the other hand, chard stood out with beta-carotene levels over twice the concentrations found in the other species. In the spring experiment, fennel and spinach had the highest average nitrate and beta-carotene levels, respectively. These characteristics varied across species, ranging from 1733.33 mg L^{-1} to 2178 mg g^{-1} (Table 3).

Except for P content, significant differences existed in the mineralogical compositions of the species used in the autumn experiment. In this context, kale microgreens contained higher Fe, Mg, and Ca levels, while chard showed higher Na levels, surpassing only radish (Table 4). Among the species in the spring experiment, mizuna microgreens exhibited the highest Fe, Mg, and Ca concentrations, while spinach had the highest average Na content. The variations in Fe, Mg, Ca, Na, and P content in the spring experiment were 45.46 mg kg^{-1}, 84.3 mg kg^{-1}, 594.2 mg kg^{-1}, and 25.55 mg kg^{-1}, respectively, and the number of sample repetitions did not allow for a comparison of means. (Table 4).

Table 4. Nutrient values of different species used for the production of microgreens.

Commercial Name	Fe (mg kg^{-1})	Mg (mg kg^{-1})	Ca (mg kg^{-1})	Na (mg kg^{-1})	P (mg kg^{-1})
			Autumn experiment [1]		
Chard	70.6 bc	207.4 bc	308.9 c	32.9 a	1.1 a
Carrot	75.8 bc	203.3 bc	313.6 c	26.3 ab	1.1 a
Radish	79.9 b	202.2 c	410.9 b	23.5 b	0.9 a
Lettuce	156.8 a	268.5 ab	353.1 c	26.2 ab	1.0 a
Kale	80.6 b	300.0 a	552.8 a	26.2 ab	1.2 a
Arugula	69.1 c	202.6 bc	380.7 bc	25.7 ab	0.8 a
p-value	0.000	0.001	0.000	0.015	n.s.
			Spring experiment [2]		
Mizuna	79.2	292.1	846.3	25.7	0.9
Fennel	54.5	279.4	587.6	25.9	0.4
Spinach	33.8	207.8	281.6	31.1	0.4
Broccoli	68.4	253.7	252.1	5.6	0.4

n.s.: not significant. [1] For the autumn experiment, means followed by the same letter in the column do not differ using the Tukey test at a 5% probability. [2] For the spring experiment, no statistical analyses were performed.

2.4. Organoleptic Properties and Preference for Purchasing Vegetables

Significant differences were observed in the various species' appearance in the autumn and spring experiments. In the autumn experiment, chard was rated as having an extremely pleasant appearance, while carrot and arugula received the worst evaluations. Similarly, mizuna had the best appearance in the spring experiment, while spinach had the worst. Despite these differences, none of the species received scores equal to or greater than 5, which could indicate negative evaluations.

In both experiments, color differences were observed between species. According to the respondents, chard and Mizuna consistently had the best color perception.

For texture, there were no differences between species in the autumn experiment. However, significant differences were observed in the spring experiment, with spinach having the most pleasant texture and receiving the highest evaluation (Table 5).

Table 5. Average values of sensory evaluations regarding appearance, color, texture, and flavor of different species used to produce microgreens.

Commercial Name	Appearance *	Color *	Texture *	Flavor *
		Autumn experiment		
Chard	2.3 a	2.1 a	3.0 a	4.3 ab
Kale	2.7 ab	2.3 ab	2.8 a	3.2 a
Lettuce	2.9 ab	3.1 c	2.6 a	3.5 ab
Radish	3.0 ab	2.9 bc	3.2 a	4.5 b
Arugula	3.1 b	2.6 abc	3.2 a	4.5 b
Carrot	3.2 b	2.6 abc	3.2 a	3.7 ab
p-value	0.000	0.000	n.s.	0.000
		Spring experiment		
Mizuna	2.2 a	2.0 b	2.6 ab	3.4 ab
Fennel	2.7 ab	2.7 ab	3.1 a	4.4 a
Spinach	3.2 b	3.3 a	2.2 b	2.7 b
p-value	0.05	0.02	0.05	0.01

n.s.: not significant. * The means followed by the same letter in the column do not differ from each other using the Tukey test at 5% probability. Each attribute was evaluated from 1 (extremely pleasant) to 9 (extremely unpleasant) (Table 8).

Notably, for the appearance, color, and texture characteristics, all averages obtained resulted in a positive evaluation of the species since, according to the scale used, the evaluations remained very pleasant and moderately pleasant (values vary from 2.2 to 3.2). However, for the flavor characteristic, the evaluators' scores varied between moderately pleasant and slightly pleasant (values range from 3.2 to 4.5 in the autumn experiment), with radish and arugula the species that received the worst evaluations, while kale was considered the tastiest. For the spring experiment, spinach and fennel differed significantly, with spinach being considered the species with the best flavor (Table 5).

Regarding purchase intention, among the species used in the autumn experiment, there was no difference between the average scores of the evaluators. Thus, the variation between 2.18 and 2.84 implied a purchase intention result of "would probably buy" (Table 6). In addition, for the spring experiment, it was found that mizuna had a significantly lower average than fennel and spinach, receiving a purchase intention of "I would definitely buy it", while the other species had the same purchase intention as the species in the experiment autumn, "I would probably buy" (Table 6).

Table 6. Summary of consumers' purchasing preferences regarding the different species used to produce microgreens.

Commercial Name	Purchase Intention *	Result of Purchase Intention
	Autumn experiment	
Chard	2.2 a	I would probably buy it
Kale	2.4 a	I would probably buy it
Lettuce	2.5 a	I would probably buy it
Radish	2.6 a	I would probably buy it
Arugula	2.8 a	I would probably buy it
Carrot	2.8 a	I would probably buy it
p-value	n.s.	
	Spring experiment	
Mizuna	2.0 a	I would definitely buy it
Fennel	2.6 a	I would probably buy it
Spinach	2.5 a	I would probably buy it
p-value	n.s.	

n.s.: not significant. * The means followed by the same letter in the column do not differ using the Tukey test at a 5% probability.

In the autumn experiment, of the total number of respondents, 53% were men and 47% were women, with most aged between 20 and 25 (79%). Additionally, 65% of the judges declared that they frequently consumed vegetables, but 84% responded that they had never consumed microgreens. In the spring experiment, of the total number of respondents, 68% were men and 32% were women, the majority of whom were between 20 and 25 years old (68%); from the total, 68% were frequent vegetable consumers, but 55% had never consumed microgreens.

2.5. Correlation Analysis

The search for direct relationships between the quality of microgreens and consumers' acceptance of them can help define production planning. In this way, a correlation analysis was carried out to determine which characteristics could positively interfere with the choice of a certain species. It is also possible to determine which characteristics do not affect sensory evaluations or which could negatively impact the choice. While red represents significant interactions, these are related to sensory evaluations, with lower averages indicating positive evaluations. In this sense, all the interactions highlighted with thick lines seem positive, with interference between the factors of Na, Fe, beta-carotene content, and the appearance of the species.

Mg levels interfere with texture and flavor characteristics, and there is a direct relationship between total proteins and total phenols, the latter with the plants' characteristic color (Figure 1). For all the correlation groups formed, there is evidence that the variation in the composition of species, nutrients, and bromatological components significantly alters the sensory evaluations of the individuals participating in the autumn experiment.

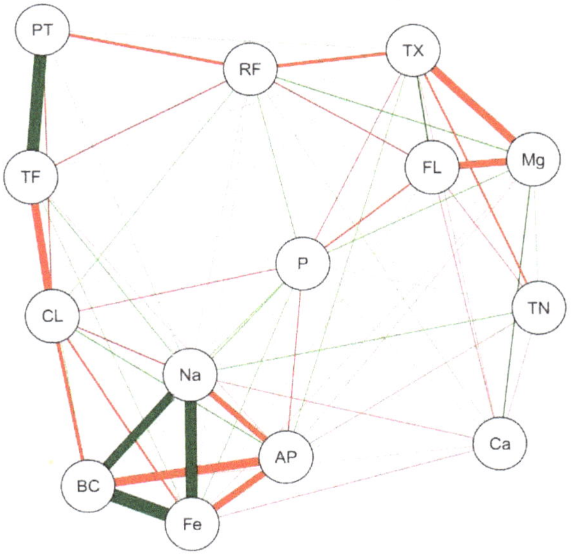

Figure 1. Pearson correlation network between the variables of total protein (TP), total phenols (TF), nitrates (TN), beta-carotene (BC), raw fiber (RF), iron (Fe), manganese (Mg), calcium (Ca), sodium (Na), phosphorus (P), appearance (AP), color (CL), texture (TX), and flavor (FL) of different species of microgreens evaluated during the autumn experiment. Positive correlations were highlighted in green, while negative correlations were highlighted in red; determining the line thickness followed a cut-off value of 0.7, corresponding to 70% reliability.

3. Discussion

3.1. Biometric and Productive Analysis

Notably, there is great variation in development among the different species used to grow microgreens (Table 2). This fact, combined with the environmental conditions of cultivation, will define a set of interactions encompassing sowing and harvesting cycles. In this sense, using unheated greenhouses, common environments for vegetable producers, creates a reliance on local environmental conditions, leading to increased variability in temperature and relative air humidity.

The variability in height development between species is related to intrinsic genetic and morpho-physiological factors. Species have different characteristics regarding the amount of seed reserves: for example, lettuce, has small seeds and low reserves, compared to radish, which has larger seeds, enabling the development of larger seedlings. However, the values obtained are based on height values previously reported in the literature for species cultivated in different environmental conditions [17–19].

During the autumn experiment, low temperatures may explain the lengthening of the production cycle, 33 days. The cultivation period was characterized by average daily temperatures below 20 °C, with around 50% of days with average temperatures below 15 °C. In this sense, harvests are reported to be carried out between 4 and 12 days after sowing for some of the species used in the present study, such as radish, chard, and spinach [20]. However, such studies are mostly carried out in controlled environmental chambers, optimizing the conditions for plant development. In addition to temperature, other environmental factors can also significantly interfere with cultivation, such as incident radiation, influencing crop productivity averages, as seen in the present study (Table 2), in which the average production obtained in autumn cultivation was 5.7 times lower than that obtained in spring. Such characteristics greatly influence the development of crops, resulting in seasonality for species such as chard [21] and arugula [10], for which it appears that autumn crops, when the temperature and incident radiation are decreasing, are less productive than crops in spring when temperature and radiation are rising.

3.2. Nutritional and Mineralogical Analysis

Plants harvested early, especially microgreens, are considered bioactive foods or superfoods due to their high levels of elements beneficial to human health [10]. In this sense, other studies have demonstrated that nutritional elements and bioactive compounds are found in higher microgreens values than conventionally harvested vegetables [11,22].

The protein levels in this research did not show differences between the species evaluated in the autumn experiment, ranging between 3.66% and 4.75%. In the spring experiment, the values ranged between 2.4% and 2.9%. The literature reports protein values like those obtained in this study, and in some cases, they exceed the ones previously reported (Table 3), as with radish, carrot [23], and arugula [24]. No variations were observed in this study's protein, phenols, and fiber contents, and the variation in the nitrate contents was confirmed (Table 3).

Among the species, significant differences were only found in the autumn experiment, with the lowest nitrate content recorded for radish (1099.55 mg L^{-1}) and the highest concentration observed for lettuce (4169.27 mg L^{-1}). For the spring experiment, the lowest value was found for spinach (3866.67 mg L^{-1}), while the highest concentration was verified for fennel (5600.00 mg L^{-1}). High variability in nitrate levels is close to those previously verified by other authors, in which different species vary between 13 and 8279 mg kg^{-1}, which are also influenced by the growing season [25]. High variability in nitrate levels can also occur in species belonging to the same botanical family, as verified in the present study (Table 3), which may be related to genetic factors and cultivation environment and inputs used, which are also influenced by the growing season [25]. In addition, due to the negative implications of consuming high nitrate levels, it is essential to find alternatives to reduce these levels in microgreens.

Beta-carotene is a red-orange pigment found in fruits and vegetables; it is a precursor of vitamin A and has an important physiological role in humans, such as vision and development. Also, it works as an important antioxidant that allows the maintenance of cellular metabolism [5]. This compound can be obtained abundantly in microgreens, implying that consuming these young plants can help prevent disorders and diseases, increasing the quality of life [9]. Therefore, an increase in beta-carotene levels is desired in microgreens. Significant variation was also observed for this compound between the species studied, with the highest concentrations of chard, fennel, and spinach (Table 4). These three species have practically double the concentration of beta-carotene compared to the other species evaluated. Just like beta-carotene, the presence of minerals in species used for producing microgreens can benefit human health, allowing the natural replacement of minerals and avoiding problems related to their absence [9,25]. These characteristics make microgreens foods that promote human health and which, in the future, can be included in specific diets to reduce problems related to inflammatory and cardiovascular diseases and obesity, among others. The variation observed between species in terms of their mineral composition (Table 4) implies that different species, or even genetic materials from the same species, can be used in a specific way for a given need. The literature shows a positive relationship between early harvesting and the amount of nutrients available, with microgreens tending to contain higher amounts of minerals [26]. Also, combining species could result in an even richer food, generally increasing the nutrients available to the final consumer.

3.3. Organoleptic Properties and Preference for Purchasing Vegetables

Sensory perception is something particular and varies between people according to their perception of what is pleasant or unpleasant. In this sense, the sensory evaluation of the species of microgreens used in this study, in both experiments, was considered positive since there were no average measurements of scores above 5 points (Table 5). Thus, it would be verified if evaluators were indifferent to characteristics and arugula, which obtained the highest averages in the flavor assessment, followed by fennel, belonging to the Apiaceae family, which has a pronounced aromatic characteristic [27].

Bitterness can also trigger an aversion reaction to the product due to evolutionary issues since it can be related to toxic compounds in nature [6]. The highest averages were verified for the flavor characteristic. This is related to the fact that some species used in producing microgreens may present marked spiciness or bitterness due to the high levels of compounds present, such as sulfur in the case of some species. This positive evaluation also resulted in positive purchase intention for all species evaluated, varying between 1.95 for mizuna and 2.84 for carrots on the 1 to 9 scale used. It was also found that purchase intention evaluations formed different groups; thus, the species that received more positive evaluations were mainly women over 30 years old. Moreover, those who claimed to consume vegetables and microgreens frequently or sometimes also evaluated two (spinach and mizuna) of the eight species used more positively (Figure 2). In this sense, it is also observed that groups of younger women who frequently consume vegetables and microgreens tend to opt for species with a more robust flavor (Table 5).

Similar to those observed for women, the evaluations of the group of men also indicated that the best evaluations were concentrated on evaluators over 30 years old and who habitually consume vegetables regularly (Figure 3). For this group, we also found that two of the three species indicated as the ones with main purchase intentions, fennel and arugula, received high evaluations regarding flavor (Table 5), indicating a preference for these evaluators over species with a strong flavor.

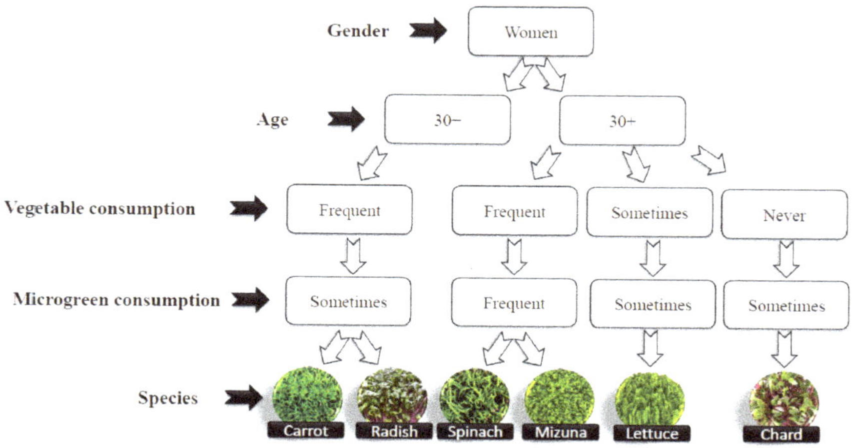

Figure 2. Map of female reviewers' purchase intention.

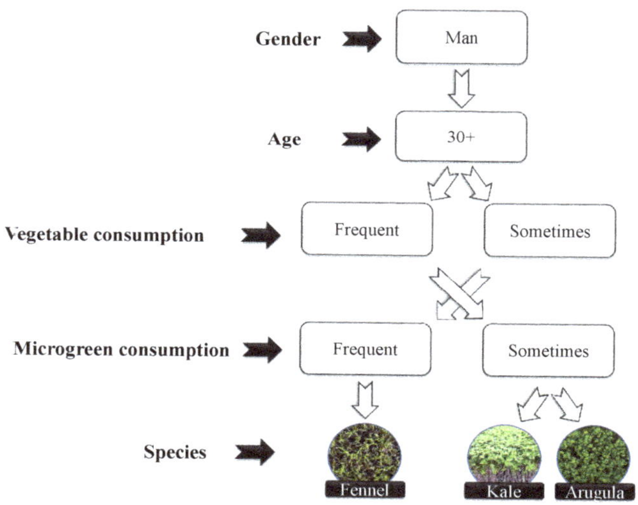

Figure 3. Map of male reviewers' purchase intention.

3.4. Correlation Analysis

Our findings enabled us to identify relationships between microgreens' composition and evaluators' sensory perception in the autumn experiment (Figure 1). In this sense, we can highlight that the presence of higher levels of beta-carotene results in better evaluations of the general appearance of microgreens. This fact may be related to the color of the microgreens, as more intense colors tend to influence consumers' decision-making [28].

For other species consumed as mature vegetables, it appears that the presence of beta carotene increases consumer acceptability, increasing evaluations regarding flavor and general sensory characteristics, such as tomatoes [29]. Consumers also gave higher scores in sensory evaluation to orange sweet potatoes, rich in beta carotene, when compared to those with purple and yellow colors [30], a result like that observed for cassava cultivars, in which biofortification with beta-carotene positively influenced greater acceptability on the part of the evaluators, who also related part of their choices to the more intense orange color [31]. This relationship between beta-carotene content and consumer acceptability can have important implications regarding the definitions of species to be cultivated, increasing commercialization potential. In addition to economic factors, using species-rich compounds in this compound can benefit consumers, improving the general characteristics of food quality [5]. Also, the positive relationships between magnesium texture and magnesium flavor may imply the choice of vegetables rich in this nutrient, which participates in a series of metabolic processes in the human body, improving performance, balancing blood sugar levels, and promoting heart health, among others [32].

In sensory terms, as demonstrated in this research (Figure 1), it can be seen in the literature that magnesium is related to the flavor and texture of a series of vegetables, as it is involved in metabolic activities that will influence the levels of sugars in vegetables, as well as on texture, increasing firmness and post-harvest life [33]. In addition, the relationship between phenol levels and the positive evaluation regarding color (Figure 1) can be attributed to the high correlation between these compounds and the existence of pigments with high antioxidant capacity, such as betacyanin, betaxanthin, and betalain [34,35]. Again, choices regarding the antioxidant capacity of the species used can be targeted, increasing the quality of the product sold to the final consumer.

4. Materials and Methods

In 2023, two experiments were carried out at the University of Talca's Soilless Cultivation and Vertical Agriculture Unit, located at the Talca Campus in the Maule Region, South Central of Chile (35°24′26″ SL and 71°38′10″ WL). The experiments took place from April to May (autumn) and August to September (spring) and involved cultivating eleven different horticultural species as microgreens in rigid plastic trays (55 cm long × 28.5 cm wide) filled with moistened peat substrate up to 2 cm high (Table 7). These eleven species were selected because they are among the most common vegetables grown in the South-Central region of Chile. The rationale behind using a diverse range of species was to observe variations in the production process, product quality, and consumer acceptance, providing a comprehensive understanding of how different microgreens perform under controlled soilless cultivation conditions.

Table 7. General characteristics of the species used to produce microgreens.

	Commercial Name	Scientific Name	Commercial Variety
Autumn experiment	Chard	*Beta vulgaris* L. var. *cicla*	Penca roja
	Carrot	*Daucus carota* L.	Chantenay Royal 2
	Radish	*Raphanus sativus* L.	Pablo
	Lettuce	*Lactuca Sativa* L.	Tango roja
	Kale	*Brassica oleracea* L. var. *sabellica*	Verde rizado
	Arugula	*Eruca sativa* Mill.	Astro
Spring experiment	Mizuna	*Brassica rapa* L. subsp. *nipposinica*	Verde
	Fennel	*Foeniculum vulgare* L.	Etrusco
	Spinach	*Spinacia oleracea* L.	Woodduck
	Anise	*Pimpinella anisum* L.	-
	Broccoli	*Brassica oleracea* L. var. *italica*	Italian Green Sprouting

The seeds were sown manually in the trays and then covered until they sprouted. They received daily watering, and a nutrient solution with a low EC of 1.3 dS m^{-1} was applied every three waterings. Temperature and relative humidity in the greenhouse were measured using sensors; these variables ranged from 3.7 °C to 35 °C and 34% to 95% in the autumn experiment and 3.6 °C to 36.6 °C and 23% to 95% in the spring experiment, respectively (Figures 4 and 5).

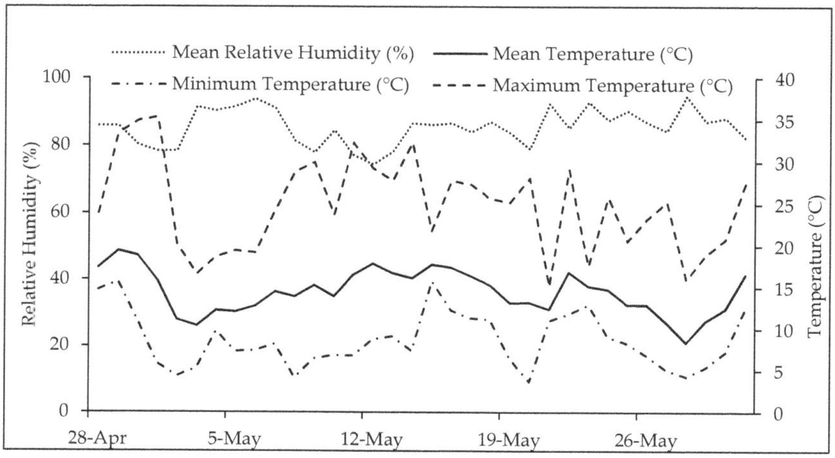

Figure 4. Relative humidity values and maximum, mean, and minimum temperatures during the autumn experiment.

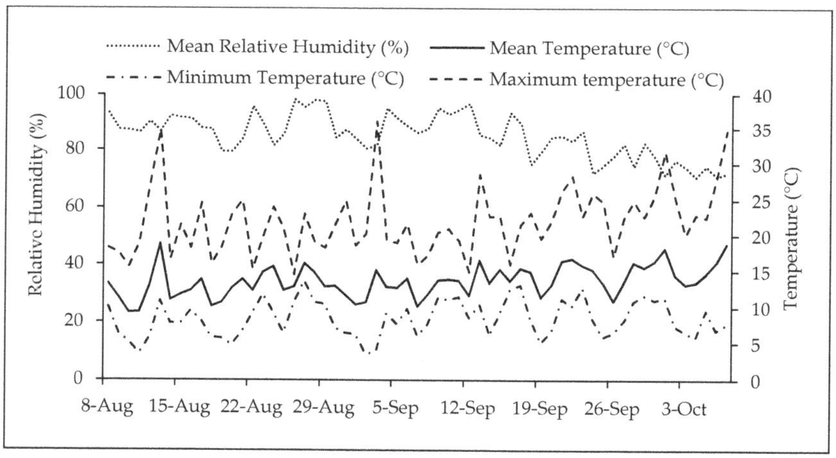

Figure 5. Relative humidity values and maximum, mean, and minimum temperatures during the spring experiment.

In each experiment, a completely randomized design with six replicates was implemented, with one tray assigned to each species as the experimental unit. After 33 days from sowing, the microgreens were harvested, and detailed plant height (cm) and fresh weight (g) measurements were taken as part of the agronomic evaluations. Following these measurements, the samples underwent chemical analysis to determine their nutritional content.

Total protein content (%) and nitrate levels (g L^{-1}) were assessed using the Kjeldahl method, following the A.O.A.C. Official Methods of Analysis [36]. Total phenols (ppm of GAE) were quantified using the Folin–Ciocalteu method as outlined in the A.O.A.C. Official Methods of Analysis [36]. For β-carotene content (mg equiv 100 g^{-1}), an organic solvent miscible in water, was employed to extract carotenoids, ensuring accurate quantification. Crude fiber content (%) was determined by the gravimetric method using a Velp fiber extractor [36–38].

Additionally, mineral contents—including iron (Fe), magnesium (Mg), calcium (Ca), sodium (Na), and phosphorus (P)—were measured in mg kg^{-1} using the electrothermal atomic absorption (EAA) method with Analytik Jena NOVAA 800 equipment. However, due to the small sample size in the spring experiment, only one repetition per species was conducted for chemical determination, which prevented statistical analysis of these variables.

To assess consumer acceptance, a sensory panel was conducted with 57 evaluators for the autumn experiment and 22 for the spring experiment. The panels evaluated all species based on flavor, texture, general appearance, and color using a nine-point hedonic scale, where one indicated "extremely pleasant" and nine indicated "extremely unpleasant" [39] (Table 8). In addition, consumers' purchase intentions and habitual consumption patterns were evaluated using a five-point scale (Table 8).

Table 8. Code and definition of the sensory scale and purchase preference for microgreens.

Code	Sensory Scale	Purchase Preference Scale
1	Extremely pleasant	I would definitely buy it
2	Very pleasant	I would probably buy it
3	Moderately pleasant	May be maybe not
4	Sightly pleasant	I probably would not buy it
5	It does not matter to me	I definitely would not buy it
6	Slightly unpleasant	-
7	Moderately unpleasant	-
8	Very unpleasant	-
9	Extremely unpleasant	-

Statistical analyses of the parametric variables were carried out using analysis of variance (ANOVA), while non-parametric variables were subjected to the Kruskal–Wallis test. When significance was found, mean separation was performed using the Tukey Honest Significant Difference (HSD) test (p = 0.05). Additionally, a correlation network analysis was conducted using the Rbio program to explore relationships between variables [40].

5. Conclusions

This study demonstrates that cultivating various microgreen species in unheated greenhouses in South Central Chile is a viable strategy to enhance both nutritional quality and consumer acceptance of fresh produce. The findings confirm that microgreens produced under these conditions consistently exhibit high levels of essential nutrients and bioactive compounds, such as proteins, phenols, and beta-carotene, which are critical for promoting human health. Notably, the variation in nutritional composition among species, such as the high phenol and β-carotene content observed in Swiss chard, carrots, and species from the Brassicaceae family, underscores the importance of selecting specific cultivars to target desired nutritional and functional outcomes. These compounds are particularly valued for their antioxidant activity, with β-carotene also contributing anti-inflammatory effects and a high vitamin A content, making these microgreens a potent functional food. The agronomic evaluation revealed that radish had the highest yield among the evaluated species, further highlighting its potential for commercial production. Additionally, the sensory evaluation indicated a strong consumer preference for microgreens, demonstrating a strong relationship between β-carotene content and appearance, as well as magnesium content with taste and texture. This interaction between organoleptic perception and nutritional composition reinforces the appeal of microgreens as a health-promoting food choice. From a broader perspective, this research validates the potential of microgreens as a healthy, nutritious, productive, and sustainable food option, offering an alternative solution to global food security challenges. The high acceptance of these products among local consumers suggests a significant opportunity for their expansion in regional markets. However, future research should address challenges related to exploring additional species, optimizing post-harvest handling, and further elucidating their direct benefits to human health. Finally, the successful cultivation of microgreens in unheated greenhouses not only enhances food diversity and nutritional quality but also presents a sustainable agricultural practice that aligns with local market demands. Ongoing research and development should focus on refining cultivation practices, expanding species diversity, and enhancing post-harvest strategies to maximize microgreens' health benefits and market potential, thereby contributing to more resilient and sustainable food systems.

Author Contributions: Conceptualization, P.R., G.C. and E.P.V.; methodology, P.R., G.C., C.M., P.G., G.R.S. and E.P.V.; software, P.R., P.G., G.R.S. and E.P.V.; validation, P.R., G.C., M.U., C.M., E.P.V. and F.F.-P.; formal analysis, G.C. and E.P.V.; investigation, P.R., G.C. and E.P.V.; resources, P.R., G.C. and E.P.V.; data curation, P.R., G.C., C.M. and E.P.V.; writing—original draft preparation, P.R., G.C., E.P.V. and C.M.; writing—review and editing, P.R., G.C., M.U., C.M., F.F.-P. and E.P.V.; visualization, G.C. and F.F.-P.; supervision, G.C., F.F.-P. and E.P.V.; project administration, G.C.; funding acquisition, G.C. All authors have read and agreed to the published version of the manuscript.

Funding: This work was supported through the Chilean projects FOVI-ANID 220031 and FIC Agricultura Vertical Hortícola No. BIP 40.036.334-0. Also, the Institutional Financial Support Program for National and International Mobility (PIAF mob) of the State University of Mato Grosso do Sul, Brazil.

Data Availability Statement: This article was written within the framework of confidential projects, which mandate the non-disclosure of data generated and analyzed. Consequently, the data supporting the reported results are not publicly available due to privacy restrictions.

Conflicts of Interest: The authors declare no conflicts of interest. The funders had no role in the design of the study, in the collection, analyses, or interpretation of data, in the writing of the manuscript, or in the decision to publish the results.

References

1. Xiao, Z.; Codling, E.E.; Luo, Y.; Nou, X.; Lester, G.E.; Wang, Q. Microgreens of Brassicaceae: Mineral Composition and Content of 30 Varieties. *J. Food Compos. Anal.* **2016**, *49*, 87–93. [CrossRef]
2. Partap, M.; Sharma, D.; Hn, D.; Thakur, M.; Verma, V.; Ujala, B.; Bhargava, B. Microgreen: A Tiny Plant with Superfood Potential. *J. Funct. Foods* **2023**, *107*, 105697. [CrossRef]
3. Carrasco, G.; Manríquez, P.; Galleguillos, F.; Fuentes-Peñailillo, F.; Urrestarazu, M. Evolution of Soilless Culture in Chile. In Proceedings of the Acta Horticulturae, International Society for Horticultural Science (ISHS). Leuven, Belgium, 9 September 2021; pp. 267–274.
4. Xiao, Z.; Lester, G.E.; Luo, Y.; Wang, Q. Assessment of Vitamin and Carotenoid Concentrations of Emerging Food Products: Edible Microgreens. *J. Agric. Food Chem.* **2012**, *60*, 7644–7651. [CrossRef] [PubMed]
5. Sharma, S.; Katoch, V.; Kumar, S.; Chatterjee, S. Functional Relationship of Vegetable Colors and Bioactive Compounds: Implications in Human Health. *J. Nutr. Biochem.* **2021**, *92*, 108615. [CrossRef] [PubMed]
6. Caracciolo, F.; El-Nakhel, C.; Raimondo, M.; Kyriacou, M.C.; Cembalo, L.; de Pascale, S.; Rouphael, Y. Sensory Attributes and Consumer Acceptability of 12 Microgreens Species. *Agronomy* **2020**, *10*, 1043. [CrossRef]
7. Zhang, X.; Bian, Z.; Yuan, X.; Chen, X.; Lu, C. A Review on the Effects of Light-Emitting Diode (LED) Light on the Nutrients of Sprouts and Microgreens. *Trends Food Sci. Technol.* **2020**, *99*, 203–216. [CrossRef]
8. Kyriacou, M.C.; El-Nakhel, C.; Graziani, G.; Pannico, A.; Soteriou, G.A.; Giordano, M.; Ritieni, A.; De Pascale, S.; Rouphael, Y. Functional Quality in Novel Food Sources: Genotypic Variation in the Nutritive and Phytochemical Composition of Thirteen Microgreens Species. *Food Chem.* **2019**, *277*, 107–118. [CrossRef]
9. Choe, U.; Yu, L.L.; Wang, T.T.Y. The Science behind Microgreens as an Exciting New Food for the 21st Century. *J. Agric. Food Chem.* **2018**, *66*, 11519–11530. [CrossRef]
10. Caruso, G.; Stoleru, V.; De Pascale, S.; Cozzolino, E.; Pannico, A.; Giordano, M.; Teliban, G.; Cuciniello, A.; Rouphael, Y. Production, Leaf Quality and Antioxidants of Perennial Wall Rocket as Affected by Crop Cycle and Mulching Type. *Agronomy* **2019**, *9*, 194. [CrossRef]
11. Bhaswant, M.; Shanmugam, D.K.; Miyazawa, T.; Abe, C.; Miyazawa, T. Microgreens—A Comprehensive Review of Bioactive Molecules and Health Benefits. *Molecules* **2023**, *28*, 867. [CrossRef]
12. Ferrón-Carrillo, F.; Guil-Guerrero, J.L.; González-Fernández, M.J.; Lyashenko, S.; Battafarano, F.; da Cunha-Chiamolera, T.P.L.; Urrestarazu, M. LED Enhances Plant Performance and Both Carotenoids and Nitrates Profiles in Lettuce. *Plant Foods Hum. Nutr.* **2021**, *76*, 210–218. [CrossRef] [PubMed]
13. Fuentes-Peñailillo, F.; Gutter, K.; Vega, R.; Silva, G.C. New Generation Sustainable Technologies for Soilless Vegetable Production. *Horticulturae* **2024**, *10*, 49. [CrossRef]
14. Parkes, M.G.; Azevedo, D.L.; Cavallo, A.C.; Domingos, T.; Teixeira, R.F.M. Life Cycle Assessment of Microgreen Production: Effects of Indoor Vertical Farm Management on Yield and Environmental Performance. *Sci. Rep.* **2023**, *13*, 11324. [CrossRef] [PubMed]

15. Carrasco, G.; Fuentes-Peñailillo, F.; Pérez, R.; Rebolledo, P.; Manríquez, P. An Approach to a Vertical Farming Low-Cost to Reach Sustainable Vegetable Crops. In Proceedings of the 2022 IEEE International Conference on Automation/XXV Congress of the Chilean Association of Automatic Control (ICA-ACCA), Curico, Chile, 24–28 October 2022; pp. 1–6.
16. Maluin, F.N.; Hussein, M.Z.; Nik Ibrahim, N.N.L.; Wayayok, A.; Hashim, N. Some Emerging Opportunities of Nanotechnology Development for Soilless and Microgreen Farming. *Agronomy* **2021**, *11*, 1213. [CrossRef]
17. Ying, Q.; Kong, Y.; Zheng, Y. Growth and Appearance Quality of Four Microgreen Species under Light-Emitting Diode Lights with Different Spectral Combinations. *HortScience* **2020**, *55*, 1399–1405. [CrossRef]
18. Frąszczak, B.; Kleiber, T. Microgreens Biometric and Fluorescence Response to Iron (Fe) Biofortification. *Int. J. Mol. Sci.* **2022**, *23*, 14553. [CrossRef]
19. Hoang, G.M.; Vu, T.T. Selection of Suitable Growing Substrates and Quality Assessment of Brassica Microgreens Cultivated in Greenhouse. *Acad. J. Biol.* **2022**, *44*, 133–142. [CrossRef]
20. Sharma, A.; Hazarika, M.; Heisnam, P.; Pandey, H.; Devadas, V.S.; Wangsu, M.; Kartha, B.D. Factors Affecting Production, Nutrient Translocation Mechanisms, and LED Emitted Light in Growth of Microgreen Plants in Soilless Culture. *ACS Agric. Sci. Technol.* **2023**, *3*, 701–719. [CrossRef]
21. Kołota, E.; Adamczewska-Sowińska, K.; Czerniak, K. Yield and Nutritional Value of Swiss Chard Grown for Summer and Autumn Harvest. *J. Agric. Sci.* **2010**, *2*, 120–124.
22. Abaajeh, A.R.; Kingston, C.E.; Harty, M. Environmental Factors Influencing the Growth and Pathogenicity of Microgreens Bound for the Market: A Review. *Renew. Agric. Food Syst.* **2023**, *38*, e12. [CrossRef]
23. Ghoora, M.D.; Babu, D.R.; Srividya, N. Nutrient Composition, Oxalate Content and Nutritional Ranking of Ten Culinary Microgreens. *J. Food Compos. Anal.* **2020**, *91*, 103495. [CrossRef]
24. Kommeroski, M.R.; Portal, K.A.; Comiotto, J.; Klug, T.V.; Flores, S.H.; Rios, A.D.O. Nutritional Quality and Bioactive Compounds of Arugula (*Eruca Sativa* L.) Sprouts and Microgreens. *Int. J. Food Sci. Technol.* **2023**, *58*, 5089–5096. [CrossRef]
25. Kyriacou, M.C.; Soteriou, G.A.; Colla, G.; Rouphael, Y. The Occurrence of Nitrate and Nitrite in Mediterranean Fresh Salad Vegetables and Its Modulation by Preharvest Practices and Postharvest Conditions. *Food Chem.* **2019**, *285*, 468–477. [CrossRef] [PubMed]
26. Turner, E.R.; Luo, Y.; Buchanan, R.L. Microgreen Nutrition, Food Safety, and Shelf Life: A Review. *J. Food Sci.* **2020**, *85*, 870–882. [CrossRef]
27. Fabek Uher, S.; Radman, S.; Opačić, N.; Dujmović, M.; Benko, B.; Lagundžija, D.; Mijić, V.; Prša, L.; Babac, S.; Šic Žlabur, J. Alfalfa, Cabbage, Beet and Fennel Microgreens in Floating Hydroponics—Perspective Nutritious Food? *Plants* **2023**, *12*, 2098. [CrossRef]
28. Rahman, S.M.E.; Mele, M.A.; Lee, Y.-T.; Islam, M.Z. Consumer Preference, Quality, and Safety of Organic and Conventional Fresh Fruits, Vegetables, and Cereals. *Foods* **2021**, *10*, 105. [CrossRef]
29. Vogel, J.T.; Tieman, D.M.; Sims, C.A.; Odabasi, A.Z.; Clark, D.G.; Klee, H.J. Carotenoid Content Impacts Flavor Acceptability in Tomato (*Solanum Lycopersicum*). *J. Sci. Food Agric.* **2010**, *90*, 2233–2240. [CrossRef]
30. Leksrisompong, P.P.; Whitson, M.E.; Truong, V.D.; Drake, M.A. Sensory Attributes and Consumer Acceptance of Sweet Potato Cultivars with Varying Flesh Colors. *J. Sens. Stud.* **2012**, *27*, 59–69. [CrossRef]
31. Tagliapietra, B.L.; Zanon, A.J.; Tironi, L.F.; Streck, N.A.; Richards, N.S.P.D.S. Nutritional Quality and Sensory Acceptance of Biofortified Cassava. *Braz. J. Food Technol.* **2021**, *24*, e2020247. [CrossRef]
32. Olusanya, R.N.; Kolanisi, U.; Ngobese, N.Z. Mineral Composition and Consumer Acceptability of Amaranthus Leaf Powder Supplemented Ujeqe for Improved Nutrition Security. *Foods* **2023**, *12*, 2182. [CrossRef]
33. Ahmed, N.; Zhang, B.; Bozdar, B.; Chachar, S.; Rai, M.; Li, J.; Li, Y.; Hayat, F.; Chachar, Z.; Tu, P. The Power of Magnesium: Unlocking the Potential for Increased Yield, Quality, and Stress Tolerance of Horticultural Crops. *Front. Plant Sci.* **2023**, *14*, 1285512. [CrossRef] [PubMed]
34. Sarker, U.; Hossain, M.M.; Oba, S. Nutritional and Antioxidant Components and Antioxidant Capacity in Green Morph *Amaranthus* Leafy Vegetable. *Sci. Rep.* **2020**, *10*, 1336. [CrossRef] [PubMed]
35. Fayezizadeh, M.R.; Ansari, N.A.; Sourestani, M.M.; Hasanuzzaman, M. Biochemical Compounds, Antioxidant Capacity, Leaf Color Profile and Yield of Basil (*Ocimum* sp.) Microgreens in Floating System. *Plants* **2023**, *12*, 2652. [CrossRef] [PubMed]
36. AOAC International. *Official Methods of Analysis*, 15th ed.; Helrich, K., Ed.; Oxford University Press: Arlington, VA, USA, 1990; ISBN 0-935584-42-0.
37. Food and Agriculture Organization. *FAO Food and Nutrition Papers: 14/7 Food Analysis: General Techniques, Additives, Contaminants and Composition*; FAO: Rome, Italy, 1986.
38. Martínez Guijarro, M.R. *Análisis Instrumental. Espectrometría de Absorción Atómica (EAA)*; Universitat Politècnica de València: Valencia, Spain, 2020.

39. Meilgaard, M.C.; Carr, B.T.; Civille, G.V. *Sensory Evaluation Techniques*, 3rd ed.; CRC Press: Boca Raton, FL, USA, 1999; ISBN 9781003040729
40. Lopes Bhering, L. Rbio: A Tool for Biometric and Statistical Analysis Using the R Platform. *Crop Breed. Appl. Biotechnol.* **2017**, *17*, 187–190. [CrossRef]

plants

MDPI

Article

Nutrient Use Efficiency and Cucumber Productivity as a Function of the Nitrogen Fertilization Rate and the Wood Fiber Content in Growing Media

Rita Čepulienė, Lina Marija Butkevičienė and Vaida Steponavičienė *

Department of Agroecosystems and Soil Sciences, Vytautas Magnus University, K. Donelaičio Street 58, LT-44248 Kaunas, Lithuania; rita.cepuliene@vdu.lt (R.Č.); lina.butkeviciene@vdu.lt (L.M.B.)

* Correspondence: vaida.steponaviciene@vdu.lt

Abstract: A peat substrate is made from peat from drained peatlands, which is a limited resource. A realistic estimate is that 50% of the world's wetlands have been lost. Peat is used in horticulture, especially for the cultivation of vegetables in greenhouses. The consequences of peatland exploitation are an increase in the greenhouse effect and a decrease in carbon stocks. Wood fiber can be used as an alternative to peat. The chemical properties of growing media interact and change continuously due to the small volume of growing media, which is limited by the growing container. This study aims to gain new knowledge on the impact of nutrient changes in the microbial degradation of carbon compounds in wood fiber and mixtures with a peat substrate on the content and uptake of nutrients required by plants. The cucumber (*Cucumis sativus* L.) variety 'Dirigent H' developed in the Netherlands was cultivated in growing media of a peat substrate and wood fiber: (1) peat substrate (PS); (2) wood fiber (WF); (3) wood fiber and peat substrate 50/50 *v*/*v* (WF/PS 50/50); (4) wood fiber and peat substrate 25/75 *v*/*v* (WF/PS 25/75). The rates of fertilization were the following: (1) conventional fertilization (CF); (2) 13 g N per plant (N_{13}); (3) 23 g N per plant (N_{23}); (4) 30 g N per plant (N_{30}). The experiment was carried out with three replications. As the amount of wood fiber increased, the humidity and pH of the growing media increased. The fertilization of the cucumbers with different quantities of nitrogen influenced the nutrient uptake. The plants grown in the 50/50 and 25/75 growing media had the best Cu uptake when fertilized with N_{23}. When the plants grown in the wood fiber media and the 50/50 media were fertilized with N_{13}, N_{23}, and N_{30}, the Mn content in the growing media at the end of the growing season was significantly lower than the Mn content in the media with conventional fertilization. Thus, nitrogen improved the uptake of Mn by the plants grown not only in the wood fiber, but also in the combinations with a peat substrate. Growing plants in wood fiber and fertilizing them with N_{13} can result in the optimum uptake of micronutrients. The number and biomass of cucumber fruits per plant were influenced by the amount of wood fiber in the growing media and the application of nitrogen fertilizer. The highest number of fruits and biomass of fruits per plant obtained were significantly higher when the cucumbers were grown in WF/PS 50/50 growing media with additional N_{13} fertilization.

Keywords: wood fiber; cucumber productivity; nutrient uptake; nitrogen fertilization; physicochemical properties of growing media

1. Introduction

The decline in arable land is putting pressure on agricultural producers. This is not only linked to changes in the ecosystem—soil degradation and climate change—but also to increasing urbanization. These changes are forcing farmers to look for alternatives to produce more from less land per unit area. These global changes have led to a significant increase in soilless crop production in recent years [1].

Soilless crop production can be 2–5 times more productive, with 10 times less water consumption compared to that of field production. This results in year-round production, a better taste, and a higher nutritional value [2]. Switching from an outdoor soil-based crop production system to a soilless system can improve water use efficiency, especially in closed-loop systems with a recirculating solution of water and nutrients that is collected in a closed drainage water tank for reuse [3,4]. Another reason is related to ecology and environmental protection; the intensive destruction of peatland areas that are important for the ecosystem and the use of artificial substrates cause waste accumulation problems and environmental pollution. Peat has been the dominant organic component of growing media in many parts of the world for the last 50 years due to its excellent physical properties, such as excellent air and water retention, low pH and salinity, and protection against pests and diseases [5]. Due to its ease of use and beneficial properties, peat has, over time, become the key component of growing media and still dominates the greenhouse sector. Worldwide, around 40 million m^3 of peat is used annually in horticulture [6], with Germany (8.5 million m^3) and the UK (2.5 million m^3) being the main consumers of peat in horticulture [7]. Other resources, such as treated and untreated waste wood or its bark, can also be used as growing media or as the components of growing media, and renewable raw materials have a lot of potential for their use in agriculture for growing plants in closed systems without soil. The preliminary studies by Dutch researchers have shown that the performance of five peat–pine bark mixtures in the region is comparable to that of perlite at the vegetative and early fruiting stages of cucumbers [4,8,9]. Waste management strategies to reduce, reuse, and recycle waste should be further and more rigorously applied in soilless cropping systems [2,4,9]. Wood-based substrates, including pine bark and wood fiber, have good air permeability and high saturated hydraulic conductivity [10]. Wood-based substrates are more sustainable than peat and more readily available and cheaper than coir [8]. Wood fiber or sawdust is a by-product of the wood industry, produced from clean, broken wood. Wood substrates are typically produced by chopping and grinding a tree, as well as from the leaves, twigs, branches, and needles removed during forest thinning. Sawdust is a by-product of wood that remains after production operations and has been used as an ingredient in other growing media materials, especially where peat is scarce or expensive. Wood sawdust can be used as a component of growing media with peat [4,11]. Wood fiber is produced using mechanical defibrillation or, most commonly, water vapor by squeezing clean wood chips through a thermoplastic press. Wood fiber can be compressed for ease of transport [12]. Using wood fiber as a growing medium dates back to 2000 in Spain, France, and Germany [13]. New wood waste recycling plants have been established in the USA, Ireland, Germany, the UK, and Latvia [14]. Materials that are easy to produce, financially viable, environmentally friendly, and capable of providing quality growing media will become the future substitutes for rockwool and peat.

Climate change, soil degradation, and other ecological problems will drive the development of soilless cropping systems and the choice of growing media in the near future. Growing crops in soilless culture systems involves high energy consumption for heating in the cold season and artificial lighting, expensive greenhouse construction [15,16], fertilizer use [3], post-harvest transportation, and packaging [17]. On the other hand, SCSs (soilless culture systems) help to alleviate many of the problems associated with conventional in situ cultivation, such as soil-borne diseases and pests, and the precise control of water and fertilizer requirements [18,19]. The resulting high-quality production and higher yields in a shorter time justify the production costs in these systems [20].

Innovative, sustainable, and renewable materials, such as compost, bark, wood, coir, and other organic wastes, with low production costs and short transport routes may become the growing media of the future and dominate the market for growing media raw materials. In the future, waste and organic materials will be highly competitive materials, similar to wood and bark today [21]. As an alternative to synthetic and peat-based substrates, wood fiber from wood processing can be used, but further research is needed.

Nutrient availability is one of the main factors determining the suitability of organic substrates for plant growth [22], which may depend on their elemental composition and other factors affecting nutrient forms and dynamics, including the presence of dissolved organic compounds, the biological stability of the growing media, and the adsorption capacity [23]. The availability of nutrient elements, including magnesium and calcium, directly influences the rate of synthesis of certain metabolites in plants [24,25]. Wood fiber in growing media can influence the uptake of nutrients by a plant, as it has a nutrient storage function [26]. Nutrient cycling in wood fiber growing media is a complex and dynamic process influenced by various environmental factors and plant species, making it difficult to predict nutrient availability and uptake [27]. However, wood fiber is a suitable material for horticultural growing media due to its fibrous structure, hydrophilic nature, and low nutrient or growth-inhibiting content compared to those of other components such as bark [28].

Given the above, this study was carried out to identify a suitable soilless media made of wood fiber and peat, select optimum fertilization to compensate for nitrogen immobilization, and ensure optimum nutrient uptake and physicochemical properties for the growing media.

2. Materials and Methods

2.1. Site Description and Experiment Design

This experiment was conducted in Lithuania at Vytautas Magnus University Agriculture Academy, Joint Research Centre of Agriculture and Forestry, in a regulated-climate greenhouse and the Soil and Crop Ecology Laboratory of Experimental Station on 2021. The cucumbers were grown for 97 days. The cucumbers selected for the experiment with different growing media under controlled climate conditions were common short cucumbers (*Cucumis sativus* L.), variety 'Dirigent H', developed in the Netherlands. The cucumber seeds were sown in 28 × 38 cm plastic tray with 24 holes of 5.5 × 6 × 6 cm on 21 May 2021. The tray was filled with peat substrate for seed germination, consisting of a peat fraction of up to 7 mm. The substrate was enriched with 2.0 g L^{-1} fertilizer $N_{14}P_{16}K_{18}$. The pH of the germination substrate was 6. The seedlings were grown from seeds in a climate-controlled chamber with lighting, RUMED1303. The temperature of the germination chamber was 25 °C, and the humidity was 80%. As the seedlings produced their second true leaf, they were transferred to the greenhouse and planted one plant at a time in 5 L plastic growing containers (D—23 cm (SBH)) filled with test growing media. Four growing media (four treatments of the experiment (factor A)) were studied: (1) 100% peat substrate (PS) (control substrate); (2) 100% wood fiber (WF); (3) 50/50% wood fiber/peat by volume (WF/PS 50/50); and (4) 25/75% wood fiber/peat by volume (WF/PS 25/75). The peat substrate was produced from a fraction of *Sphagnum* peat with a particle size <20 mm by the company Rekyva (Lithuania). Wood fiber was produced by Lithuanian company Nereta only from softwood highest-quality chips (residues of the lumber industry). Wood fiber is RAL-certified. Both the mixtures of wood fiber and peat were prepared just before the start of the experiment (company Rekyva, Lithuania). The wood fiber density was 0.16 g cm^{-3}, the air dry moisture content was 41%, and the water absorption total was 2.1 g g^{-1}. The chemical composition of the wood fiber and the peat substrate was analyzed in a laboratory of the Lithuanian Agricultural Advisory Service and in an agrochemical research laboratory of the Lithuanian Research Centre for Agriculture and Forestry. The characteristics of the materials used for the cucumber-growing media are presented in Table 1.

The growing containers were arranged in rows on either side of the irrigation pipe. Each row contained 12 growing containers spaced 30 cm apart. The whole experiment consisted of 16 rows, i.e., 192 growing containers. One irrigation capillary (drip sprinkler) was introduced into each growing container (Figure 1).

Each treatment consisted of 48 growing containers, with one cucumber plant each. Each treatment of the growing media (substrate) was divided into four groups with four fertilization backgrounds (factor B): (1) conventional fertilization (control); (2) N_{13} as an additional fertilization—13 g N per plant; (3) N_{23} as an additional fertilization—23 g N per plant; and (4) N_{30} as an additional fertilization—30 g N per plant.

The experiment was carried out with three replications. The growing containers were arranged in a completely randomized design.

The average daily temperature in the greenhouse during the experiment was 25 °C, the humidity was 65%, and the daily light integral was 21.8 mol·m^{-2} d^{-1}. The cucumbers were watered every two hours. Water was supplied to each plant using a capillary irrigation system. The growing containers were protected with polyethylene film to prevent excess water from draining into the drainage system. The plants were grown as a single stem throughout the study. The cucumbers grew upwards using polypropylene ropes.

The experiment was carried out for 80 days after transplanting the seedlings into the greenhouse.

Table 1. Characteristics of wood fiber and *Sphagnum* peat used in experiment as growing media and their parts.

Parameter	Wood Fiber	*Sphagnum* Peat
pH	4.6	5.3
EC mS m^{-1}	58	76
C:N	435	80
N total % DW *	0.22	1.08
P mg kg^{-1} DW	57	181
K mg kg^{-1} DW	472	674
Mg mg kg^{-1} DW	29.3	23.2
Fe mg kg^{-1} DW	113.0	86.2
Cu mg kg^{-1} DW	2.9	3.0
Zn mg kg^{-1} DW	10.4	12.4
B mg kg^{-1} DW	15.3	14.0
Cl mg kg^{-1} DW	117.0	29.9

*—DW, these parameters were determined as dry weight.

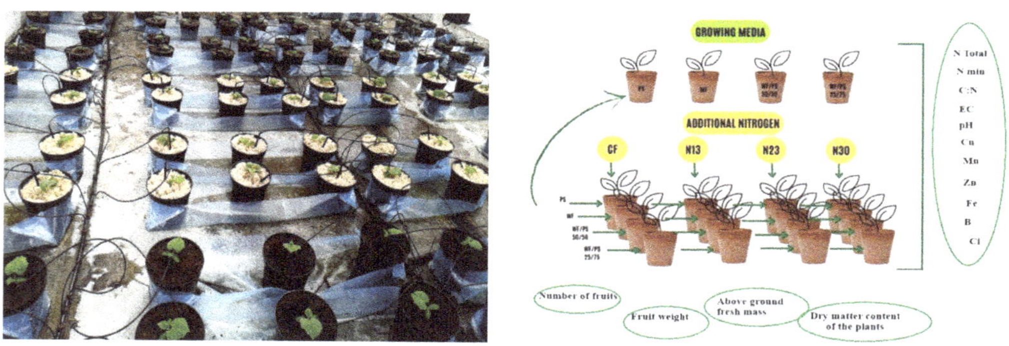

Figure 1. Experiment after transplanting cucumbers into growing containers and experimental design.

2.2. Plant Fertilization

In the experiment, conventional fertilization was carried out with the mineral fertilizer YaraMila® COMPLEX NPK 12-11-18 with microelements. This complex fertilizer contains all the essential nutrients needed for the plants. The chemical composition of the fertilizer is as follows: nitrogen (N): 12%; phosphorus (P_2O_5/P): 11/4.8%; potassium (K_2O/K): 18/14.9%; magnesium (MgO/Mg): 2.7/1.8%; sulfur (SO_3^-/S): 20/8%; boron (B): 0.015%; iron (Fe): 0.20%; manganese (Mn): 0.02%; and zinc (Zn): 0.02%. A total of 1.60 kg of complex fertilizer was dissolved in 1 m^3 of water. Until they flowered, the cucumbers were fertilized once a week with 500 mL of the nutrient solution per plant. The cucumbers were fertilized twice a week with this fertilizer during the later stages of growth until the end of the growing season. For additional nitrogen fertilization, water-soluble calcium nitrate ($Ca(NO_3)_2$) was used. The chemical composition of the fertilizer was as follows: total nitrogen (N)—15.5% ((nitrate nitrogen ($N–NO_3$)—14.4%) and ammonia nitrogen ($N–NH_4$)—1.1%); and calcium oxide (CaO)—26.3% (calcium—18.8%). Calcium nitrate was applied once a week (11 times every 7 days) to the plants during the growing season to provide them with the additional nitrogen required in the experiment (Figure 2).

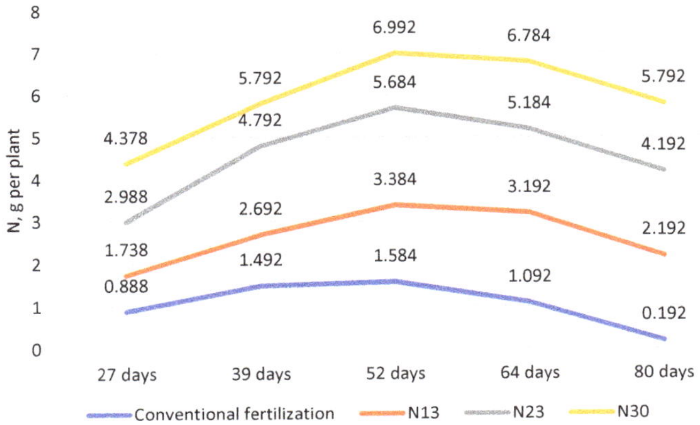

Figure 2. Nitrogen fertilization during cucumber growing season (days after transplanting). N13, N23, and N30—additional nitrogen fertilization.

2.3. Measurement Properties of Growing Media

The agrochemical parameters of the growing media were determined at the end of the experiment. Samples of the growing media were taken from the two growing containers of each treatment after cutting the aboveground part of the plant for agrochemical analyses.

Organic carbon and nitrogen in the growing media were determined according to standard methods. Organic carbon (C) was determined using the Tyurin spectrophotometric method [24]. The sample was heated with a solution of potassium bichromate in sulfuric acid (+160 °C). The final measurement of C at 590 nm was carried out on a Cary 60 spectrophotometer (Varian, Palo Alto, CA, USA) using glucose standard solutions. The total nitrogen (N_{total}) content of the soil was determined by the Kjeldahl method (the Kjeldahl apparatus Vapodest), with a spectrophotometric measurement at 655 nm. The mineral nitrogen (N_{min}) content was determined spectrometrically using sulfosalicylic acid [29].

Copper (Cu) and zinc (Zn) contents were determined according to EN 13650:2003 and EN 8288:2002. Manganese (Mn) and total iron (Fe) were determined according to EN 13650:2003, AOAC 974.27. Boron (B) and chlorine (Cl) were determined using LST EN 13650:2003, LST EN ISO 11885:2009. Chemical analysis of nutrients was carried out at the Agrochemical Research Laboratory of the Lithuanian Research Centre for Agriculture and Forestry. All nutrients were determined in three replications.

The electrical conductivity (EC), density (D), and moisture content (M) of the growing media were measured with a Delta-T Device HH2 Moisture Meter (Cambridge, UK) at a depth of 10 cm in each growing container six times per growing season. The acidity (pH) of the growing media was determined by potentiometric determination in a 1 N KCl extract according to ISO 10390:2005.

The variation (%) in micronutrient content in the growing media during the growing season was calculated by summing the nutrient content of the growing media (dry matter) at the start of the experiment with the nutrient content of the fertilizer. This amount is equivalent to 100%. The micronutrient content (%) of the growing media (dry matter) at the end of the experiment, minus 100, was taken as nutrient uptake.

2.4. Plant Growth and Yield Measurements

Cucumber productivity was determined by picking, weighing, and counting the marketable fruit (17–20 cm) from each cucumber treatment separately. The cucumber fruits were harvested twice a week. Total cucumber productivity during the growing season was determined at the end of the experiment (80 days after transplanting) by converting the results obtained into productivity in kg per plant.

At the end of each stage of the study, the plants were cut from the two growing containers of each treatment, and the green biomass of the aboveground part was determined. The plants with all their morphological parts were weighed and crushed. The dry biomass of the aboveground part of the plant was determined by drying at 105 °C in a drying oven Memmert (Schwabach, Germany) to a constant biomass.

2.5 Statistical Analysis

All data were statistically analyzed using the computer program package STATISTICA10 using the method of two-way (growing media × fertilization rate) analysis of variance (ANOVA). Statistically significant differences in the data were determined by Fisher's and the least significant difference (LSD) tests at the probability levels of 95% ($p < 0.05$) and 99% ($p < 0.01$). Principal component analysis (PCA) was performed using principal component and classification methods. PCA is a widely used statistical technique that simplifies complex multidimensional data into a smaller set of interpretable variables called principal components.

3. Results

3.1. Cucumber Productivity and Physicochemical Properties of Growing Media

The moisture content of the cucumber growing media depended on the amount of wood fiber and additional N, but significant differences were found in the non-factor-related interaction (growing media × fertilization rate) (Table 2). Increasing the amount of wood fiber in the media increased the moisture content from 9.8 to 35.0% compared to the moisture content of the peat substrate. Additional N fertilization significantly reduced the moisture content of the cucumber growing media by up to 37.6% compared to that of the growing media without additional fertilization. The density of the growing media was influenced only by its composition, decreasing with an increasing wood fiber content. Electrical conductivity (EC) depended on the interaction of the factors, on the proportion of wood fiber in the growing media, and on the addition of N fertilizer. The highest EC value was found in the growing media mixture WF/PS 25/75. The index decreased with an increasing fiber content in the media. The EC value in the wood fiber media was higher than that in the peat substrate, but lower compared to the EC value in the WF/PS 50/50 media. The additional fertilization with N_{30} increased the EC value by a factor from 1.1 to 3.3 compared to the EC value in the media without additional fertilization. A higher proportion of the wood fiber in the growing media had an alkalizing effect by increasing the pH, while the additional fertilizer $N_{13–30}$ significantly decreased the pH of the media.

The dry biomass and green biomass of the cucumber plants were significantly influenced by the interaction of factors. The higher proportion of wood fiber in the media (WF/PS 50/50 and WF) and the additional fertilization with higher rates of $N_{23,30}$ significantly reduced the dry biomass content of the aboveground biomass of the plants. The aboveground green biomass was significantly higher only when the cucumbers were grown on the peat substrate without additional N fertilization. Increasing the proportion of wood fiber and N fertilization in the growing media resulted in a significant decrease in the average aboveground green biomass of the plants. The amount of fruit and biomass of cucumber fruits on the plant were influenced by the amount of wood fiber in the growing media and additional N fertilization. The highest number of fruits was produced on the plant when they were grown in a 50/50 mixture of peat substrate and wood fiber WF/PS, with additional fertilization with the lowest rate of nitrogen N_{13}. The cucumbers were also significantly higher in fruit biomass when grown in a WF/PS 50/50 mixture without additional fertilizer and with the lowest rate of additional N_{13}.

Table 2. The indicators of cucumber productivity and the physicochemical properties of the growing media as a function of the nitrogen fertilization rate and the wood fiber content in the growing media.

Factors		Physicochemical Properties of Growing Media				Indicators of Cucumber Productivity			
Growing Media (A)	N Fertilization Rate (B)	Moisture, %	Density, g cm^{-2}	EC, mS m^{-1}	pH	Dry Biomass Content of the Plants, %	Aboveground Fresh Biomass, g per Plant	Number of Fruits, per Plant	Fruit Biomass, kg per Plant
				Growing media					
PS		41.48 c	0.39 a	399 c	5.18 c	18.95 a	195.6 a	35.15 b	5.61 b
WF/PS 25/75		45.55 c	0.39 a	513 a	5.85 b	17.43 b	164.1 d	34.85 b	5.48 b
WF/PS 50/50		50.98 b	0.30 b	370 d	6.55 a	16.48 c	187.6 b	41.18 a	6.22 a
WF		55.99 a	0.23 b	420 b	5.98 b	14.30 d	172.0 c	36.80 b	5.68 b
				Fertilization					
	CF	55.40 a	0.33 a	172 d	6.18 a	17.13 ab	212.9 a	36.61 b	6.06 a
	N_{13}	54.90 a	0.33 a	434 c	5.88 b	17.25 a	204.0 a	40.00 a	6.03 a
	N_{23}	49.14 b	0.33 a	522 b	5.80 b	16.13 c	157.5 b	34.55 b	5.40 b
	N_{30}	34.55 c	0.33 a	575 a	5.70 b	16.65 b	144.9 c	36.82 b	5.51 b
				Influence and interactions					
A		**	*	**	**	**	**	**	**
B		**	ns	**	**	**	**	**	**
A × B		ns	ns	**	ns	**	**	**	**

Note: peat substrate (PS); wood fiber/peat by volume (WF/PS 50/50); wood fiber/peat by volume (WF/PS 25/75); wood fiber (WF); conventional fertilization (CF). Data with the same letters for Factor A and Factor B are not significantly different at $p < 0.05$. * and **—the level of statistical significance at $p < 0.05$ and $p < 0.01$, respectively; ns—not significant.

Principle component analysis (PCA) showed that the physicochemical properties of the growing media were altered by increasing the proportion of wood fiber and the quantity of nitrogen fertilizer and that the productivity of cucumbers depended on them (Figure 3). Only the pH of the growing media had no significant effect on the cucumbers' performance, irrespective of the rate of nitrogen fertilization. The other physicochemical properties of the growing media were correlated in distinct groups. When the cucumbers were grown without supplementary fertilization (CF), the green aboveground biomass and the fruit content of the plants correlated strongly with the moisture content of the growing media, while the dry biomass content of the aboveground biomass was influenced by the density of the media. There was a strong correlation between the EC of the growing media and the biomass of fruit on the plants, but a negative correlation was found between this and the moisture of the growing media.

With additional N_{13} fertilization, a strong correlation was obtained in the group between the EC of the growing media and the biomass of the fruit on the plants, with a weaker correlation with the moisture content of the media. The other strongly correlated group was the plant aboveground dry biomass and the density of the growing media. The cucumber fruit content and plant aboveground green biomass were negatively correlated with the EC of the growing media.

When the nitrogen fertilizer rate was increased to N_{23}, only the dry biomass content of the aboveground plant correlated with the density of the growing media. The other cucumber performance parameters were not affected by the physical characteristics. When the cucumbers were fertilized at the maximum rate of N_{30}, the largest group of correlated parameters was the aboveground dry biomass of the plant with the electrical conductivity and density of the growing media. In another group, there was a strong correlation between the cucumber fruit biomass and the moisture content of the growing media.

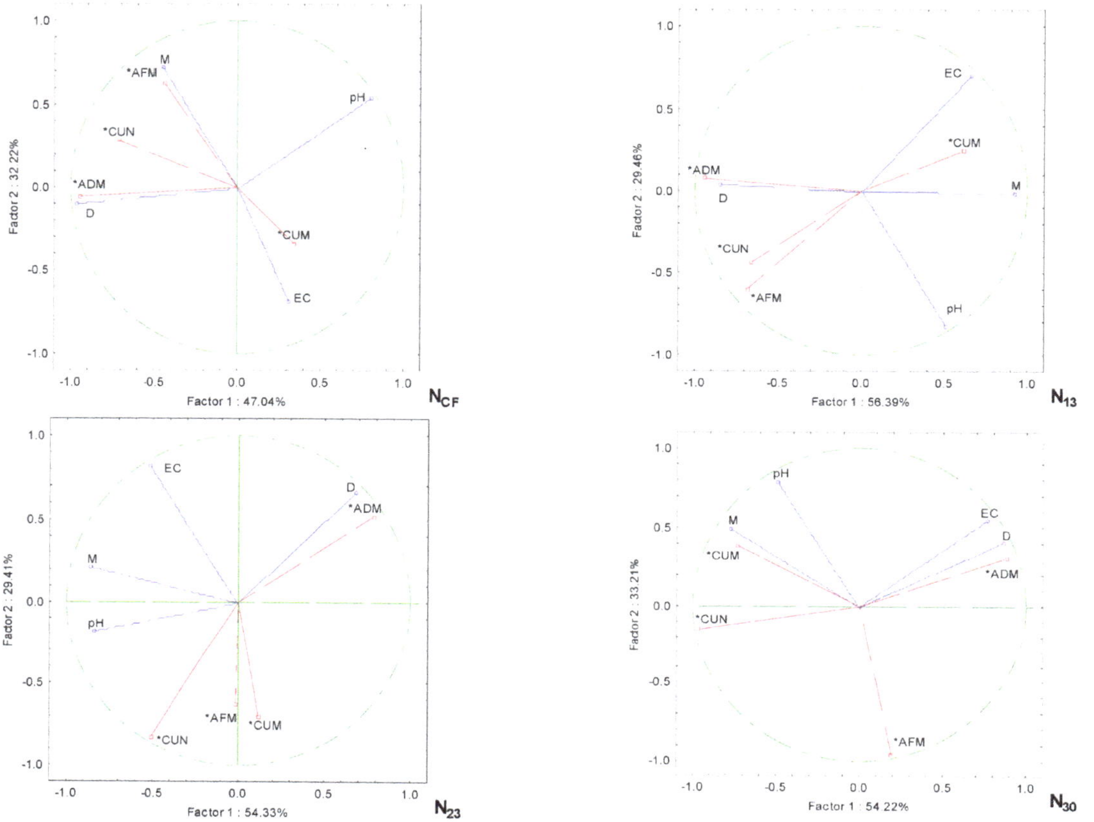

Figure 3. PCA analysis shows the relationship between the physicochemical properties of the growing media with wood fiber and the productivity parameters for the cucumbers grown under different nitrogen fertilization rates. Note: the density of the growing media (D), the electrical conductivity of the growing media (EC), the moisture of the growing media (M), the aboveground dry biomass content of the plants (ADM), the aboveground fresh biomass content of the plants (AFM), the number of cucumbers per plant (CUN), the cucumber biomass per plant (CUM), conventional fertilization (CF), the fertilization rates (N_{13}, N_{23} and N_{30}), and active variables (*).

3.2. Micronutrient Content, Nitrogen Content, and C:N at the End of the Growing Season in the Growing Media

This study found that the highest uptake of micronutrients during the growing season was observed in the plants grown in the wood fiber (Table 3). The plants grown in this medium had a 37 percentage point higher copper uptake than the plants grown in the peat substrate with conventional fertilization. The plants grown in wood fiber showed the best uptake of zinc and manganese when fertilized with the highest N fertilization rate of N_{30}, which was 13 and 28 percentage points, respectively, compared to that of conventional fertilization in wood fiber. Only the iron uptake of the plants grown in wood fiber decreased with increasing nitrogen levels. The plants fertilized with N_{30} nitrogen absorbed 15 percentage less iron than the plants grown in the wood fiber with conventional fertilization, but 6 percentage more than the plants grown on the peat substrate with conventional fertilization.

Table 3. The variation (%) in micronutrient content in the growing media during the growing season, taking into consideration the amount obtained with fertilizer.

Growing Media	Fertilization Rate	Cu	Zn	Mn	Fe	B	Cl
PS	CF	−2	−67	−18	−65	1	853
	N_{13}	−4	−66	−37	−80	21	586
	N_{23}	16	−42	−36	−78	76	870
	N_{30}	−11	−28	−26	−64	44	1044
WF/PS 25/75	CF	13	−45	−1	−59	31	14
	N_{13}	7	−33	−9	−74	50	97
	N_{23}	−3	−36	−2	−77	56	91
	N_{30}	35	−24	−7	−59	102	120
WF/PS 50/50	CF	−31	−47	−14	−47	1	−9
	N_{13}	−25	−52	−22	−49	26	14
	N_{23}	−33	−48	−22	−63	36	119
	N_{30}	−20	−40	−21	−63	81	144
WF	CF	−39	−62	−60	−86	−18	118
	N_{13}	−39	−70	−77	−82	27	150
	N_{23}	−39	−53	−71	−62	25	232
	N_{30}	−39	−75	−78	−71	12	269

Note: a negative value indicates that the micronutrients were taken up by the plant. PS—peat substrate; WF/PS 50/50—wood fiber/peat by volume; WF/PS 25/75—wood fiber/peat by volume; WF—wood fiber; CF—conventional fertilization.

A higher nitrogen content N_{30} interfered with the uptake of copper and zinc by the plants grown on the peat substrate and on the WF/PS 50/5 and WF/PS 25/75 wood fiber media. The maximum N_{30} fertilization on the peat substrate reduced the zinc uptake by 39 percentage points, by 7 percentage points on the WF/PS 50/50 growing media, and by 21 percentage points on the WF/PS 25/75 growing media compared with the zinc uptake by the plants given the conventional fertilization. Thus, the plants grown in these growing media showed the best uptake of zinc under conventional fertilization. The plants grown in WF/PS 50/50 and WF/PS 25/75 had the best uptake of copper when fertilized with N_{23}. Copper uptake increased by 2 and 16 percentage points, respectively, compared to that under conventional fertilization.

Manganese and iron uptake by the plants was better when fertilized with increased nitrogen. The plants grown on the peat substrate showed the best uptake of these trace elements when fertilized with N_{13} nitrogen, with increases of 19 and 15 percentage points, respectively, compared to that of conventional fertilization. The plants grown on the WF/PS 50/50 media showed the best uptake of manganese and iron when fertilized with N_{23} nitrogen. The plants grown on WF/PS 25/75 had the best uptake of manganese when fertilized with N_{13} nitrogen and iron when fertilized with N_{23} nitrogen. The cucumbers grown on the WF media showed a better uptake of Fe and Mn than the cucumbers grown on the other growing media studied. The cucumbers fertilized with conventional fertilizer (CF) in WF were 32 percentage points more efficient in Fe uptake than those in PS.

The trends of the boron and chlorine contents in the growing media were similar. The smallest changes in these elements during the growing season were observed in the media where the plants were fertilized with conventional fertilization. When the plants were fertilized with N_{13} nitrogen, the boron content of the growing media increased from 19 percentage points in the WF/PS 25/75 growing media to 45 percentage points in the wood fiber. Additionally the chlorine content increased from 23 percentage points in the WF/PS 50/50 growing media to 73 percentage points in the WF/PS 25/75 growing media compared with the levels of these elements in the media where the plants were treated with conventional fertilization. As the nitrogen supply to the plants was further increased up to the maximum rate of N_{30}, the levels of these chemical elements increased in all the growing media studied, and at the end of the cucumber growing season, the levels of boron increased from 30 percentage points in WF to 80 percentage points in WF/PS 50/50.

Additionally the level of chlorine increased from 106 percentage points in WF/PS 25/75 to 191 percentage points in PS compared with the level of the growing media where the plants were treated with conventional fertilization.

At the end of the cucumber growing season, the micronutrient content of the growing media was determined to assess how the plant nutrient uptake depended on the fertilization with different nitrogen rates and the amount of wood fiber in the growing media. The comparison of the copper content of all the media at all the fertilizer rates showed that at the end of the growing season, the WF media contained significantly less copper, by as much as 99%, and the WF/PS 25/75 media contained significantly more copper than the PS media under conventional fertilization (CF).

This study showed that the highest uptake of micronutrients during the growing season was observed for plants grown on WF. The plants grown on this medium had a 37-percentage-point-higher copper uptake than the plants grown on PS. The plants grown in the wood fiber had the best uptake of zinc and manganese when fertilized with the highest nitrogen rate N_{30}, which was 13 and 28 percentage points, respectively, compared to that of the conventional fertilizer WF. Only the uptake of iron by WF plants decreased with increasing nitrogen levels. The plants fertilized with N_{30} absorbed 15 percentage less of it than those subjected to WF and conventional fertilization, but 6 percentage points more than those subjected to PS and conventional fertilization (CF). The boron content was more influenced by the composition of the media than by nitrogen fertilization. Boron uptake by the plants was better from PS and WF without additional nitrogen fertilizer. The higher nitrogen rate made the uptake of boron from the growing media more difficult.

At the end of the cucumber growing season, the highest levels of total and mineral nitrogen were found in the WF/PS 25/75 growing media, but the carbon/nitrogen ratio was found in the WF media (Table 4). PS had the lowest Fe content compared to those of the WF growing medium and the mixed medium, with an average of 1.9 times more of the content of this trace element. The total and mineral nitrogen contents of the growing media increased with increasing nitrogen fertilization. The C:N ratio was more dependent on the proportion of wood fiber in the growing media and decreased with an increasing nitrogen fertilizer rate up to N_{30}.

Table 4. The influence of the N fertilization rate and the wood fiber content of the growing media on the micronutrient content, the nitrogen content, and C:N at the end of the growing season.

Factor		Micronutrient Content of the Growing Media mg kg^{-1}						N_{total} %	N_{min} mg kg^{-1}	C:N
Growing Media (A)	N Fertilization Rate (B)	Cu	Mn	Zn	Fe	B	Cl			
					Growing media					
PS		2.73 b	17.11 c	6.59 ab	27.25 c	18.98 b	281 b	1.28 b	136.4 bc	52.50 b
WF/PS 25/75		2.92 b	22.28 b	5.32 b	45.50 b	20.93 ab	312 ab	1.35 a	116.6 c	48.25 c
WF/PS 50/50		10.61 a	36.39 a	7.44 a	76.00 a	23.95 a	292 b	1.26 b	154.1 ab	47.92 c
WF		2.52 b	11.70 d	6.78 ab	34.58 c	17.08 b	342 a	0.88 c	186.6 a	94.75 a
					Fertilization					
	CF	4.37 b	24.80 a	6.51 b	50.25 a	15.43 b	225 d	0.90 c	74.9 c	79.17 a
	N_{13}	4.65 ab	20.38 b	5.70 c	41.75 b	19.58 a	262 c	1.05 b	155.4 b	62.75 b
	N_{23}	4.25 b	21.63 b	7.14 a	42.33 b	22.05 a	344 b	1.37 a	224.5 a	51.75 c
	N_{30}	5.52 a	20.68 b	6.78 ab	49.00 a	23.88 a	395 a	1.45 a	239.0 a	49.75 d
					Influence and interactions					
A		**	**	*	**	*	*	**	**	**
B		*	*	*	*	*	**	*	**	**
A × B		*	*	*	*	ns	*	*	**	**

Note: peat substrate (PS); wood fiber/peat by volume (WF/PS 50/50); wood fiber/peat by volume (WF/PS 25/75); wood fiber (WF); conventional fertilization (CF). Data with the same letters for Factor A and Factor B are not significantly different at $p < 0.05$. * and **—the level of statistical significance at $p < 0.05$ and $p < 0.01$, respectively; ns—not significant.

Principle component analysis showed that increasing the proportions of wood fiber and nitrogen fertilizer in the growing media resulted in a change in the trace element content of the media, which, in turn, influenced the cucumbers' performance. These parameters were correlated in separate groups (Figure 4). The green biomass of the cucumber plants in the substrate without additional fertilization (NCF) was strongly dependent in the group on B and Fe, and to a lesser extent on Mn and Cu. The cucumber fruit sets in pcs m^{-2} were more dependent on B than on the Fe content in the growing media and significantly less on Mn and Cu. The cucumber fruit biomass per plant was strongly correlated with Zn, but there was a negative correlation between this trace element and the plant dry biomass.

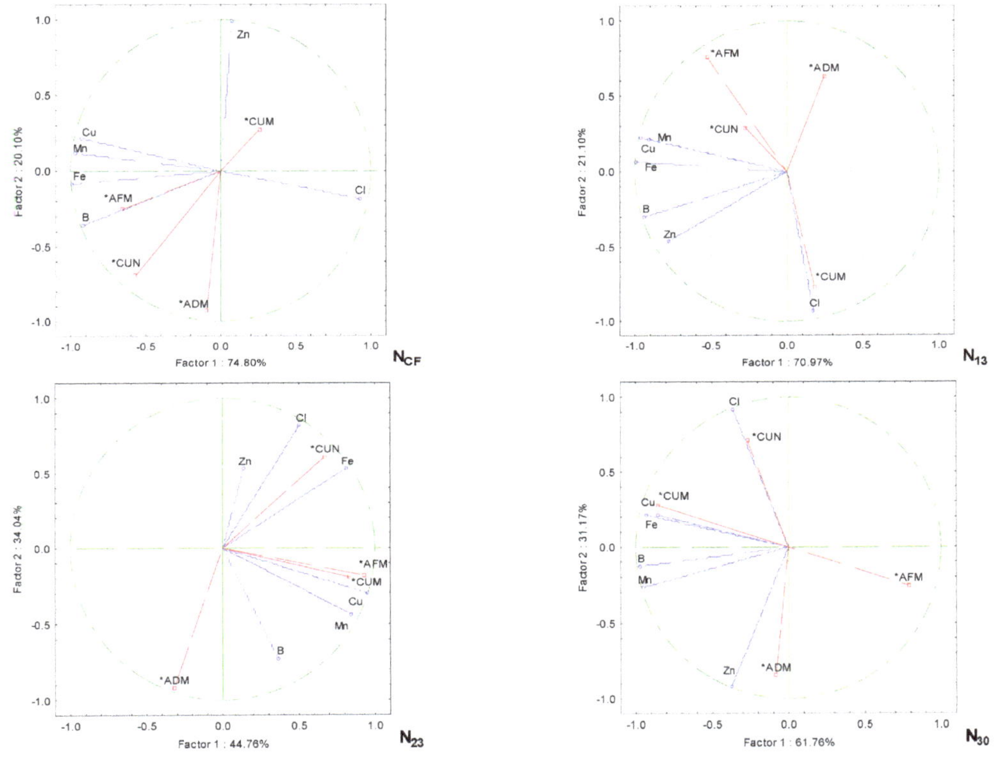

Figure 4. PCA analysis shows the relationship between the micronutrients on the growing media with wood fiber and the productivity parameters for the cucumbers under different nitrogen fertilization rates. Note: the density of the growing media (D), the electrical conductivity of the growing media (EC), the moisture of the growing media (M), the aboveground dry biomass content of the plants (ADM), the aboveground fresh biomass content of the plants (AFM), the number of cucumber fruits per plant (CUN), the number of cucumber fruits per plant (CUM), conventional fertilization (CF), the fertilization rates (N_{13}, N_{23} and N_{30}), and active variables (*).

In the growing media, increasing the nitrogen rate (N_{13}), the number of cucumber fruit pieces m^{-2}, and the green biomass of the plants in kg m^{-2} correlated strongly with Mn and Cu and less strongly with the Fe content of the substrate. The number of cucumber fruits per plant (CUM) was strongly correlated with the Cl content of the growing media.

When the nitrogen fertilizer level was increased to N_{23}, a correlation was observed in two groups. The cucumber fruit biomass in units m^{-2} was strongly correlated with Zn, Cl, and Fe, while in the other group, the cucumber plant green biomass and the fruit biomass were strongly correlated with Cu and Mn, with a weaker correlation with B. A negative correlation was found between the plant dry biomass and Zn.

When the fertilizer rate was increased to a maximum of N_{30}, there was a very strong correlation between the fruit content and the Cl content, and the fruit biomass was strongly dependent on the Cu and Fe contents of the growing media. The dry biomass content of the aboveground part of cucumber plants at N_{30} was already positively correlated with the Zn content of the growing media. A negative correlation was found between the green aboveground biomass of the plants and the Cu and Fe contents of the media.

Principle component analysis showed that increasing the proportion of wood fiber and nitrogen fertilizer in the growing media resulted in changes in the total and mineral nitrogen contents and the carbon/nitrogen ratio in the growing media and that these changes were related to the cucumber performance parameters (Figure 5). The fruit biomass was significantly influenced by the C:N ratio, while the other performance parameters were not affected by higher nitrogen rates and C:N. In the substrate without supplementary nitrogen (N_{CF}), the number of cucumber fruits in pcs m^{-2} was strongly influenced by N_{total}, while the dry biomass content of the aboveground part of the plants correlated not only with N_{total}, but also with N_{min}. When N fertilization was increased with N_{13}, the cucumber fruit biomass depended on the N_{min} content, while the fruit content and the aboveground dry biomass correlated strongly with N_{total} in the growing media. The fruit content of the cucumbers on the N_{min} growing media was significantly influenced by the addition of more intensive N fertilization with N_{23}. This was also strongly correlated with the C:N ratio. The cucumber green aboveground biomass was strongly correlated with the N_{total} content. A negative correlation was found between the biomass of fruit on the plant and the N_{min} in the substrate.

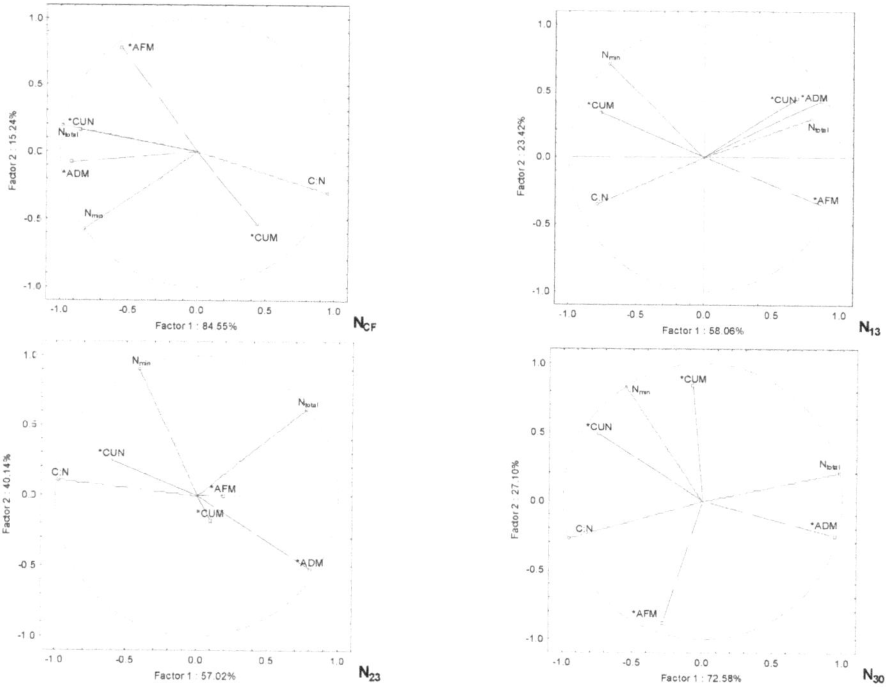

Figure 5. PCA analysis shows the relationship between the N_{total}, N_{min}, and C:N in the growing media with wood fiber on the productivity parameters for the cucumbers with different nitrogen fertilization rates. The density of the growing media (D), the electrical conductivity of the growing media (EC), the moisture of the growing media (M), the aboveground dry biomass content of the plants (ADM), the aboveground fresh biomass content of the plants (AFM), the number of cucumber fruits per plant (CUN), the number of cucumber fruits per plant (CUM), conventional fertilization (CF), the fertilization rates (N_{13}, N_{23} and N_{30}), and active variables (*).

This research shows that the composition of the growing medium and the amount of fertilizer need to be carefully balanced to ensure optimum growing conditions for cucumbers. Combinations of wood fiber and peat with moderate rates of nitrogen can be effective for cucumber cultivation, providing better access to moisture and nutrients, while too much nitrogen can interfere with the uptake of nutrients and lead to problems of balanced plant nutrition.

4. Discussion

Plants grown under climate-controlled conditions take in more nutrients from the growing media than those grown outdoors. This high nutrient demand is due to the shallow penetration of the root system (limited by the volume of the growing container), the weak suction power of the roots, and the large difference between the biomass of the aboveground part of the plant and the root system. To provide the plants grown with wood fiber under controlled climatic conditions in different media with the required micronutrient content, the micronutrient content of the media at the beginning and the end of the study was determined. Wood fiber contains a significant amount of nutrients, and the concentration of these nutrients in wood varies between tree species and depends on the nutrient supply of the soil [29]. Fe use efficiency can contribute to improving plants' physiology by playing the key role as a component of the catalytic centers of various bio-protein redox enzymes, including cytochromes, peroxidases, and catalases. This trace element is vital for nitrogen assimilation and photosynthesis [30].

Wood fiber, wood shavings, and sawdust are wastes from the wood industry. All these products have good water and air retention. Wood fiber can be used in crop production as a stand-alone substrate or in combination with other components such as peat [15,31]. Typically, wood fiber-enriched substrates are used for growing ornamental plants, shrubs, or seedlings in pots [32,33], but are also used in vegetable production [34]. The sorption capacity depends on many factors: the amount of wood fiber in the substrate, the physical and chemical properties of the fiber, and its type and origin [35,36]. In this study, the moisture absorption of the wood fiber/peat blend WF/PS 50/50 was significantly higher compared to that of PS and WF/PS 25/75 (Table 2). Fertilization with higher nitrogen rates of N_{23} and N_{30} reduced the moisture absorption of the growing media, but there was no interaction between media composition and fertilization. Principle component analysis showed that by increasing the proportion of fiber in the growing media, the green biomass of the cucumber plants correlated very strongly with the moisture content of the growing media without additional fertilization with NCF (Figure 3). The biomass of cucumber fruits on the plant correlated strongly with the additional fertilization at the highest rate of N_{30} [37], pointing out that the addition of 40% and 50% wood fiber to peat reduces the water content by 16% vv^{-1} and increases the air content of the substrate by the same amount. This indicates that it is more difficult to overwater, but also requires the more accurate monitoring of the water content and the state of the nutrients. Wood fiber media need to be watered more frequently than peat. *Pelargonium* grown in wood fiber was also found to flower earlier than the ones grown in other media or mixtures with wood fiber [11,14]. In our experiment, the cucumbers were irrigated equally in all the media. However, the moisture content was significantly highest in WF, but was reduced by additional fertilization with N_{23} and N_{30}. Wood fiber has a high level of total porosity and, in most cases, a very high number of air-filled pores and a relatively low level of readily available water. It was found also to have a higher air diffusion rate compared to that of peat [11,38]. Therefore, wood fiber is used to optimize the physical properties of other components in growing media, e.g., by reducing the bulk density, increasing the air capacity, and improving the re-wetting capacity [16,39]. The density of the growing media was highly dependent on their composition. The density of the WF media and the WF/PS 50/50 mixture was significantly lower compared to that of PS. The lower proportion of wood fiber in the WF/PS 25/75 mixture was equivalent to the density of PS. Fertilization with N did not affect the density of the growing media. The dry biomass of the cucumber plants correlated very strongly with the density of the growing media as the proportion of wood fiber in the growing media increased (Figure 3).

In their studies on growing media with wood fiber [40], the authors found that strawberries grown in a mixture of 75% wood fiber and 25% peat gave the best yields. This suggests that a mixture of growing media with wood fiber can improve plants' performance. The composition of the growing media did not affect the chemical composition of the berries, but the berries of the plants grown in this mixture had a lower firmness than those grown in coir fiber. The most productive plants were in mixtures containing up to 75% wood fiber, even without a starter fertilizer [40]. Cucumbers were the most productive in a medium made of wood fiber 50% and peat 50% [29]. Other researchers' studies on ornamental plants have shown that in growing media with 10% and 20% wood fiber, the leaf greenness index, the flower biomass, and the visual score did not differ from those grown on 100% peat substrate. A total of 40% wood fiber harmed all the growth parameters and the macro- and micronutrient contents of the leaves. The plants grown on a peat substrate enriched with 20% wood fiber and fertilized with nitrogen had the highest leaf greenness index, the highest number of flowers, and the highest N, P, Ca, Na, Fe, Mn, and Cu contents. This study showed that high-quality ornamental plants can be successfully grown in a growing media with 20% wood fiber and additional nitrogen fertilization [9]. In our experiment, the most productive cucumbers were grown in a WF/PS 50/50 mixture with additional N_{13} fertilizer. Higher nitrogen rates resulted in a lower fruit number and biomass per plant (Table 2).

Electrical conductivity (EC) is an indicator of the concentrations of soluble salts (Na^+, Mg^{2+}, Ca^{2+}, Cl^-, $SO_4{}^{2-}$, HCO^{3-}, K^+, NO^{3-}, etc.) [41]. High EC has negative effects on germination, photosynthesis, plant vigor, yield, and the nutritional and economic values [42]. Salinity is detrimental to plant growth and development due to water stress, cytotoxicity due to excessive levels of ions such as sodium (Na^+) and chloride (Cl^-), and nutritional imbalances. In our study, it was found that at the highest N_{30} fertilizer rate, the dry biomass content of the aerial part of the plant was dependent on electrical conductivity (Figure 3). On the other hand, a very low EC value of the media indicates nutrient deficiency. During the anion exchange process, excess sodium ions can replace calcium and magnesium, leading to changes in the structure and fertility of the media. This results in the depletion of available nutrients. That different plants differ in their salt sensitivity-tolerance range [43]. Cucumber is a salt-sensitive species, and the EC of the growing media must be <2.7 dS/m, otherwise significant yield losses are inevitable. Our results also showed that EC was influenced by the biomass of fruit grown without additional nitrogen fertilization and at the lowest N_{13} rate (Figure 3). Zucchini (moderately salt-sensitive) grow well in a media with an EC of 2.6–2.8 dS/m, while tomatoes can tolerate salt concentrations of up to 2.9 dS/m with no yield loss [43]. An EC < 2.5 dS/m is considered optimal for creating an ideal substrate [44], but it has been reported that an EC < 3.5 dS/m is the most suitable for growing media and substrates with high EC have higher nutrient sorption and buffering capacities [39]. This study points out that the cucumber fruit biomass with increasing WF in the growing media was strongly dependent on EC, but only without additional fertilizer (NCF) and with the lowest N_{13} rate (Figure 3). When the nitrogen rate was increased to N_{30}, a greater influence of EC was observed on the aboveground dry biomass than on the fruit biomass. The replacement of growing media with wood fiber for *Zinnia hibrida* and *Tagetes erecta* has been investigated [45]. The authors suggest that in media with wood fiber, fertilization with 100–200 mg L^{-1} N can maintain an adequate pH and EC levels of the substrate solution and plant growth without the need for additional N fertilization at the later growth stages.

In our experiment, increasing the proportion of wood fiber in the growing media increased the pH from 5.2 to 6.6, while additional N fertilization reduced the pH value by 7.8%. It has been pointed out that wood waste materials also have a naturally higher pH compared to that of peat, which requires less limestone for initial pH adjustment, but the effect on the substrate's ability to protect itself against pH changes is unknown [46]. Another study also shows that the pH of the substrate solution decreases with an increasing

N concentration [45]. Principle component analysis showed that the pH had no significant effect on the cucumber biometric parameters by increasing the proportion of wood fiber in the growth media (Figure 3). A moderately strong relationship was found between the fruit biomass on the plant and the pH when the plants were fertilized at the maximum rate of N_{30}.

Inducing moderate salinity and/or nutritional stress and the biofortification of vegetables with micronutrients beneficial to human health (iodine, iron, molybdenum, selenium, silicon, and zinc) are well-known methods that have been successfully used to enhance the content of health-promoting phytochemicals in vegetables [47]. It has been shown experimentally that EC and the pH vary with the growing media, depending on the wood components, and can modify the chemical properties of the substrate differently. Other work has shown that increasing the ratios of some wood components increases the pH of the substrate and inhibits EC during the growing season [12]. Some researchers provided this conclusion after three seasons of tomato cultivation in three different substrates—wood fiber, coir, and perlite. The growing media did not show significant differences in tomato yields, but the yields were obtained slightly earlier in wood fiber [13,48].

Plants grown under controlled climatic conditions take up more nutrients from the growing media than those grown outdoors. The high nutrient requirement is due to the shallow penetration of the root system (limited by the volume of the vegetative receptacle), the weak suction power of the roots, and the large difference between the biomass of the aboveground part of the plant and the root system. The nitrogen concentration in wood fiber is generally low, so additional nitrogen is added either at the production stage or by the consumer. Therefore, the nutrient levels, especially nitrogen, can vary considerably depending on the production process. The addition of nitrogen during production can increase the level of water-soluble N. The levels of K, Na, Mg, and Ca can also vary considerably. Some wood fibers have been found to have slightly elevated P levels. The level of extractable (plant-available) nutrients is high, indicating that decomposition processes and nutrient exchange take place in wood fiber [49]. Wood fiber products used in combination with peat, the main component of the growing media, have been found to have a high water-holding capacity, which may have an impact on the retention of nutrients in the growing media [50]. Another aspect to consider is the efficiency with which these mineral nutrients can be used to achieve good production without using too much fertilizer, which can cause environmental pollution. In our experiment, at the end of the cucumber growing season, the levels of Cu, Mn, Zn, Fe, and B were found to be significantly higher in the WF/PS 50/50 mixture than those in the PS, the WF, and the WF/PS 25/75 mixture (Table 4). Only Cl was detected at higher levels in the WF growing media than those in the other media. Other researchers suggest that the accumulation of chloride should be attributed to unfavorable nutrients in the root environment [51]. Additional fertilization with N_{30} increased the levels of Cu, Zn, Fe, B, and Cl in the WF growth media. This may be related to the increase in EC, as the increase in N rate blocked the uptake of nutrients from the media, resulting in the highest micronutrient levels in the root zone. Sometimes, the Mn and Zn levels can vary over a wide range, but generally, the concentrations and availability of the other trace elements are very low. In general, the levels of trace elements in the WF media are higher than those of the water-soluble trace elements [11]. Scientists have investigated the chemical composition of nutrient solutions in the root environment of tomatoes grown in wood fiber and rockwool under different levels of nitrate nitrogen in nutrient solutions at 200, 220, and 240 mg $N–NO_3·dm^{-3}$. Despite the wide range of carbon-to-nitrogen ratios (C:N) in wood fiber, no significant variation in nitrate content in the root environment was observed. This was the result of an appropriate frequency of nutrient solution application per day [51]. Our results show that the N_{min} and the N_{total} increased significantly in the growing media (cucumber root zone), while C:N decreased significantly when the N fertilization rate was increased to N_{23} (Table 4). Although the N_{min} content in the WF media was highest at the end of the cucumber growing season, the C:N content was not as high as that in the WF media. The addition of wood fiber to the growing media may improve macronutrient uptake in the short term, but may reduce micronutrient uptake in the medium term, highlighting the complex effects of wood fiber on nutrient availability and plant growth [52]. The decreasing effect of $N\text{-}NH_4$ and $N\text{-}NO_3$ contents in wood fiber on cucumber growth can be attributed to biosorption due to the high C:N = 125 ratio in these media [53].

In our experiment, similar results were obtained; the highest C:N ratio was found in the wood fiber growing media compared to those of the peat substrate and the media mixtures, but additional N fertilization reduced this ratio. The N_{total} content was low in the WF media at the end of the growing season. The WF media had the highest N_{min}, but there was no significant difference compared to that of the WF/PS 50/50 growing media. It can be concluded that for the cucumbers grown on WF, WF/PS 50/50 nitrogen uptake was more difficult than it was for the cucumbers grown on the PS and WF/PS 25/75 growing media. When fertilization was increased to N_{23} and N_{30}, mineral N remained in the growing media (Table 4). Wright et al. (2008) found that plants grown on soils containing woody material require about 100 mg L^{-1} N more fertilizer to achieve a similar result to that of the plants grown on peat substrates. Different types of wood component also differ in their physical properties and their ability to immobilize N. The differences in air and water porosity, total water-holding capacity, and bulk density can also affect microbial activity and plant growth. The main cause of N immobilization is biosorption by microorganisms that use the N available in plants [54]. This can lead to nutritional deficiencies in ornamental plants and a reduction in their aesthetic value [50]. The negative effects of the sorption of N and other components on plant growth and flowering can be eliminated by appropriate fertilization. To overcome N immobilization in growing media with wood fiber, it is necessary to provide additional nutrient solutions from the beginning of plant growth [11,39,55].

Wood fiber-based growing media can be produced with the desired physical and chemical properties for the optimal growth of a wide range of plants and vegetables. Other studies on the yield of plants grown on wood fiber media show that additional fertilizers are needed, but growth may depend on the nature of the wood fiber and its content in the growing media. Wood fiber can be a reliable, consistent, renewable, and cost-effective alternative to the traditional peat substrate in greenhouses. The use of wood fiber in growing media is in line with sustainable development policies and can help reduce the carbon footprint of the product.

5. Conclusions

The cucumber fruit biomass and quantity were highest when the cucumbers were grown on the WF/PS 50/50 media. Additional fertilization with N_{13} significantly increased the cucumber yield. Higher nitrogen rates of N_{23} and N_{30} reduced the fruit biomass and the number of cucumbers per plant. Wood fiber reduced the density and increased the moisture content of the growing media. Increasing the nitrogen content decreased the moisture content of the growing media. Wood fiber reduced the acidity of the media. At the end of this study, significantly higher levels of Cu, Mn, Zn, Fe, and B were found in WF/PS 50/50 than those in the PS, WF, and WF/PS 25/75 media. The application of N_{30} fertilizer increased the levels of the trace elements Cu, Zn, Fe, B, and Cl in the EC growing media. It can be assumed that the excessive nitrogen applied blocked the uptake of nutrients from the media. At the end of the growing season, the micronutrient content in the root zone of cucumbers fertilized with N_{23} and N_{30} was highest. At the end of the growing season, the N_{min} content was higher in the growing media with a higher proportion of wood fiber, but the C:N content remained highest in the WF medium. The optimum medium with wood fiber for cucumber production in a greenhouse was found to be WF/PS 50/50 with additional fertilization with N_{13}. Thus, wood fiber-based substrates can be produced with the desired physical and chemical properties for the optimal growth of a wide range of plants and vegetables.

Further research is needed to evaluate wood fiber-based soilless substrates containing different types of wood to assess not only N immobilization, but also the availability of other nutrients and the effect on the plants. A comparison of the range of wood production residues available from using different methods for the development of growing media is needed to show the advantages and disadvantages of wood products as substrate components. Cultivation technologies need to be developed based on nutrient management strategies that depend on the type of wood and the quantity used in the growing media. The cultivation technologies should be tested before making science-based recommendations to growers.

Author Contributions: R.Č. and L.M.B. designed the research framework and contributed to the application of the study methodology and the analysis of the results; R.Č., L.M.B. and V.S. played an active role in writing, reviewing, and editing this manuscript. All authors have read and agreed to the published version of the manuscript.

Funding: This research received no external funding.

Data Availability Statement: The data are contained within the article.

Conflicts of Interest: The authors declare no conflicts of interest.

References

1. Raviv, M. Can compost improve sustainability of plant production in growing media? *Acta Hortic.* **2017**, *1168*, 119–134. [CrossRef]
2. Gruda, N.S. Increasing Sustainability of Growing Media Constituents and Stand-Alone Substrates in Soilless Culture Systems. *Agronomy* **2019**, *9*, 298. [CrossRef]
3. Chu, Y.-T.; Brown, P.B. Sustainable marine aquaponics: Effects of shrimp to plant ratios and C/N ratios. *Front. Mar. Sci.* **2021**, *8*, 1825. [CrossRef]
4. Asaduz-Zaman, M.; Saifullah, M.; Mollick, A.S.R.; Hossain, M.M.; Halim, G.M.A.; Asao, T. Influence of Soilless Culture Substrate on Improvement of Yield and Produce Quality of Horticultural Crops. In *Soilless Culture-Use of Substrates for the Production of Quality Horticultural Crops*; IntechOpen: London, UK, 2015; pp. 1–32. [CrossRef]
5. Zawadzińska, A.; Salachna, P.; Nowak, J.S.; Kowalczyk, W. Response of Interspecific Geraniums to Waste Wood Fiber Substrates and Additional Fertilization. *Agriculture* **2021**, *11*, 119. [CrossRef]
6. Yang, T.; Altland, J.; Samarakoon, U. Evaluation of organic substrates as an alternative to perlite for cucumber production in the Dutch bucket hydroponic system. *Acta Hortic.* **2021**, *1317*, 319–326. [CrossRef]
7. Schmilewski, G. Growing medium constituents used in the EU. *Acta Hortic.* **2009**, *819*, 33–46. [CrossRef]
8. Kuisma, E.; Palonen, P.; Yli-Halla, M. Reed canary grass straw as a substrate in soilless cultivation of strawberry. *Sci. Hortic.* **2014**, *178*, 217–223. [CrossRef]
9. Dorais, M.; Ménard, C.; Bégin, G. Risk of phytotoxicity of sawdust substrates for greenhouse vegetables. *Acta Hortic.* **2007**, *761*, 589–595. [CrossRef]
10. Carlile, W.R.; Cattivello, C.; Zaccheo, P. Organic growing media: Constituents and properties. *Vadose Zone J.* **2015**, *14*, 1–13. [CrossRef]
11. Maucieri, C.; Nicoletto, C.; Van Os, E.; Anseeuw, D.; Van Havermaet, R.; Junge, R. Hydroponic technologies. In *Aquaponics Food Production Systems*; Springer: Berlin/Heidelberg, Germany, 2019; Volume 10, ISBN 978-3-030-15943-6.
12. Jackson, B.E.; Wright, R.D.; Alley, M.M. Comparison of fertilizer nitrogen availability, nitrogen immobilization, substrate carbon dioxide efflux, and nutrient leaching in peat-lite, pine bark, and pine tree substrates. *HortScience* **2009**, *44*, 781–790. [CrossRef]
13. Domeño, I.; Irigoyen, I.; Muro, J. New Wood Fibre Substrates Characterization in Hydroponic Tomato Culture. *Eur. J. Hortic. Sci.* **2010**, *75*, 89–94.
14. Carlile, W.; Raviv, M.; Prasad, M. Organic soilless media components. In *Soilless Culture Theory and Practice*; Elsevier: Amsterdam, The Netherlands, 2019; p. 691. [CrossRef]
15. Theurl, M.C.; Hörtenhuber, S.J.; Lindenthal, T.; Palme, W. Unheated soil-grown winter vegetables in Austria:Greenhouse gas emissions and socio-economic factors of diffusion potential. *J. Clean. Prod.* **2017**, *151*, 134–144. [CrossRef]
16. Barrett, G.E.; Alexander, P.D.; Robinson, J.S.; Bragg, N.C. Achieving environmentally sustainable growing media for soilless plant cultivation systems—A review. *Sci. Hortic.* **2016**, *212*, 220–234. [CrossRef]
17. Antón, A.; Torrellas, M.; Montero, J.I.; Ruijs, M.; Vermeulen, P.; Stanghellini, C. Environmental impact assessment of Dutch tomato crop production in a Venlo glasshouse. *Acta Hortic.* **2012**, *927*, 781–791. [CrossRef]
18. Tzortzakis, N.; Nicola, S.; Savvas, D.; Voogt, W. Editorial: Soilless Cultivation through an Intensive Crop Production Scheme. Management Strategies, Challenges and Future Directions. *Front. Plant Sci.* **2020**, *11*, 363. [CrossRef]
19. Beretta, D.; Ripamonti, M. Evaluation of wood fiber as component of substrates for container nursery crops. *Acta Hortic.* **2019**, *1305*, 77–82. [CrossRef]

20. Sivaranjani, S.; Panwar, V.P. Soil nutrient dynamics under mountainous landscape: Issues and challenges. In *Understanding Soils of Mountainous Landscapes: Sustainable Use of Soil Ecosystem Services and Management*; Elsevier: Amsterdam, The Netherlands, 2023; pp. 131–149. [CrossRef]
21. Sindhu, S.S.; Gholap, D.B.; Singh, M.C.; Dhiman, M.R. Effect of medium amendments on growth and flowering in gerbera. *Indian J. Hortic.* **2010**, *67*, 391–394.
22. Lira-Martins, D.; Quesada, C.A.; Strekopytov, S.; Humphreys-Williams, E.; Herault, B.; Lloyd, J. Wood Nutrient-Water-Density Linkages Are Influenced by Both Species and Environment. *Front. Plant Sci.* **2022**, *13*, 778403. [CrossRef]
23. Krug, D.; Wagenführ, A.; Weber, A.; Wenderdel, C. *Fiber-Based Materials*; Springer: Berlin/Heidelberg, Germany, 2023; pp. 1491–1561 [CrossRef]
24. Heineman, K.D.; Turner, B.L.; Dalling, J.W. Variation in wood nutrients along a tropical soil fertility gradient. *New Phytol.* **2016**, *211*, 440–454. [CrossRef]
25. Gull, M.; Kausar, A. Comparative Analysis of Various Growth, Ions Uptake and other Physiological Performance of Two Rice (*Oryza sativa* L.) Genotypes Under Altering Saline Conditions. *Pharmacophore* **2018**, *9*, 73–81.
26. Gruda, N.; Savvas, D.; Colla, G.; Rouphael, Y. Impacts of genetic material and current technologies on product quality of selected greenhouse vegetables—A review. *Eur. J. Hortic. Sci.* **2018**, *83*, 319–328. [CrossRef]
27. Gruda, N. Current and future perspective of growing media in Europe. *Acta Hortic.* **2012**, *960*, 37–43. [CrossRef]
28. Caballero, R.; Pajuelo, P.; Ordovás, J.; Carmona, E.; Delgado, A. Evaluation and correction of nutrient availability to Gerbera jamesonii H. Bolus in various compost-based growing media. *Sci. Hortic.* **2009**, *122*, 244–250. [CrossRef]
29. Čepulienė, R.; Butkevičienė, L.M.; Skinulienė, L.; Steponavičienė, V. Response of Cucumbers (*Cucumis sativus* L.) to Waste Wood Fiber Substrates and Additional Nitrogen Fertilization. *Plants* **2022**, *11*, 3464. [CrossRef]
30. Harris, C.N.; Dickson, R.W.; Fisher, P.R.; Jackson, B.E.; Poleatewich, A.M. Evaluating Peat Substrates Amended with Pine Wood Fiber for Nitrogen Immobilization and Effects on Plant Performance with Container-grown Petunia. *HortTechnology* **2020**, *30*, 107–116. [CrossRef]
31. Gruda, N.; Bisbis, M.; Tanny, J. Impacts of protected vegetable cultivation on climate change and adaptation strategies for cleaner production—A review. *J. Clean. Prod.* **2019**, *225*, 324–339. [CrossRef]
32. Zheng, J.; Guo, R.; Li, D.; Zhang, J.; Han, S. Nitrogen addition, drought and mixture effects on litter decomposition and nitrogen immobilization in a temperate forest. *Plant Soil* **2017**, *416*, 165–179. [CrossRef]
33. Jackson, B.E.; Wright, R.D. Pine tree substrate: An alternative and renewable growing media for horticulture crop production. *Acta Hortic.* **2009**, *819*, 265–272. [CrossRef]
34. Gruda, N.; Schnitzler, W. Suitability of wood fiber substrates for production of vegetable transplants II.: The effect of wood fiber substrates and their volume weights on the growth of tomato transplants. *Sci. Hortic.* **2004**, *100*, 333–340. [CrossRef]
35. Geisseler, D.; Horwath, W.R.; Joergensen, R.G.; Ludwig, B. Pathways of nitrogen utilization by soil microorganisms—A review. *Soil Biol. Biochem.* **2010**, *42*, 2058–2067. [CrossRef]
36. Fain, G.B.; Gilliam, C.H.; Sibley, J.L.; Boyer, C.R.; Witcher, A.L. Wholetree substrate and fertilizer rate in production of greenhouse-grown petunia (*Petunia* × *hybrida* Vilm.) and marigold (*Tagetes patula* L.). *HortScience* **2008**, *43*, 700–705. [CrossRef]
37. Eveleensas, B.; Van Winkelis, A.; Blokas, C. Medienos pluoštas vazoninių augalų kultūroje; durpių pakeitimas iki 50% tūrio? *ISHS Acta Hortic.* **2021**, *1317*, 165–174. [CrossRef]
38. Gruda, N.; Schnitzler, W. Wood fibers as a peat alternative substrate for vegetable production. *Eur. J. Wood Wood Prod.* **2006**, *64*, 347–350. [CrossRef]
39. Clemmensen, A.W. Physical Characteristics of Miscanthus Composts Compared to Peat and Wood Fiber Growth Substrates. *Compos. Sci. Util.* **2013**, *12*, 219–224. [CrossRef]
40. Visconti, F.; de Paz, J.M. Electrical conductivity measurements in agriculture: The assessment of soil salinity. In *New Trends and Developments in Metrology*; Cocco, L., Ed.; IntechOpen: London, UK, 2016; pp. 99–126. [CrossRef]
41. Aurdal, S.M.; Woznicki, T.L.; Haraldsen, T.K.; Kusnierek, K.; Sønsteby, A.; Remberg, S.F. Wood Fiber-Based Growing Media for Strawberry Cultivation: Effects of Incorporation of Peat and Compost. *Horticulturae* **2023**, *9*, 36. [CrossRef]
42. Savvas, D.; Gianquinto, G.P.; Tüzel, Y.; Gruda, N. *Good Agricultural Practices for Greenhouse Vegetable Crops—Principles for Mediterranean Climate Areas*; FAO: Rome, Italy, 2013; p. 217. Available online: http://www.fao.org/3/i3284e/i3284e.pdf (accessed on 10 October 2024).
43. Agarwal, P.; Saha, S.; Hariprasad, P. Agro-industrial-residues as potting media: Physicochemical and biological characters and their influence on plant growth. *Biomass-Convers. Biorefinery* **2023**, *13*, 9601–9624. [CrossRef]
44. Owen, W.G.; Jackson, B.E.; Whipker, B.E.; Fonteno, W.C. Pine Wood Chips as an Alternative to Perlite in Greenhouse Substrates: Nitrogen Requirements. *HortTechnology* **2016**, *26*, 199–205. [CrossRef]
45. Nguyen, V.T.; Wang, C.-H. Use of organic materials as growing media for honeydew melon seedlings in organic agriculture. *Commun. Soil Sci. Plant Anal.* **2017**, *48*, 2137–2147. [CrossRef]
46. Rouphael, Y.; Kyriacou, M.C. Enhancing quality of fresh vegetables through salinity eustress and biofortification applications facilitated by soilless cultivation. *Front. Plant Sci.* **2018**, *9*, 1254. [CrossRef]
47. Harris, C. Evaluating Wood Fiber Soilless Substrates for Effects on Plant Performance and Nutrient Management in Container Crops. Master's Thesis, University of New Hampshire, Durham, NH, USA, 2019. Available online: https://scholars.unh.edu/thesis/1293 (accessed on 10 October 2024).

48. Parkinson, J.A.; Allen, S.E. A wet oxidation procedure suitable for the determination of nitrogen and mineral nutrients in biological material. *Commun. Soil Sci. Plant Anal.* **2008**, *6*, 1–11. [CrossRef]
49. Muro, J.; Irigoyen, I.; Samitier, P.; Mazuela, P.; Salas, M.; Soler, J.; Urrestarazu, M. Wood fiber as growing medium in hydroponic crop. *Acta Hortic.* **2005**, *697*, 179–185. [CrossRef]
50. Wright, R.D.; Jackson, B.E.; Browder, J.F.; Latimer, J.G. Growth of chrysanthemum in a pine tree substrate requires additional fertilizer. *HortTechnology* **2008**, *18*, 111–115. [CrossRef]
51. Komosa, A.; Kleiber, T.; Pirog, J. Contents of macro- and microelements in root environment of greenhouse tomato grown in rockwool and wood fiber depending on nitrogen levels in nutrient solustions. *Acta Sci. Pol. Hortorum Cultus* **2010**, *9*, 59–68.
52. Fontana, M.; Johannes, A.; Zaccone, C.; Weisskopf, P.; Guillaume, T.; Bragazza, L.; Elfouki, S.; Charles, R.; Sinaj, S. Improving crop nutrition, soil carbon storage and soil physical fertility using ramial wood chips. *Environ. Technol. Innov.* **2023**, *31*, 103143. [CrossRef]
53. Vandecasteele, B.; Muylle, H.; De Windt, I.; Van Acker, J.; Ameloot, N.; Moreaux, K.; Coucke, P.; Debode, J. Plant fibers for renewable growing media: Potential of defibration, acidification or inoculation with biocontrol fungi to reduce the N drawdown and plant pathogens. *J. Clean. Prod.* **2018**, *203*, 1143–1154. [CrossRef]
54. Piróg, J.; Komosa, A.; Markiewicz, B. The effect of wood fiber density on the content of macro and microelements in the root environment of greenhouse cucumber. *J. Fruit Ornam. Plant Res.* **2009**, *70*, 81–89. [CrossRef]
55. Depardieu, C.; Prémont, V.; Boily, C.; Caron, J. Sawdust and Bark-Based Substrates for Soilless Strawberry Production: Irrigation and Electrical Conductivity Management. *PLoS ONE* **2016**, *11*, e0154104. [CrossRef]

plants

MDPI

Article

Nitrogen Immobilization by Wood Fiber Substrates Strongly Affects the Photosynthetic Performance of Lettuce

Lingyi Wu [1], Ruohan Li [1], Juncheng Liu [2], Wenzhong Cui [2], Zhiyong Qi [1,*] and Wanlai Zhou [1,*]

[1] Institute of Urban Agriculture, Chinese Academy of Agricultural Sciences, Chengdu 610213, China; 82101225393@caas.cn (L.W.); 82101235419@caas.cn (R.L.)

[2] School of Mechanical Engineering, Chengdu University, Chengdu 610100, China; ljceng@icloud.com (J.L.); 17784240198@163.com (W.C.)

* Correspondence: qizhiyong@caas.cn (Z.Q.); zhouwanlai@caas.cn (W.Z.)

Abstract: Wood fiber substrates are widely used as peat substitutes in horticulture, but the impact of their high carbon-to-nitrogen ratio on nitrogen immobilization and crop photosynthetic performance remains unclear. This study systematically examined the effects of wood fiber substrates on lettuce photosynthetic performance and underlying physiological mechanisms using pot experiments. Two substrate treatments—peat (control) and wood fiber—were combined with three nitrogen levels: low, medium, and high (63, 127, and 210 mg $N \cdot L^{-1}$). Results indicated that wood fiber substrates significantly reduced the availability of fast-acting nitrogen, leading to a substantial decrease in lettuce biomass (39.0–56.8%), total nitrogen content (7.2–39.9%), and chlorophyll content (13.7–36.2%). Chlorophyll fluorescence kinetics analysis revealed that wood fiber substrates impair photosystem function through multiple pathways. At the early stage (15 days), key effects included structural damage to the donor side of PSII(Photosystem II), indicated by L and K peaks, and inhibited electron transfer on the PSI(Photosystem I) acceptor side (δRo decreased by 15.08–27.90%, along with a reduction in W_{OI} amplitude). The findings provide an important theoretical basis for optimising nitrogen management strategies for wood fibre substrates.

Keywords: chlorophyll fluorescence; JIP test; PSI; PSII

1. Introduction

Peat, the most widely used substrate [1,2], is non-renewable and poses significant ecological concerns [3,4]. To address this issue, the horticultural industry is actively seeking sustainable alternative materials [5–7]. Among these alternatives, wood fiber has emerged as a promising peat substitute due to its unique physical properties—low bulk density, high porosity, and good rewettability—as well as its eco-friendliness [8,9].

It is now gaining widespread global adoption. Despite its advantages in soilless cultivation, the high carbon-to-nitrogen (C/N) ratio of wood fiber can induce nitrogen immobilization, which may negatively affect plant growth [10]. Nitrogen immobilization of the cultivation substrate refers to the phenomenon in which inorganic nitrogen from fertiliser sources is fixed in the substrate due to the unique physicochemical properties of the substrate's organic materials and is difficult to be absorbed and utilised by the crop [3,11,12]. Because the exogenous nitrogen is partially immobilised by the substrate material, the amount of nitrogen that the plant can take up becomes less. Studies have shown that microbial nitrogen immobilization significantly increases when the substrate's C/N ratio exceeds 30:1, thereby reducing plant-available nitrogen [13–15]. However, wood fiber substrates typically exhibit C/N ratios of 100:1 or higher, making nitrogen immobilization a significant concern [16]. Numerous studies have shown that Woody Substrate (WS), such as wood fibre, wood chip substrate, bark substrate, etc., tends to have severe nitrogen fixation during crop cultivation, leading to crop nitrogen deficiency, which greatly increases the difficulty of nitrogen and nutrient management in WS cultivation [17,18]. Gruda et al. [19] reported that complete wood fiber substrates exhibited significant nitrogen immobilization, leading to tomato growth inhibition. Wright et al. [12] found that plants grown in wood-fractionated substrates required an additional 100 mg·L^{-1} of nitrogen to achieve growth comparable to peat-based cultivation. These findings indicate that nitrogen immobilization in wood fiber substrates may severely restrict their horticultural applications, highlighting the urgent need for optimization through scientific nitrogen management strategies.

Nitrogen availability plays a crucial role in shaping the structure and function of the photosynthetic system, while nitrogen deficiency severely disrupts the stability and efficiency of the photosynthetic apparatus [20]. Chlorophyll a fluorescence measurement is a widely used technique for evaluating the impact of stress factors on photosynthesis [21]. The JIP test, a robust mathematical model developed by Strasser et al. [22], quantifies energy uptake by the PSII antenna complex, energy capture in the PSII reaction center, electron transport from PSII to PSI, and the final reduction of the PSI electron acceptor. The JIP test has been widely employed to assess crop responses to nitrogen deficiency [23–25]. Nitrogen deficiency decreases chlorophyll a and accessory pigments while altering the structure of chloroplast-like vesicle membranes [23]. This typically results in impaired PSII reaction center function, a reduced pool of electron acceptors, disrupted electron transfer processes, and diminished photochemical efficiency [20,24].

Nitrogen immobilization in woody substrates reduces plant-available nitrogen, thereby significantly impacting plant growth and physiological functions. While numerous studies have explored the effects of nitrogen deficiency on plant photosynthesis, research on plant photosynthetic responses to nitrogen immobilization in wood substrates remains limited. Moreover, whether nitrogen deficiency caused by immobilization in wood-fiber substrates induces physiological effects similar to low-nitrogen stress lacks systematic investigation.

Based on this, this study systematically examined how nitrogen immobilization in wood fiber substrates affects lettuce photosynthesis and its underlying mechanisms. Potting experiments were conducted with two substrate types (peat and wood fiber) and three nitrogen levels (low, medium, and high). Chlorophyll fluorescence kinetics and photosynthetic parameters were analyzed to elucidate these effects. In this study, it was hypothesised that nitrogen fixation by wood fibre substrates would have a negative effect on the photosynthetic performance of lettuce. The findings will contribute to optimizing nitrogen management strategies for wood fiber substrates and facilitating their efficient use in soilless cultivation.

2. Results

2.1. Fast-Acting Nitrogen Supply in the Substrate

N immobilization by wood fiber substantially reduced fast-acting nitrogen supply in the substrate (Figure 1). At day 15, the ammonia nitrogen content in the wood fiber substrate was reduced by 22.0%, 21.2% and 20.2% compared to the control at high, medium and low nitrogen application levels, respectively, with all differences being statistically significant ($p < 0.05$); At day 30, that reductions were 23.1%, 13.4% and 13.3%, respectively. Nutrient measurements were taken 2–3 days after applying the nutrient solution, reflecting the residual nitrogen state in the substrate following both lettuce uptake and nitrogen immobilization by the substrate. Although the total nitrogen applied to the wood fiber substrate was equivalent to that of the control substrate, the biomass and total nitrogen content of the lettuce grown in it were significantly lower, suggesting reduced nitrogen uptake by plants in the wood fiber substrate (Figure 2). Theoretically, the remaining quick nitrogen concentration in the wood fiber substrate should be higher than that in the control; however, actual measurements revealed a significantly lower concentration of quick nitrogen. This paradox suggests that the wood fiber substrate significantly reduces the availability of quick-acting nitrogen by strongly fixing nitrogen and converting exogenous inorganic nitrogen into organic forms, which are less accessible to plants. Table 1 shows that the wood fiber substrate has a high nitrogen immobilization potential, reaching 115 mg·L^{-1} within 4 days, equivalent to 18.40 mg per pot of substrate. Over the 30-day experimental period, the total N input per tray of substrate was approximately 67.30 mg, 40.32 mg and 20.00 mg for high, medium and low N applications, respectively, whereas wood fiber fixes between 27.3% and 92.0% of the applied N in 4 days, which is sufficient to result in a significant reduction in the supply of fast-acting N to the substrate.

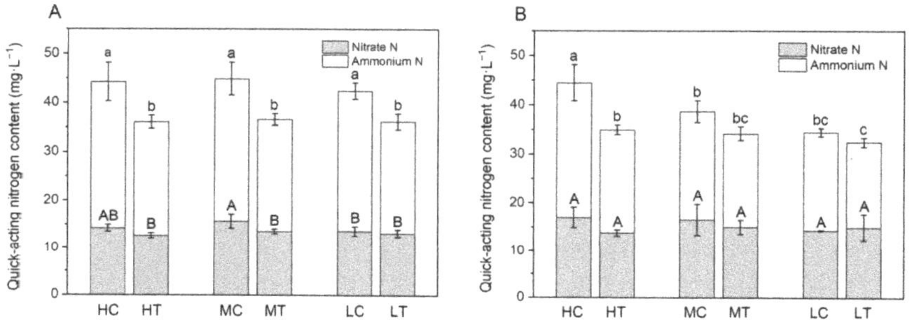

Figure 1. Concentration of quick nitrogen in the substrate under different treatments ((**A**) is for Day 15; (**B**) is for Day 30). High nitrogen & control substrate (HC), high nitrogen & wood fiber substrate (HT), medium nitrogen & control substrate (MC), medium nitrogen & wood fiber substrate (MT), low nitrogen & control substrate (LC), and low nitrogen & wood fiber substrate (LT). Within each organic acid or germination index, values that do not share a letter are significantly different according to ANOVA and Duncan's multiple range test ($p < 0.05$).

Table 1. Nitrogen immobilization potential of wood fibers. High nitrogen (100% N, HN), medium nitrogen (60% N, MN), and low nitrogen (30% N, LN).

Form	HN	MN	LN
Nutrient solution nitrogen concentration (mg N·L^{-1})	210.30	127.60	62.50
Nutrient solution input per pot (L)	0.32	0.32	0.32
Total nitrogen input to single pot (mg)	67.30	40.32	20.00
Wood fiber nitrogen immobilization capacity (mg N·L^{-1})		115.00	
Volume of wood fiber in single pot substrate (L)		0.16	
4-day nitrogen immobilization potential of single-pot substrates (mg)		18.40	

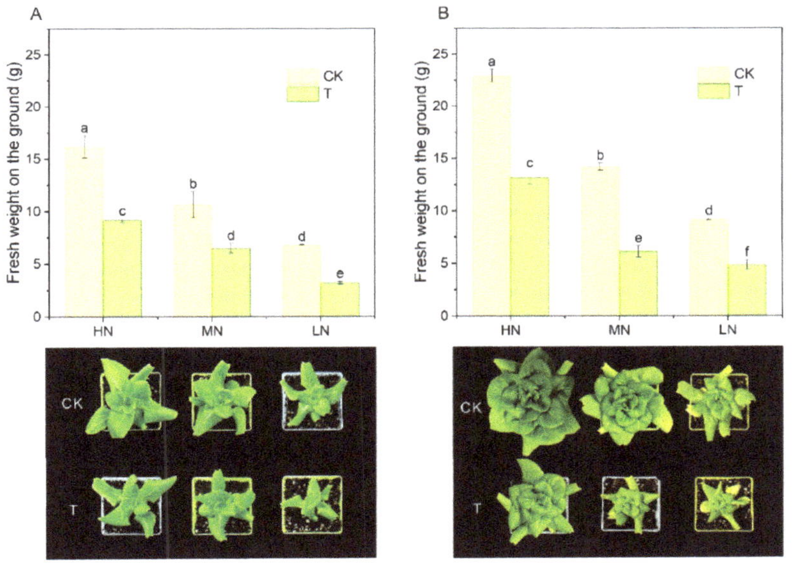

Figure 2. Above ground fresh weight and morphological characteristics of plants under different treatments ((**A**) for Day 15 and (**B**) for Day 30). High nitrogen (100% N, HN), medium nitrogen (60% N, MN), and low nitrogen (30% N, LN). Control substrate (CK) and wood fiber substrate (T). Within each organic acid or germination index, values that do not share a letter are significantly different according to ANOVA and Duncan's multiple range test ($p < 0.05$).

2.2. Lettuce Growth Performance and Nutrient Accumulation

Wood fiber appeared to have a significant negative effect on the growth performance of lettuce (Figure 2).The aboveground fresh weight of lettuce in the wood fiber substrate was reduced by 43.5%, 39.0% and 52.9% at day 15 compared to the control at high, medium and low nitrogen application levels, respectively; at day 30, this reduction was 42.6%, 56.8% and 47.2%, with the difference reaching the level of significance on both occasions ($p < 0.05$). Morphological characteristics of lettuce also showed that at the same level of nitrogen application, the control plants had larger and thicker smooth leaves with intense green colour, while the wood fiber group had shorter plants and narrower leaves. By day 30, this morphological difference was even more pronounced, with wood fiber-based lettuce showing significant nitrogen deficiency characteristics such as short plants and yellowed leaf ends.

Further, nutrient accumulation in lettuce cultivated on wood fiber substrate was also significantly lower than the control (Figure 3). Nitrogen content in lettuce gradually decreased with decreasing nitrogen application, both in wood fiber substrate and control, indicating that nitrogen availability directly affected nitrogen accumulation in lettuce leaves under the conditions of this experiment. The N content of lettuce in the wood fiber substrate was reduced by 39.9%, 24.2%, and 7.2% at day 15 compared to the control at high, medium, and low N application levels; at day 30, this reduction was 21.7%, 9.5%, and 9.9%, respectively. The accumulation of phosphorus and potassium in leaves of lettuce cultivated on wood fiber substrates at different levels of nitrogen application showed a similar pattern to that of nitrogen, being significantly lower than that of the control substrate in most cases, and decreasing with decreasing nitrogen application.

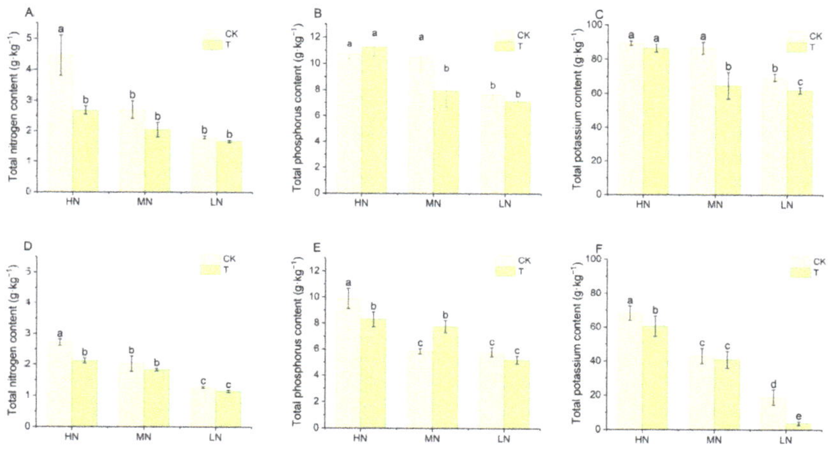

Figure 3. N, P, and K contents of plants under different treatments ((**A–C**) for Day 15 and (**D,F**) for Day 30). High nitrogen (100% N, HN), medium nitrogen (60% N, MN), and low nitrogen (30% N, LN). Control substrate (CK) and wood fiber substrate (T). Within each organic acid or germination index, values that do not share a letter are significantly different according to ANOVA and Duncan's multiple range test ($p < 0.05$).

2.3. *Chlorophyll a Fluorescence Properties*

2.3.1. OJIP Chlorophyll Fluorescence Kinetics

The OJIP chlorophyll fluorescence kinetics can well reflect the changes in primary photochemical reactions and photosynthetic functions of PSII [26]. OJIP is the four key stages of the Chlorophyll Fluorescence Induction Kinetics (CFIK) curve for studying the functional state of photosynthetic system II (PSII).It reflects the efficiency of light energy absorption, transfer and electron transfer in PSII by irradiating dark-adapted plant leaves with high-intensity light and recording changes in fluorescence intensity on millisecond time scales. The chlorophyll fluorescence induction curves for all treatments showed a typical OJIP shape (Figure 4A,D). To visualise the comparative effects of the treatments on transient kinetics, the curves were plotted as relative variable fluorescence $V_t = (F_t - F_o)/(F_M - F_o)$ and $\Delta V_t = F_{t(treated)} - F_{t(HC)}$ (Figure 4B,E). Rising O-J step in the OJIP curve correlates with the degree of closure of the PSII reaction center [24]. The J-I part of the curve corresponds to the reduction of the secondary electron acceptor Q_B, plastoquinone (PQ), cytochrome (Cyt b6f), and plastocyanin (PC) [21]. Fluorescence transient analysis showed that the V_t of lettuce under different treatments showed significant divergence at points J and I, especially at point I, indicating that the effect of wood fiber substrate on lettuce mainly occurred at the O-J and J-I periods.

To further explore the initial site of action of nitrogen nutrition, the JIP test parameter V_J/V_I was introduced (Figure 4C,F). At 15 days of age, V_J and V_I increased with decreasing nitrogen application in both wood fiber and control. Whereas, at the same level of nitrogen application, both V_J and V_I were significantly higher in the wood fiber group than in the control group. Considering that the wood fiber substrate itself reduces the effective supply of nitrogen, the above facts suggest that reduced nitrogen supply can lead to elevated V_J and V_I, reflecting the openness of the PSII reaction center in lettuce and reduced electron transfer from Q_A^- to Q_B^- on the PSII receptor side. It is noteworthy that only the wood fiber group had significantly higher V_J and V_I at both low and medium N levels, whereas only V_I was significantly higher in the other treatments, suggesting that the effect of nitrogen deficiency firstly affected V_I, and only when nitrogen deficiency was severe to a certain extent did it further lead to a significant increase in V_J. At 30 days of age, V_J and V_I

showed a significant increase in V_J and V_I in and only in the low nitrogen treatments, and there was no significant difference between wood fiber and control, suggesting that the effect of the wood fiber substrate was weakening with time, but that the effect of severe low nitrogen continued to affect the chlorophyll fluorescence performance of lettuce. In addition, at the same level of nitrogen application, both V_J/V_I were lower in the wood fiber group than in the control group, suggesting that nitrogen immobilization by wood fiber had a greater effect on the lettuce I step.

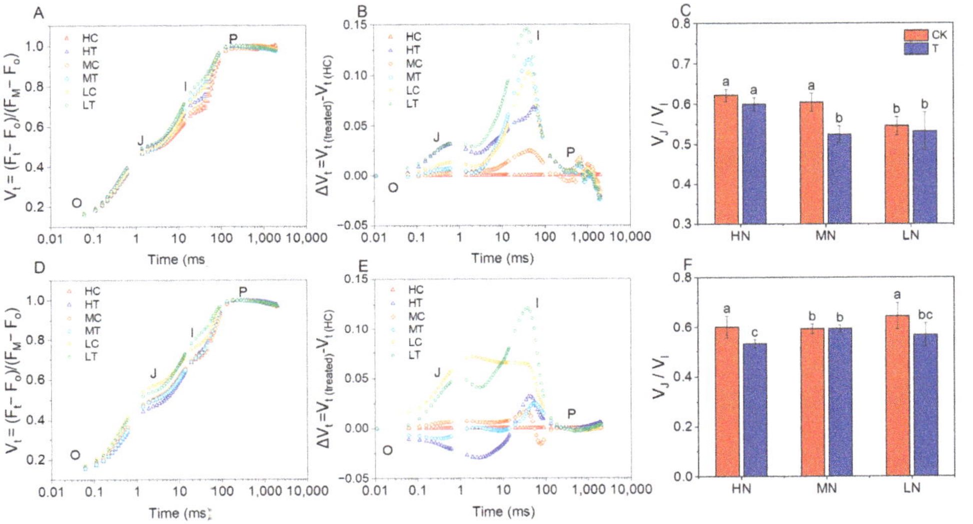

Figure 4. Relative variable fluorescence curves and V_J/V_I of lettuce under different treatments ((**A–C**) for Day 15 and (**C–F**) for Day 30). High nitrogen (100% N, HN), medium nitrogen (60% N, MN), and low nitrogen (30% N, LN). Control substrate (CK) and wood fiber substrate (T). High nitrogen & control substrate (HC), high nitrogen & wood fiber substrate (HT), medium nitrogen & control substrate (MC), medium nitrogen & wood fiber substrate (MT), low nitrogen & control substrate (LC), and low nitrogen & wood fiber substrate (LT). Within each organic acid or germination index, values that do not share a letter are significantly different according to ANOVA and Duncan's multiple range test ($p < 0.05$).

To further assess the events reflected in the OK, OJ, OI and IP phases, additional normalisations and corresponding subtractions (differential kinetics) of the fluorescence rise kinetic curves are also presented. Fluorescence rise kinetic curves for the different treatments were double normalised between O (20 µs) and K steps (300 µs) to show the L peak, and differential kinetics were plotted on a linear time scale from 0 to 300 µs, $W_{OK} = (F_t - F_O)/(F_K - F_O)$, $\Delta W_{OK} = W_{OK(treated)} - W_{OK(HC)}$ (Figure 5A,C). Simultaneous double normalisation between O (20 µs) and J-steps (2 ms) was performed to show the K-peak and to plot the differential dynamics on a linear time scale from 0 to 2 ms, $W_{OJ} = (F_t - F_O)/(F_J - F_O)$, $\Delta W_{OJ} = W_{OJ(treated)} - W_{OJ(HC)}$ (Figure 5B,D). At 15 days of age, lettuce in the wood fiber group showed distinct L and K peaks at all levels of nitrogen application, whereas the control group showed L and K peaks only at low nitrogen. Surprisingly, the wood fiber group exhibited the highest L and K peaks at high nitrogen levels, which usually implies the most intense adversity stress. at 30 days of age, significant L and K peaks were only evident in low-nitrogen wood fiber lettuce.

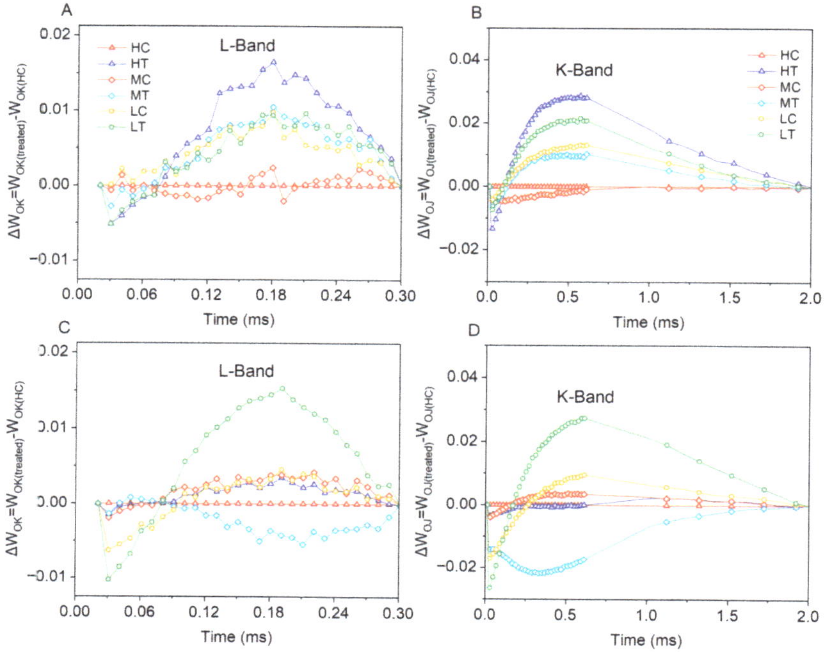

Figure 5. ΔW_{OK}, ΔW_{OJ} of lettuce under different treatments ((**A**,**B**) are for Day 15, (**C**,**D**) are for Day 30). High nitrogen & control substrate (HC), high nitrogen & wood fiber substrate (HT), medium nitrogen & control substrate (MC), medium nitrogen & wood fiber substrate (MT), low nitrogen & control substrate (LC), and low nitrogen & wood fiber substrate (LT).

As shown in Figure 6, the chlorophyll fluorescence transients of each group were double normalised between F_O and F_I and expressed as relative variable fluorescence $W_{OI} = (Ft - Fo)/(F_I - Fo)$, $\Delta W_{OI} = W_{OI(treated)} - W_{OI(HC)}$, $F_I = Ft = 30$ ms, and plotted $W_{OI} \geq 1$ over the 30–270 ms time range to show how the IP phase was affected. ΔW_{OI} was used to show the effect of different treatments on the J-step, and the results were generally consistent with ΔVt. The amplitude of the I-P phase in the $W_{OI} \geq 1$ fraction reflects the size of the terminal electron acceptor pool on the receptor side of photosystem I (PSI), with smaller amplitudes indicating a smaller terminal electron acceptor pool on the PSI receptor side [24]. The wood fiber substrate as well as the amount of nitrogen applied affected the amplitude of the W_{OI} curves to varying degrees. At 15 days of age, the amplitude of the W_{OI} curve decreased with decreasing nitrogen application, both for wood fiber substrate and control. At the same level of nitrogen application, the W_{OI} amplitude of lettuce in the wood fiber group was even lower. The W_{OI} performance of the different treatments at 30 days of age was essentially the same as at 15 days of age. Considering that the wood fiber substrate itself reduces the effective supply of nitrogen, the above facts suggest that the reduction in nitrogen supply reduces the terminal electron receptor pool on the PSI receptor side and that this inhibition persists over the 30-day cultivation period.

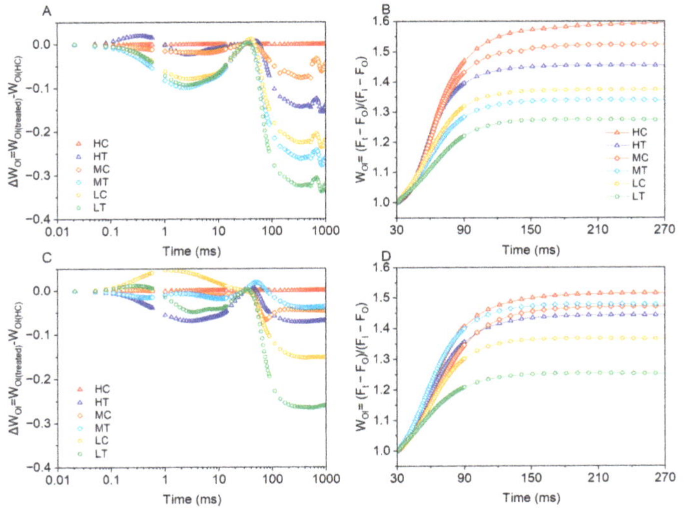

Figure 6. W_{OI}, ΔW_{OI} of lettuce under different treatments ((**A**,**B**) are for Day 15, (**C**,**D**) are for Day 30). High nitrogen & control substrate (HC), high nitrogen & wood fiber substrate (HT), medium nitrogen & control substrate (MC), medium nitrogen & wood fiber substrate (MT), low nitrogen & control substrate (LC), and low nitrogen & wood fiber substrate (LT).

2.3.2. Photosynthetic Electron Transfer

The quantum yield-related data were normalized and visualized as a spider plot to illustrate changes in lettuce quantum yield, from photon uptake to the electron acceptor at the PSI terminal, under each treatment. As shown in Figure 7, at 15 days, the maximum quantum yield for primary photochemistry (φPo), the probability of electron transfer beyond Q_A^- (ψEo), and the quantum yield for electron transport (φEo) did not differ significantly. Subsequently, electron transfer efficiency diverged significantly. δRo represents the probability of electron transport from the reduced intersystem electron acceptor (RE) to the final electron acceptor in PSI. At 15 days, δRo was significantly lower in the wood fiber group than in the CK under the same nitrogen level. The reductions were 15.08%, 27.90%, and 17.68% at high, medium, and low nitrogen levels, respectively. Similarly, φRo, the quantum yield for the reduction of end electron acceptors at the PSI acceptor side (RE), decreased by 20.22%, 27.84%, and 21.46%, respectively. At 30 days, similar to ΔV_t trends, differences in electron transfer efficiency diminished across treatments, except for the two low-nitrogen treatments (LC, LT).

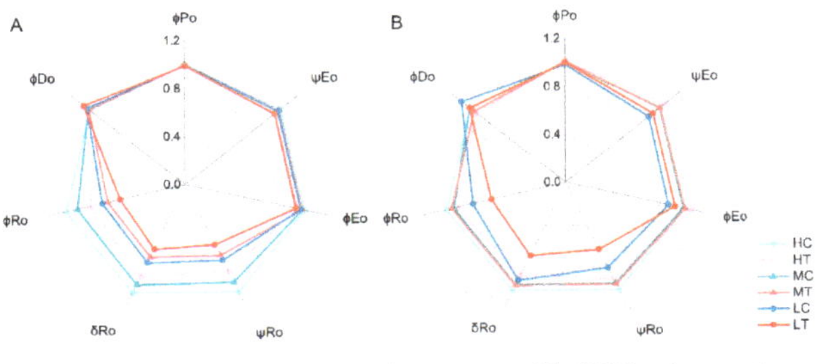

Figure 7. Normalised spider plots of quantum yield of PSII-active reaction centers in lettuce under different treatments ((**A**) for Day 15 and (**B**) for Day 30). High nitrogen & control substrate (HC), high nitrogen & wood fiber substrate (HT), medium nitrogen & control substrate (MC), medium nitrogen & wood fiber substrate (MT), low nitrogen & control substrate (LC), and low nitrogen & wood fiber substrate (LT).

Changes in electron transfer energy fluxes at the core of the active reaction were further visualized using an energy pipeline membrane model of specific fluxes in the photosynthetic apparatus (Appendix A Figure A1). The arrow sizes of ABS/RC, TR_0/RC, ET_0/RC, RE_0/RC, and DI_0/RC represent the efficiencies of light absorption, trapping, electron transport, reduction of electron acceptors, and dissipation, respectively, for each active reaction center (RC). In this study, RE_0/RC exhibited the most pronounced response to treatments, showing a significant decline in the wood fiber group compared to the control at the same nitrogen level, indicating reduced electron flux from Q_A^- to the PSI final electron acceptor. These findings indicate that nitrogen availability significantly influences electron transfer in the lettuce photosynthetic apparatus, with the strongest impact occurring between the reduced intersystem electron acceptor (RE) and the PSI final electron acceptor.

The performance index PI_{abs} quantifies PSII performance by measuring the efficiency from photon absorption to electron transport chain reduction. PI_{total} represents the efficiency of electron transfer from photon absorption to the PSI terminal electron acceptor, reflecting its performance [27]. As shown in Figure 8, at 15 days, PI_{abs} and PI_{total} were lower in the wood fiber group than in the control (C) group under the same nitrogen level. The reduction in PI_{total} was significantly greater than that in PI_{abs}, reaching statistical significance. PI_{abs} decreased by 26.02%, 3.76%, and 13.90%, while PI_{total} decreased by 51.67%, 49.35%, and 40.94% at high, medium, and low nitrogen levels, respectively. This suggests that both PSII and PSI functions were impaired in the wood fiber group, with a more pronounced reduction in the PSI terminal electron acceptor. These findings align with the aforementioned quantum yield and energy flow results. Interestingly, by 30 days, the differences between the wood fiber and control groups had disappeared, suggesting that the nitrogen immobilization-induced damage to the lettuce photosystem may be reversible.

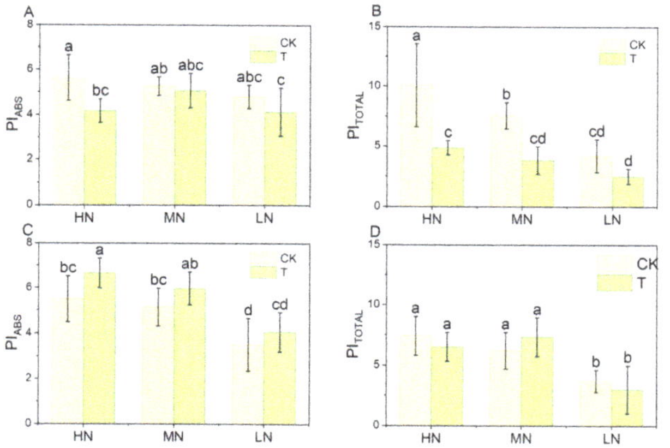

Figure 8. PI_{abs}, PI_{total} of lettuce under different treatments ((**A**,**B**) for Day 15 and (**C**,**D**) for Day 30). High nitrogen (100% N, HN), medium nitrogen (60% N, MN), and low nitrogen (30% N, LN). Control substrate (CK) and wood fiber substrate (T). Within each organic acid or germination index, values that do not share a letter are significantly different according to ANOVA and Duncan's multiple range test ($p < 0.05$).

2.4. *Lettuce Chlorophyll Content and Gas Exchange Parameters*

The chlorophyll content of lettuce cultivated on wood fiber substrate was significantly lower compared to the control (Figure 9). This was particularly evident at 30 days, where chlorophyll a and b contents of wood fiber substrate lettuce were significantly lower ($p < 0.05$) than the control at both medium and low nitrogen levels, and the total chlorophyll content decreased by 13.7%, 36.2%, and 29.6%, respectively, and with decreasing nitrogen application.

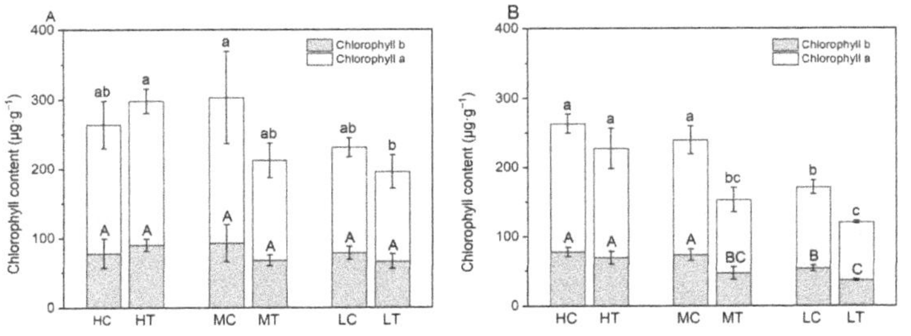

Figure 9. Chlorophyll content of lettuce under different treatments ((**A**) for Day 15 and (**B**) for Day 30). High nitrogen & control substrate (HC), high nitrogen & wood fiber substrate (HT), medium nitrogen & control substrate (MC), medium nitrogen & wood fiber substrate (MT), low nitrogen & control substrate (LC), and low nitrogen & wood fiber substrate (LT). Within each organic acid or germination index, values that do not share a letter are significantly different according to ANOVA and Duncan's multiple range test ($p < 0.05$).

Wood fiber substrates significantly reduced the net photosynthetic rate of lettuce (Figure 10). The net photosynthetic rate of lettuce gradually decreased as nitrogen levels decreased. At the same nitrogen level, the net photosynthetic rate of lettuce with wood fiber substrates was lower than that of the control. At 15 days, the net photosynthetic rate of the wood fiber group was significantly lower than that of the control group at high, medium, and low nitrogen levels, with decreases of 32.1%, 33.2%, and 27.5%, respectively ($p < 0.05$). At 30 days, the decreases were 7.2%, 34.1%, and 31.3%, respectively, and were significant ($p < 0.05$) at medium and low nitrogen levels. Interestingly, stomatal conductance, intercellular CO_2 concentration, and transpiration rate were lower in the wood fiber group compared to the control group at the same nitrogen level. However, at 30 days, these parameters were higher in the wood fiber group, indicating that CO_2 was not efficiently utilized after entering the leaves, leading to Ci accumulation. This suggests that at 15 days, photosynthesis in the wood fiber group is primarily limited by stomatal factors, while at 30 days, it is limited by non-stomatal factors [28]. This difference supports the idea that the mechanism of photosynthesis inhibition by wood fiber substrates exhibits significant stage-dependent characteristics.

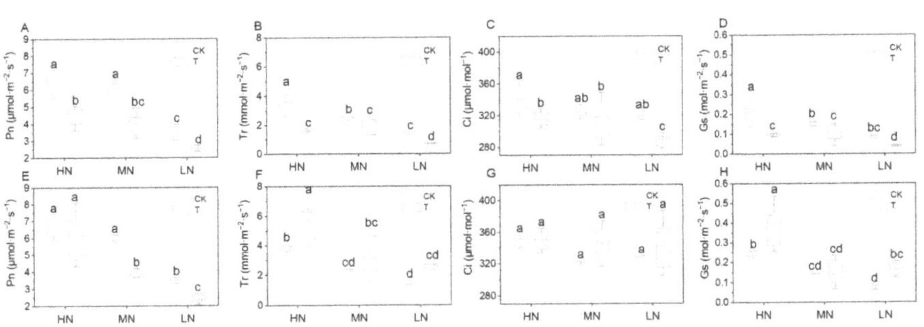

Figure 10. Photosynthetic parameters of lettuce under different treatments ((**A–D**) for Day 15 and (**E–H**) for Day 30). High nitrogen (100% N, HN), medium nitrogen (60% N, MN), and low nitrogen (30% N, LN). Control substrate (CK) and wood fiber substrate (T). Within each organic acid or germination index, values that do not share a letter are significantly different according to ANOVA and Duncan's multiple range test ($p < 0.05$).

3. Discussion

3.1. Effect of Wood Fiber Substrate on Nutrient Supply and Growth and Development of Lettuce

Wood fiber substrates with high carbon to nitrogen ratios significantly reduced the supply fast-acting nitrogen, consistent with previous findings on microbial-mediated nitrogen immobilization in organic substrates [29]. The present study showed a significant reduction in ammoniacal and nitrate nitrogen content in the wood fiber substrate compared to the control (Figure 1), which aligns with Wright et al.'s [12] findings that an additional 100 $mg \cdot L^{-1}$ nitrogen fertilizer is required for the wood substrate. Notably, despite the same total nitrogen application, lettuce biomass and total nitrogen content in the wood fiber substrate were significantly lower than in the control (Figure 2), indicating a significant inhibition of nitrogen uptake efficiency. This phenomenon may be attributed to the strong immobilization of inorganic nitrogen by microorganisms in the wood fiber substrate, converting exogenous nitrogen into organic forms and reducing effective nitrogen accumulation in plants [30]. Notably, even with 40% wood fiber in the substrate, lettuce exhibited significant nitrogen deficiency symptoms (yellowing leaves and plant dwarfism), supporting previous recommendations that wood fiber should not exceed 30% of the substrate [31]. Furthermore, nitrogen immobilization in the wood fiber substrate also affected lettuce total phosphorus and potassium content, with the gap becoming more pronounced over time. This phenomenon may be related to limited root development and down-regulation of nutrient transport protein expression due to nitrogen deficiency [32].

3.2. Effect of Wood Fiber Substrate on the Structure and Function of PSII and PSI in Lettuce

Chlorophyll a fluorescence transients provide a sensitive assay for assessing photosynthetic responses to abiotic stresses [26]. In this study, we revealed multiple mechanisms of damage to the lettuce photosystem caused by the wood fiber substrate, by analyzing the variable fluorescence curve (ΔV_t) and characteristic bands (L and K bands).

Wood fiber substrates significantly affected electron transfer in both photosystem II (PSII) and photosystem I (PSI) of lettuce. Relative variable fluorescence (V_J) at the J step reflects the efficiency of electron transfer from Q_A^- to Q_B^- on the PSII receptor side [24], And step I (V_I) characterises the redox capacity of the PSI receptor side plastoquinone (PQ) library. In the present study, V_J and V_I values were significantly higher under wood fiber treatment compared to control (Figure 3), indicating that electron transfer was blocked on both PSII donor side (OEC) and PSI acceptor side. Notably, V_J and V_I values in the wood fiber group increased simultaneously at low and medium nitrogen levels, while only V_I values were significantly higher in other treatments. This suggests that nitrogen deficiency first affects electron transfer on the PSI acceptor side, and exacerbates damage to the PSII donor side only under severe nitrogen stress. This dose-dependent effect is consistent with Guo et al.'s study on the decline of photosystem function under nitrogen stress, which may be related to reduced ferredoxin-NADP$^+$ reductase (FNR) activity and disruption of cyst-like body membrane structure caused by nitrogen deficiency [33]. At 30 days, V_J and V_I values were significantly higher only under low nitrogen treatments, with no significant difference between the wood fiber and control groups. This suggests that the negative effects of the wood fiber substrate diminished as the reproductive period progressed, but the inhibitory effects of severe low nitrogen stress on photosystem function persisted.

Structural damage on the donor side of PSII is further revealed by the kinetic features of the L and K peaks. The L peak indicates reduced connectivity of the antenna complex to reaction centers (RCs) [34,35], and the K peak reflects impaired OEC function [36]. In this study, the control group at 15 days of age showed L and K peaks only at low nitrogen levels, while the wood fiber group showed significant L and K peaks at all nitrogen levels (Figure 5), further confirming that the wood fiber substrate significantly exacerbated structural damage on the PSII donor side. At 30 days, distinct L and K peaks were observed only in the low-nitrogen wood fiber group, suggesting that the negative effects of the wood fiber substrate diminished with increased fertility, but the inhibitory effects of severe low-nitrogen stress on OEC function persisted.

A reduction in W_{OI} amplitude indicates shrinkage of the PSI-terminal electron acceptor pool (ferredoxin-NADP+ reductase) capacity [24]. Reduced nitrogen application decreased the size of the PSI receptor-side terminal electron acceptor pool, with stronger inhibition in the wood fiber group at the same nitrogen level (Figure 6). This may be related to restricted FNR synthesis due to reduced bioavailability of micronutrients (e.g., iron) in the wood fiber substrate [37]. This result, along with the 19.81–27.9% reduction in δRo in the wood fiber group (Figure 7), supports the mechanism of blocked electron transfer at the end of PSI.

Comparative analysis of chlorophyll fluorescence parameters from 15- and 30-day-old lettuce revealed a stage-specific effect of nitrogen immobilization by wood fiber substrates on photosystem function. At 15 days, the relative variable fluorescence values at the K, L, and I steps in the wood fiber group were significantly higher than those in the control group (Figure 4), indicating severe inhibition of both PSII reaction centers (particularly OEC function) and PSI receptor-side electron transfer. This is consistent with the fact that nitrogen immobilization is strongest early in planting (within 2 weeks) [30]. However, at 30 days, differences between the wood fiber and control groups were no longer significant at the K and L steps, with only the I step remaining significant, indicating superior recovery of the PSII structure compared to PSI. In performance indicator analysis, both PIabs and PItotal were significantly lower in the wood fiber group than in the control at 15 days, suggesting impaired PSII and PSI performance, particularly a significant reduction in the PSI terminal electron acceptor's reduction capacity. By 30 days, PIabs were higher in the wood fiber group than in the control, and the gap between PI_{abs} and PI_{total} narrowed, indicating compensatory restoration of photosystem function and improved synergy between PSII and PSI as nitrogen immobilization waned. These results suggest that wood fiber substrate-induced photosystem damage is reversible, with recovery potential linked to rapid PSII repair mechanisms (e.g., D1 protein turnover) and compensatory effects of PSI cyclic electron flow [38]. This finding provides a theoretical basis for optimizing the application strategy of wood fiber substrate.

3.3. *Effect of Wood Fiber Substrates on Photosynthesis*

Nitrogen immobilization by wood fiber substrates significantly inhibited chlorophyll biosynthesis in lettuce. Chlorophyll, a key pigment of the photosynthetic system, directly influences the efficiency of light energy capture and conversion [39]. Nitrogen is essential for chlorophyll porphyrin ring synthesis and also regulates pigment accumulation by influencing chloroplast developmental gene expression [25,40,41]. The reduced nitrogen availability caused by the wood fiber substrate in this study (Figure 1) directly explains the significant decline in lettuce chlorophyll content. Additionally, potassium regulates stomatal function, activates photosynthesis-related enzymes, and facilitates assimilate transport, while phosphorus plays a direct role in assimilation and photophosphorylation [42]. Thus, nitrogen deficiency, along with potassium and phosphorus shortages, further impairs lettuce photosynthetic performance.

The inhibitory effect of wood fiber substrates on photosynthesis exhibited distinct stage-dependent characteristics (Figure 10). At the early stage (15 days), stomatal limitation was the primary factor restricting photosynthesis, indicated by a 36.2–56.8% reduction in stomatal conductance (Gs) and a 2.8–10.1% decline in intercellular CO_2 concentration (Ci). This phase may be regulated by the ABA signaling pathway, triggered by acute nitrogen deficiency, which induces stomatal closure to minimize transpirational water loss [43]. By 30 days, the net photosynthetic rate (Pn) declined significantly (by 7.2–34.1%), even though Gs and Ci had recovered to control levels or higher, indicating a shift from stomatal to non-stomatal limitation [28]. Photosystem damage at this stage was characterized by: (1) impaired electron transfer at the PSI terminus, with δRo decreasing by 13.93–25.17% and RE0/RC decreasing by 8.79–19.74% (Figure 7, Supplementary Figure **A1**). (1) Indicating an imbalance in the ferredoxin redox cycle. (2) Reduced Rubisco activity, as evidenced by elevated Ci and persistently low Pn, indicating insufficient carboxylation efficiency. (3) Reduced photosynthetic complexes: chlorophyll content declined by 13.7–36.2%, leading to a lower density of photosystem reaction centers (RCs). This transition aligns with Cetner et al.'s theory that nitrogen deficiency triggers adaptive remodeling of photosynthetic organs, where plants prioritize basal metabolism over light energy conversion efficiency under prolonged nitrogen stress [23].

Wood fiber substrates likely inhibit photosynthesis through three primary mechanisms: (1) reduced chlorophyll synthesis, which limits light energy capture; (2) decreased stomatal conductance, leading to a lower net photosynthetic rate; and (3) structural impairment of the PSII and PSI systems. Notably, at 30 days, the difference between PIabs and PItotal narrowed in the wood fiber group, indicating a partial compensatory recovery of photosystem function. This suggests that wood fiber-induced damage was not entirely irreversible. These findings provide a theoretical basis for dynamically adjusting nitrogen supply during the reproductive period to mitigate its adverse effects.

4. Materials and Methods

4.1. Experimental Materials and Treatment Setup

The trial was conducted and completed in September 2023 at the Institute of Urban Agriculture, Chinese Academy of Agricultural Sciences. The substrate materials used in the trial included wood fiber, peat, vermiculite, and perlite. Wood fiber was sourced from Shanghai Meizijia Horticulture Co., Ltd. (Shanghai, China), whereas peat, vermiculite, and perlite were purchased from the market. Table 2 presents the substrate types: control substrate (CK) and wood fiber substrate (T). The control substrate was formulated as V_{peat}:$V_{perlite}$:$V_{vermiculite}$ = 9:3:1. In order to exclude the interference of other factors, we screened the wood fibre substrate for group allocation ratio before the experiment to control the physicochemical properties which were basically the same as the control substrate. The final formulation was determined as $V_{wood\ fiber}$:V_{peat}:$V_{perlite}$ = 2:2:1.

Table 2. Basic physico-chemical properties of test substrates.

	Capacity ($g{\cdot}cm^{-3}$)	Porosity (%)			pH	EC ($ms{\cdot}cm^{-1}$)	Nutrients ($g{\cdot}kg^{-1}$ Dry Weight)			
		Total Porosity	Water-Holding Porosity	Air Porosity			Quick-Acting Nitrogen	Total Nitrogen	Total Phosphorus	Total Potassium
Control substrate	0.21	0.72	0.68	0.04	6.26	0.46	0.30	0.82	0.01	2.2
Wood fiber	0.18	0.8	0.72	0.07	6.09	0.42	0.24	0.65	0	1.67

The second experimental factor was nitrogen level, which was adjusted based on Hoagland's nutrient solution [44]. Three nitrogen levels were established by varying the nitrogen concentration: low nitrogen (30% N, LN), medium nitrogen (60% N, MN), and high nitrogen (100% N, HN). The concentrations of other nutrients remained essentially unchanged, and the detailed nutrient solution formulations are presented in Table A1. Six treatment groups were included in the study: high nitrogen & control(HC), high nitrogen & treatment (HT), medium nitrogen &control (MC), medium nitrogen & treatment (MT), low nitrogen & control (LC), and low nitrogen & treatment (LT), with each treatment having 10 replicates.

4.2. Plant Cultivation and Sampling

The test crop was cream lettuce (*Lactuca sativa cv.*) and square plastic boxes (8 cm × 8 cm × 8 cm) filled with substrate were used for the cultivation experiment. Seeds were germinated at a temperature of 22 °C and 70% humidity, shaded with a shade cloth. At 21 days after lettuce sowing, when lettuce had two leaves took the seedlings with even growth and set them in the cultivation box, one plant per box. They were placed in a light incubator with the temperature set at 22 °C for 14 h during the day and 18 °C for 10 h at night. Humidity 70%, light intensity 30,000 Lux. The light quality is white LED light. An equal amount of nutrient solution was applied to each cultivation box every 5 days during crop growth.

Analytical tests were carried out by randomly taking five plants from each treatment on the 15th and 30th day after planting, respectively. The tests consisted of three categories, namely (1) chlorophyll fluorescence curve and photosynthesis analysis of lettuce were firstly carried out, (2) Lettuce was harvested and its aboveground fresh weight, nutrient content, and chlorophyll content were determined, and (3) substrates were collected and their nitrogen fixation and ammonia and nitrate-nitrogen concentrations were determined.

4.3. Determination of Physico-Chemical Properties of Cultivated Substrates

4.3.1. Determination of Physical and Chemical Properties of Substrates

Physical properties of the substrates, including bulk density, total porosity, air voids, and water-holding porosity, were measured using the ring knife method. Chemical properties were analyzed with a multi-parameter tester (Mettler Toledo Instruments, Shanghai, China): pH and electrical conductivity (EC) were determined in a 1:10 (w/v) extract of air-dried substrate and distilled water. Total nitrogen (TN) content was measured using a total organic carbon analyzer (Elemental XPERT-TOC/TN). For this, 0.1 g of freshly milled substrate was digested with 5 mL concentrated H_2SO_4 at 375 °C, followed by analysis with an inductively coupled plasma optical emission spectrometer (ICP-OES, Thermo Scientific ICAP PRO XP) [44]. Total phosphorus (TP) and total potassium (TK) were also determined by ICP-OES, while total carbon (TC) was analyzed using an elemental analyzer. Available potassium was extracted using the ammonium acetate leaching method and quantified via flame photometry. Available phosphorus was measured using the molybdenum-antimony anti-colorimetric method. The contents of cellulose, hemicellulose, and lignin were determined via the Van Soest method (a detergent fiber analysis technique).

4.3.2. Determination of Substrate Nitrogen Immobilization

According to Vandecasteele's method [18], 350 mg $N \cdot L^{-1}$ of KNO_3 was added to the material and incubated at 28 °C for 4 days. The rate of nitrogen immobilization (% N) was calculated based on the difference between the mineral nitrogen content of cultured Day0 and Day4.

The steps were as follows: firstly, KNO_3 solution with NO_3^--N content of 2000 $mg \cdot L^{-1}$ was prepared, 400 mL of each material was taken and weighed (M1), and 70 mL of KNO_3 solution was added to each material separately. Pure water was added to the material to adjust the moisture content to 70%, mixed thoroughly and the wet weight M2 was recorded. Place in a 200 mL acrylic box with a perforated lid and shake to level. On day 0 of incubation, the samples were left for half an hour and weighed 5.00 g. The samples were macerated with 25 mL 2 $mol \cdot L^{-1}$ KCL (1:5 V soil/V solution). The extract was prepared after shaking and filtration and its concentration (C1) was determined. The material was incubated at 28 °C and 70% relative humidity for 4 days, maintaining normal aeration of the cover holes. The water content was adjusted using deionised water and maintained at 70% (w/w) during incubation. On day 4 of incubation, the mineral nitrogen concentration (C2) in the material was leached using the same method as on day 0. Nitrogen immobilization rate and nitrogen immobilization were calculated from the difference between the 0-day nitrogen concentration and the 4-day nitrogen concentration. A 100% nitrogen immobilization rate means that all mineral nitrogen in the 350 $mg \cdot L^{-1}$ N material is fixed.

$$\text{Nitrogen immobilization rate} = (C1 - C2)/C1 * 100\%$$

$$\text{Nitrogen immobilization} = \frac{140 - C2 * \frac{25}{5} * M2}{0.4} (mg \cdot L^{-1})$$

4.3.3. Determination of Ammonia and Nitrate Nitrogen in Substrates

The ammonia and nitrate nitrogen contents of the extracts were determined using a continuous flow analyser (SEAL Analytical, AA3, Mequon, WI, USA) to characterise the fast-acting nitrogen content of the substrate, using a 1:5 extract of fresh substrate and 2 $mol \cdot L^{-1}$ KCl solution (w/v).

4.4. Determination of Physiological Indicators in Lettuce

4.4.1. Determination of Chlorophyll Content and Plant Nutrients in Lettuce

On the day of harvest for each crop, take the first leaf of lettuce that is fully expanded and photosynthetic pigments were determined according to the Lichtenthaler spectrophotometric method, including chlorophyll a, chlorophyll b [24]. Nitrogen, phosphorus and potassium were determined as previously described for nitrogen, phosphorus and potassium in planting substrates.

4.4.2. Determination of Chlorophyll a Fluorescence Parameters in Lettuce

Chlorophyll fluorescence parameters were measured on the day of harvest for each crop using a FluorPen FP110 handheld chlorophyll fluorometer, and leaves were dark-adapted for 30 min prior to measurement to ensure that all PSIIs were dark-adapted. The chlorophyll a fluorescence transient was then triggered with a beam of saturated red light (~3000 $\mu mol \cdot m^{-2} \cdot s^{-1}$) with a peak wavelength of 650 nm [45].The chlorophyll a fluorescence parameters involved in this study are shown in Table A2.

4.4.3. Determination of Gas Exchange Data for Lettuce

Gas exchange data were measured on the day of harvest for each crop. The first fully expanded leaf was selected from five plants with uniform growth in each treatment, and the photosynthetic rate (Pn), intercellular CO_2 concentration (Ci), and transpiration rate (Tr) were measured using a LI-6400 portable photosynthesizer on a sunny day between 9:30 and 11:00 a.m. The results showed that the photosynthesis rate (Pn), inter-cellular CO_2 concentration (Ci), and transpiration rate (Tr) were all measured at the same time and in the same manner as in the other treatments. Stomatal limiting values were judged according to the formula Ls = 1 − Ci/Ca, with Ca being the ambient CO_2 concentration of Berry et al. [46].

4.5. Statistical Analysis

This study was analysed by one-way ANOVA using IBM SPSS Statistics 27, significant differences between treatment means were detected using Duncan's multiple range test ($p < 0.05$) and graphing was done using Origin 2022.

5. Conclusions

This study systematically examined the physiological mechanisms by which wood fiber substrates influence lettuce photosynthesis via nitrogen immobilization. The high carbon-to-nitrogen ratio of the wood fiber substrate significantly reduced the supply of fast-acting nitrogen (Figure 1), decreasing lettuce nitrogen uptake efficiency (Figure 3), triggering typical nitrogen deficiency symptoms (Figure 2), and reducing chlorophyll content (Figure 9). Chlorophyll fluorescence kinetics analysis suggests that wood fiber substrates impair photosystem function through multiple pathways: In the early stage (15 days), structural damage occurred on the donor side (OEC) of PSII, and electron transfer was blocked on the acceptor side of PSI (Figure 5). In the later stage (30 days), PSII function was partially restored, but PSI damage persisted, indicating that nitrogen deficiency caused by wood fiber nitrogen immobilization affects the photosystem in stages (Figure 6). Notably, some of the photosystem damage induced by wood fibers appears reversible. The results provide a theoretical basis for optimizing nitrogen management in wood fiber substrates and suggest that negative effects can be mitigated by dynamically regulating nitrogen supply during the reproductive period and supplementing trace elements (e.g., iron and magnesium), promoting the efficient use of wood fiber substrates in horticulture.

Author Contributions: Conceptualization, L.W. and W.Z.; methodology, W.Z.; software, J.L. and R.L.; validation, W.C.; formal analysis, L.W.; investigation, R.L.; resources, Z.Q.; data curation, L.W.; writing—original draft preparation, L.W.; writing—review and editing, W.Z.; visualization, L.W.; supervision, Z.Q.; project administration, Z.Q.; funding acquisition, Z.Q. All authors have read and agreed to the published version of the manuscript.

Funding: This research was funded by the Sichuan Science and Technology Program: No. 2025ZNS-FSC0439; the Chengdu Agricultural Science and Technology Center Local Finance Special Funds Project: NASC2023ST09, NASC2024TD01; and the major task project of the Institute of Urban Agriculture, Chinese Academy of Agricultural Sciences: SZ202402.

Data Availability Statement: Data is contained within the article.

Conflicts of Interest: The authors declare no conflicts of interest.

Abbreviations

The following abbreviations are used in this manuscript:

N	Nitrogen
PSII	Photosystem system II
PSI	Photosystem system I
OEC	Oxygen evolving complex
Pn	Photosynthesis rate
Ci	Inter-cellular CO_2 concentration
Tr	Transpiration rate
Gs	Stomatal conductance
Ls	Stomatal limiting values
CK	Control substrate
T	Wood fiber substrate
L	Low nitrogen
M	Medium nitrogen
H	High nitrogen

Appendix A

Table A1. Nutrient solution formulations used in this study.

Formulation Name		Elemental (mg·L^{-1})				Nitrogen Level	Concentration of Each Element (mg·L^{-1})				
		$Ca(NO_3)_2·4H_2O$	KNO_3	$NH_4H_2PO_4$	$MgSO_4·7H_2O$		N	P	K	Ca	Mg
Hogeland and Anon Universal Nutrient Solution (modified)	HN	945	607	115	493	100%	210.3	31.0	234.4	160.2	48.1
	MN	240	607	115	493	60%	126.6	31.0	234.4	40.7	48.1
	LN	0	350	115	493	30%	62.5	31.0	135.1	0.0	48.1

Table A2. The chlorophyll fluorescence parameters involved in this study.

Terms and Formulas	Illustrations
$F_O = F$ (20 μs)	F_O (Minimum fluorescence): the fluorescence value in the dark-adapted state when all reaction centres are open, i.e., when Q_A^-, the primary electron acceptor of the reaction centre, is in the oxidation state
$F_L = F$ (150 μs)	Fluorescence value at 150 μs
$F_K = F$ (300 μs)	Fluorescence value at 300 μs
$F_J = F$ (3 ms)	Fluorescence value at 3 ms (J level of OJIP)
$F_I = F$ (30 ms)	Fluorescence value at 30 ms (I level of OJIP)
$F_M = F_P = F\ (t)_{max}$	F_M (maximum fluorescence): the fluorescence value when all reaction centres are switched off in the dark-adapted state
$F_v = F_M - F_O$	Maximum variable fluorescence
$V_L = (F_L - F_O)/\ (F_M - F_O)$	Relative variable fluorescence in the L phase of the fluorescence induction curve
$V_K = (F_K - F_O)/\ (F_M - F_O)$	Relative variable fluorescence in the K phase of the fluorescence induction curve
$V_J = (F_J - F_O)/\ (F_M - F_O)$	Relative variable fluorescence in the J phase of the fluorescence induction curve
$V_I = (F_I - F_O)/\ (F_M - F_O)$	Relative variable fluorescence in stage I of the fluorescence induction curve
$M_0 = (\Delta V/\Delta t)_0 = 4\ (F_{300\mu s} - F_O)/\ (F_M - F_O) = TR_0/RC - ET_0/RC$	Approximate initial slope (per millisecond) of transient fluorescence V = f(t).This parameter reflects the rate of closure of PS II active reaction centres (RCs).
$ABS/RC = [4\ (F_{300\mu s} - F_O)/\ (F_{2ms} - F_O)]\ (F_M/F_V) = M_0\ (1/V_J)\ (1/\phi_{Po})$	Energy flux absorbed per unit PS II active reaction centre
$TR_0/RC = \phi_{Po}\ (ABS/RC)$	Energy flux captured per unit PS II active reaction centre at t = 0
$ET_0/RC = \phi_{Eo}\ (ABS/RC)$	Energy flux transferred per unit PS II active reaction centre at t = 0
$RE_0/RC = \phi_{Ro}\ (ABS/RC) = (1 - V_I) \times (M_0/V_J)$	Specific electron flux per unit of PS II active reaction centre reducing PS I terminal electron acceptor at t = 0
$DI_0/RC = (ABS/RC) - (TR_0/RC)$	Energy flux per unit PS II active reaction centre heat dissipation at t = 0
$\phi_{Po} = TR_0/ABS = F_V/F_M = 1 - F_O/F_M$	The maximum quantum yield of the primary photochemical reaction at t = 0 refers to the quantum yield of photons captured by the PSII reaction centre and also indicates the maximum photochemical efficiency of the PSII reaction centre in the dark-adapted state
$\psi_{Eo} = ET_0/TR_0 = (1 - V_J)$	Ratio of electrons captured by the active reaction centre of PS II being transferred at t = 0
$\phi_{Eo} = ET_0/ABS = (1 - F_O/F_M)\ \psi_{Eo} = \phi_{Po}\psi_{Eo}$	Quantum efficiency of electron transfer from Q_A^- to electron transfer chain at t = 0
$\psi_{Ro} = RE_0/TR_0 = 1 - V_I$	Ratio of electrons captured by the active reaction centre of PSII being transferred to PSI at t = 0
$\delta_{Ro} = RE_0/ET_0 = (1 - V_I)/\ (1 - V_J)$	Ratio of electrons in the reduced PSI receptor measuring terminal electron acceptor to electrons in the electron transport chain
$\phi_{Ro} = RE_0/ABS = \phi_{Po}\psi_{Ro} = \phi_{Po}\psi_{Eo}\delta_{Ro}$	Quantum yields of electron acceptors at the lateral end of reduced PSI receptors
$\phi_{Do} = F_O/F_M = 1 - \phi_{Po}$	Thermal dissipation quantum yield at t = 0, reflecting the proportion of absorbed light energy dissipated by non-photochemical quenching processes in PSII

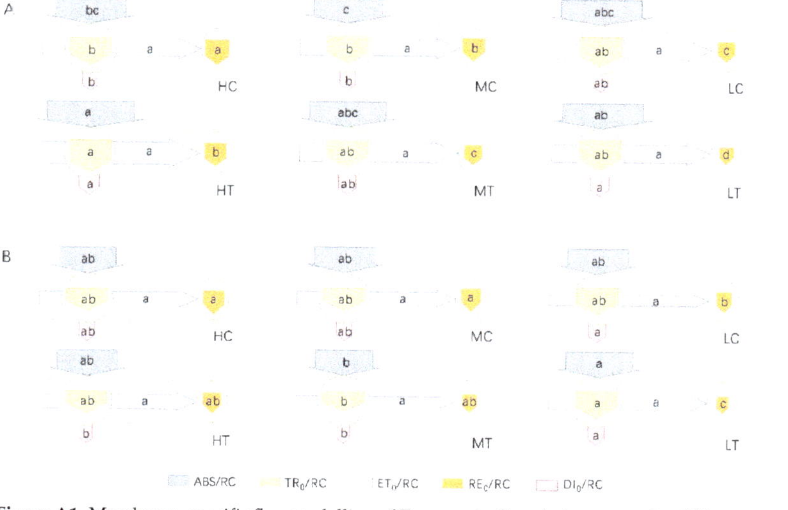

Figure A1. Membrane-specific flux modelling of Energy pipelines in lettuce under different treatments ((**A**), day 15, (**B**), day 30).

References

1. Schmilewski, G. Growing medium constituents used in the eu. *Acta Hortic.* **2009**, *819*, 33–46. [CrossRef]
2. Gruda, N.; Bisbis, M.; Tanny, J. Impacts of protected vegetable cultivation on climate change and adaptation strategies for cleaner production—A review. *J. Clean. Prod.* **2019**, *225*, 324–339. [CrossRef]

3. Jackson, B.E.; Wright, R.D.; Barnes, M.C. Pine Tree Substrate, Nitrogen Rate, Particle Size, and Peat Amendment Affect Poinsettia Growth and Substrate Physical Properties. *HortScience* **2008**, *43*, 2155–2161. [CrossRef]
4. Barrett, G.E.; Alexander, P.D.; Robinson, J.S.; Bragg, N.C. Achieving environmentally sustainable growing media for soilless plant cultivation systems—A review. *Sci. Hortic.* **2016**, *212*, 220–234. [CrossRef]
5. Gruda, N.S. Current and Future Perspective of Growing Media in Europe. 2011. Available online: https://www.actahort.org/books/960/960_3.htm (accessed on 24 January 2025).
6. Agarwal, P.; Saha, S.; Hariprasad, P. Agro-industrial-residues as potting media: Physicochemical and biological characters and their influence on plant growth. *Biomass Convers. Biorefinery* **2023**, *13*, 9601–9624. [CrossRef] [PubMed]
7. Nocentini, M.; Panettieri, M.; García de Castro Barragán, J.M.; Mastrolonardo, G.; Knicker, H. Recycling pyrolyzed organic waste from plant nurseries, rice production and shrimp industry as peat substitute in potting substrates. *J. Environ. Manag.* **2021**, *277*, 111436. [CrossRef]
8. Gruda, N.; Schnitzler, W.H. Suitability of wood fiber substrate for production of vegetable transplants. *Sci. Hortic.* **2004**, *100*, 309–322. [CrossRef]
9. Schmilewski, G. The role of peat in assuring the quality of growing media. *Mires Peat* **2008**, *3*, 1–8.
10. Dickson, R.W.; Helms, K.M.; Jackson, B.E.; Machesney, L.M.; Lee, J.A. Evaluation of Peat Blended with Pine Wood Components for Effects on Substrate Physical Properties, Nitrogen Immobilization, and Growth of Petunia (Petunia ×hybrida Vilm.-Andr.). *HortScience* **2022**, *57*, 304–311. [CrossRef]
11. Handreck, K. Use of the nitrogen drawdown index to predict fertilizer nitrogen requirements in soilless potting media. *Commun. Soil Sci. Plant Anal. Commun. Soil Sci. Plant Anal.* **1993**, *24*, 2137–2151. [CrossRef]
12. Wright, R.D.; Jackson, B.E.; Browder, J.F.; Latimer, J.G. Growth of Chrysanthemum in a Pine Tree Substrate Requires Additional Fertilizer. *HortTechnology* **2008**, *18*, 111–115. [CrossRef]
13. Bunt. A.C. *Media and Mixes for Container-Grown Plants*; Springer: Berlin/Heidelberg, Germany, 1988. [CrossRef]
14. Prasad, M. Physical, chemical and biological properties of coir dust. *Acta Hortic.* **1997**, *450*, 21–30. [CrossRef]
15. Handreck, K.A. Rapid assessment of the rate of nitrogen immobilisation in organic components of potting media: II. Nitrogen drawdown index and plant growth. *Commun. Soil Sci. Plant Anal.* **1992**, *23*, 217–230. [CrossRef]
16. Maher, M.; Prasad, M.; Raviv, M. Organic Soilless Media Components. *Soil. Cult.* **2019**, *6*, 303–378.
17. Owen, W.; Jackson, B.; Whipker, B.; Fonteno, W. Pine Wood Chips as an Alternative to Perlite in Greenhouse Substrates: Nitrogen Requirements. *HortTechnology* **2016**, *26*, 199–205. [CrossRef]
18. Vandecasteele, B.; Viaene, J.; Debode, J.; Muylle, H.; Perneel, M. Selecting raw materials for sustainable growing media: A protocol to assess the risk for N immobilization. In Proceedings of the International Society for Horticultural Science (ISHS), Leuven, Belgium, 2017; pp. 149–160. Available online: https://www.actahort.org/books/1168/1168_20.htm (accessed on 24 January 2025).
19. Gruda, N.; Tucher, S.; Schnitzler, W. N-Immobilization of wood fiber substrates in the production of tomato transplants (*Lycopersicon lycopersicum* (L.) Karst. Ex Farw.). *J. Appl. Bot. Food Qual.* **2000**, *74*, 32–37.
20. Redillas, M.C.F.R.; Jeong, J.S.; Strasser, R.J.; Kim, Y.S.; Kim, J.-K. JIP analysis on rice (*Oryza sativa* cv Nipponbare) grown under limited nitrogen conditions. *J. Korean Soc. Appl. Biol. Chem.* **2011**, *54*, 827–832. [CrossRef]
21. Kalaji, H.M.; Jajoo, A.; Oukarroum, A.; Brestic, M.; Zivcak, M.; Samborska, I.A.; Cetner, M.D.; Łukasik, I.; Goltsev, V.; Ladle, R.J.; et al. The Use of Chlorophyll Fluorescence Kinetics Analysis to Study the Performance of Photosynthetic Machinery in Plants. In *Emerging Technologies and Management of Crop Stress Tolerance*; Ahmad, P., Rasool, S., Eds.; Academic Press: New York, NY, USA, 2014; pp. 347–384.
22. Strasser, R.J.; Tsimilli-Michael, M.; Srivastava, A. Analysis of the Chlorophyll a Fluorescence Transient. In *Chlorophyll a Fluorescence: A Signature of Photosynthesis*; Papageorgiou, G.C., Govindjee, Eds.; Springer: Dordrecht, The Netherlands, 2004; pp. 321–362.
23. Cetner, M.D.; Kalaji, H.M.; Goltsev, V.; Aleksandrov, V.; Kowalczyk, K.; Borucki, W.; Jajoo, A. Effects of nitrogen-deficiency on efficiency of light-harvesting apparatus in radish. *Plant Physiol. Biochem.* **2017**, *119*, 81–92. [CrossRef]
24. Guo, P.; Ren, J.; Shi, X.; Xu, A.; Zhang, P.; Guo, F.; Feng, Y.; Zhao, X.; Yu, H.; Jiang, C. Optimized nitrogen application ameliorates the photosynthetic performance and yield potential in peanuts as revealed by OJIP chlorophyll fluorescence kinetics. *BMC Plant Biol.* **2024**, *24*, 774. [CrossRef]
25. Peng, J.; Feng, Y.; Wang, X.; Li, J.; Xu, G.; Phonenasay, S.; Luo, Q.; Han, Z.; Lu, W. Effects of nitrogen application rate on the photosynthetic pigment, leaf fluorescence characteristics, and yield of indica hybrid rice and their interrelations. *Sci. Rep.* **2021**, *11*, 7485. [CrossRef]
26. Kalaji, H.M.; Jajoo, A.; Oukarroum, A.; Brestic, M.; Zivcak, M.; Samborska, I.A.; Cetner, M.D.; Łukasik, I.; Goltsev, V.; Ladle, R.J. Chlorophyll a fluorescence as a tool to monitor physiological status of plants under abiotic stress conditions. *Acta Physiol. Plant.* **2016**, *38*, 102. [CrossRef]

27. Kalaji, H.M.; Oukarroum, A.; Alexandrov, V.; Kouzmanova, M.; Brestic, M.; Zivcak, M.; Samborska, I.A.; Cetner, M.D.; Allakhverdiev, S.I.; Goltsev, V. Identification of nutrient deficiency in maize and tomato plants by in vivo chlorophyll a fluorescence measurements. *Plant Physiol. Biochem.* **2014**, *81*, 16–25. [CrossRef]
28. Gao, G.; Feng, Q.; Liu, X. Stomatal and no-stomatal limitations to photosynthesis of Populus euphratica leaves under natural conditions. *J. Arid Land Resour. Environ.* **2020**, *34*, 182–188.
29. Debode, J.; De Tender, C.; Cremelie, P.; Lee, A.S.; Kyndt, T.; Muylle, H.; De Swaef, T.; Vandecasteele, B. Trichoderma-Inoculated Miscanthus Straw Can Replace Peat in Strawberry Cultivation, with Beneficial Effects on Disease Control. *Front. Plant Sci.* **2018**, *9*, 213. [CrossRef] [PubMed]
30. Jackson, B.E.; Wright, R.D.; Alley, M.M. Comparison of Fertilizer Nitrogen Availability, Nitrogen Immobilization, Substrate Carbon Dioxide Efflux, and Nutrient Leaching in Peat-lite, Pine Bark, and Pine Tree Substrates. *HortScience Horts* **2009**, *44*, 781–790. [CrossRef]
31. Harris, C.N.; Dickson, R.W.; Fisher, P.R.; Jackson, B.E.; Poleatewich, A.M. Evaluating Peat Substrates Amended with Pine Wood Fiber for Nitrogen Immobilization and Effects on Plant Performance with Container-grown Petunia. *HortTechnology* **2020**, *30*, 107–116. [CrossRef]
32. Zhou, H.; Liu, Y.; Mu, B.; Wang, F.; Feng, N.; Zheng, D. Nitrogen limitation affects carbon and nitrogen metabolism in mung bean (*Vigna radiata* L.). *J. Plant Physiol.* **2023**, *290*, 154105. [CrossRef]
33. Mulo, P. Chloroplast-targeted ferredoxin-NADP+ oxidoreductase (FNR): Structure, function and location. *Biochim. Biophys. Acta (BBA) Bioenerg.* **2011**, *1807*, 927–934. [CrossRef]
34. Tsimilli-Michael, M.; Strasser, R.J. The energy flux theory 35 years later: Formulations and applications. *Photosynth. Res.* **2013**, *117*, 289–320. [CrossRef]
35. Kalaji, H.M.; Bąba, W.; Gediga, K.; Goltsev, V.; Samborska, I.A.; Cetner, M.D.; Dimitrova, S.; Piszcz, U.; Bielecki, K.; Karmowska, K.; et al. Chlorophyll fluorescence as a tool for nutrient status identification in rapeseed plants. *Photosynth. Res.* **2018**, *136*, 329–343. [CrossRef]
36. Strasser, R.J.; Srivastava, A.; Tsimilli-Michael, M. The fluorescence transient as a tool to characterize and screen photosynthetic samples. *Probing Photosynth. Mech. Regul. Adapt.* **2000**, *25*, 445–483.
37. Schansker, G.; Tóth, S.Z.; Strasser, R.J. Methylviologen and dibromothymoquinone treatments of pea leaves reveal the role of photosystem I in the Chl a fluorescence rise OJIP. *Biochim. Biophys. Acta (BBA) Bioenerg.* **2005**, *1706*, 250–261. [CrossRef]
38. Takahashi, S.; Milward, S.E.; Fan, D.Y.; Chow, W.S.; Badger, M.R. How does cyclic electron flow alleviate photoinhibition in Arabidopsis? *Plant Physiol.* **2009**, *149*, 1560–1567. [CrossRef]
39. Mu, X.; Chen, Y. The physiological response of photosynthesis to nitrogen deficiency. *Plant Physiol. Biochem.* **2021**, *158*, 76–82. [CrossRef]
40. Poorter, H.; Evans, J.R. Photosynthetic nitrogen-use efficiency of species that differ inherently in specific leaf area. *Oecologia* **1998**, *116*, 26–37. [CrossRef] [PubMed]
41. Qiang, B.; Zhou, W.; Zhong, X.; Fu, C.; Cao, L.; Zhang, Y.; Jin, X. Effect of nitrogen application levels on photosynthetic nitrogen distribution and use efficiency in soybean seedling leaves. *J. Plant Physiol.* **2023**, *287*, 154051. [CrossRef] [PubMed]
42. Hasanuzzaman, M.; Bhuyan, M.H.M.B.; Nahar, K.; Hossain, M.S.; Mahmud, J.A.; Hossen, M.S.; Masud, A.A.C.; Moumita; Fujita, M. Potassium: A Vital Regulator of Plant Responses and Tolerance to Abiotic Stresses. *Agronomy* **2018**, *8*, 31. [CrossRef]
43. Ologundudu, A.F.; Adelusi, A.A. Effect of nitrogen nutritional stress on some mineral nutrients and photosynthetic apparatus of *Zea mays* L. and *Vigna unguiculata* L. *Not. Sci. Biol.* **2013**, *5*, 376–382. [CrossRef]
44. Liang, X.; Wang, H.; Zhang, Y.; Yang, R.; Zhang, D.; Zhou, W.; Qi, Z.; Lin, W. The Substitution of Sphagnum for Peat as a Culture Substrate Reduces N2O Emissions from Vegetable Production Systems. *Agronomy* **2024**, *14*, 369. [CrossRef]
45. Li, P.; Zhu, Y.; Song, X.; Song, F. Negative effects of long-term moderate salinity and short-term drought stress on the photosynthetic performance of Hybrid Pennisetum. *Plant Physiol. Biochem.* **2020**, *155*, 93–104. [CrossRef]
46. Berry, J.A.; Downton, W.J.S. Environmental Regulation of Photosynthesis. In *Photosynthesis*; Academic Press: New York, NY, USA, 1982.

plants

MDPI

Review

Nitrogen Immobilization in Organic Media: A Double-Edged Sword Affecting the Utilization of Green Waste as Growing Media

Ruohan Li, Wenzhong Cui, Min Zhang, Zhiyong Qi * and Wanlai Zhou *

Institute of Urban Agriculture, Chinese Academy of Agricultural Sciences, Chengdu 610213, China; liruohan000120@163.com (R.L.); 17784240198@163.com (W.C.); zhangmin10062023@163.com (M.Z.)
* Correspondence: qizhiyong@caas.cn (Z.Q.); zhouwanlai@caas.cn (W.Z.)

Abstract

This review proposes a "phenomenon–mechanism–regulation" framework for understanding nitrogen immobilization during the conversion of green waste into growing media. Nitrogen immobilization acts as a double-edged sword: intense short-term immobilization, typically occurring within the first 1–2 weeks after substrate establishment, can rapidly deplete mineral nitrogen and induce plant nitrogen deficiency, whereas the immobilized nitrogen is subsequently incorporated into microbial biomass and lignin-associated organic pools, forming a slow-release reservoir that enhances nitrogen retention and reduces leaching losses. Owing to its extremely high C/N ratio (often >100) and the coexistence of labile carbon fractions and recalcitrant compounds (e.g., lignin and phenolics), green waste exhibits substantially stronger immobilization potential than conventional media. Empirical evidence indicates that nitrogen immobilization can reach 10–115 mg $N \cdot L^{-1}$ within a few days in wood-derived substrates, and additional fertilization of up to 100 mg $N \cdot L^{-1}$ may be required to maintain crop growth. Mechanistically, nitrogen immobilization is governed by the coupling of microbial assimilation—driven by stoichiometric C/N imbalance (typically triggered when C/N > 20–25)—and abiotic chemical fixation, including reactions between NH_4^+/NO_2^- and lignin-derived phenolics forming stable organic nitrogen compounds. The relative dominance of these pathways is jointly regulated by carbon quality, nitrogen form, and pH. Based on these mechanisms, regulatory strategies are summarized at multiple scales, including feedstock pretreatment to reduce labile carbon availability, substrate formulation to optimize C/N balance, and model-assisted intelligent fertigation to synchronize nitrogen supply with crop demand. Overall, this study provides a theoretical basis for improving green waste valorization and promoting sustainable horticultural production.

Keywords: organic residues; soilless substrates; Biotic N immobilization; abiotic N immobilization

1. Introduction

Growing media serve as the physical carrier and nutrient interface in soilless culture systems and therefore strongly influence the rhizosphere environment and crop productivity [1]. Peat has long been the preferred medium because of its favorable physicochemical properties; however, its non-renewability and the ecological damage caused by its extraction have made the search for sustainable, locally available alternatives an urgent priority in horticulture [2–5].

Among various candidates, green waste is considered a promising peat alternative because it is abundant, renewable, and relatively inexpensive. Globally, the generation of municipal solid waste continues to increase and is projected to rise from approximately 2.1 billion t in 2023 to 3.8 billion t by 2050, indicating that the management pressure associated with organic wastes, including green waste, will further intensify [6]. In China, green waste has become the second largest category of organic solid waste in urban areas after municipal household waste, with an annual generation of approximately 40–70 million t. Moreover, its production has increased rapidly, rising from 10.7134 million t in 2003 to 60.9705 million t in 2022, and is projected to reach 83.35–91.48 million t by 2030 [7]. In contrast, countries in Europe and the United States incorporated garden waste into systems for the separate collection and resource utilization of biowaste at an earlier stage. For example, in the United States, yard trimmings amounted to approximately 35.4 million t in 2018, accounting for 12.1% of total municipal solid waste. In the European Union, municipal waste generation reached 511 kg per capita in 2023, of which 48% was recycled through material recovery, composting, and related pathways [8,9]. However, despite this vast resource potential and decades of exploration, the high-value utilization of green waste in professional horticulture remains disproportionately low. This discrepancy highlights a persistent gap between the theoretical availability of biomass and the practical stability of substrate performance.

Predominantly composed of plant fibers, green waste can be processed into growing media with excellent water-holding capacity and aeration performance, while providing essential nutrients such as N, P, and K [2,10,11]. Its local availability may also reduce the carbon footprint of substrate production and transport, aligning with the principles of circular agriculture. However, its high C/N ratio often induces nitrogen immobilization, which remains a critical bottleneck restricting the large-scale application of green waste as growing media [12–14]. While nitrogen immobilization is traditionally viewed as a simple biological competition for nutrients, emerging evidence suggests a more complex interplay between microbial consumption and abiotic chemical immobilization. Nevertheless, the relative contributions of these biotic and abiotic pathways, particularly in high-lignin substrates, remain a subject of ongoing debate and lack quantitative clarity.

This process (nitrogen immobilization) has a dual role in green-waste-based media. In the short term, strong immobilization can induce nitrogen deficiency in plants. In the longer term, however, the immobilized nitrogen may contribute to a stable organic nitrogen pool that is gradually remineralized, thereby supporting sustained nutrient supply and reducing the risks of nitrate leaching and ammonia volatilization. Current research, however, tends to focus heavily on the detrimental short-term effects, often overlooking the long-term ecological benefits and the potential to steer this process via precise management. Because nitrogen supply, transformation, and retention are central to agricultural productivity and environmental sustainability [15], a comprehensive re-evaluation of nitrogen immobilization in green-waste-based media is essential. However, current knowledge is fragmented: microbial immobilization has been extensively studied, whereas abiotic pathways, particularly those involving lignin-derived compounds, remain poorly characterized, and their relative contributions are rarely integrated into existing frameworks. Moreover, most current studies are case-specific and lack a unified conceptual framework linking mechanisms, controlling factors, and management strategies. This review therefore examines its biochemical mechanisms and major controlling factors and proposes a regulatory framework from feedstock pretreatment to cultivation management, with the aim of supporting the conversion of green waste into a more efficient and manageable horticultural resource (Figure 1).

https://doi.org/10.3390/plants15091298

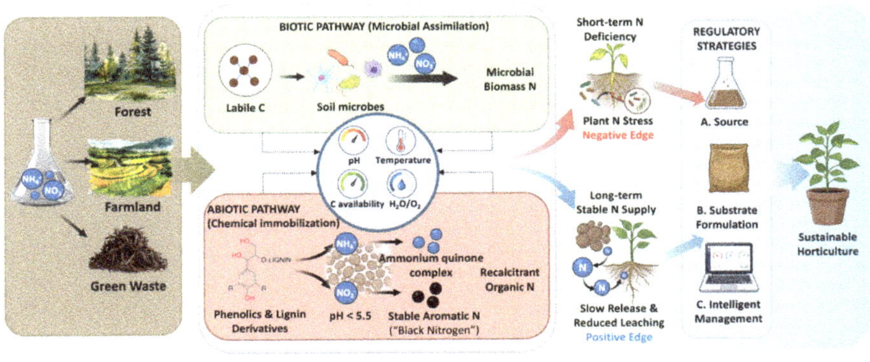

Figure 1. Nitrogen immobilization in organic media: mechanisms, effects, and regulatory strategies.

2. Nitrogen Immobilization in Organic Media

2.1. Nitrogen Immobilization in Natural Soil

Nitrogen immobilization is a fundamental biogeochemical process in natural soils. By converting reactive inorganic N—primarily ammonium (NH_4^+) and nitrate (NO_3^-)—into organic forms, this process plays a key role in regulating nitrogen retention, primary productivity, and environmental risk in terrestrial ecosystems. The intensity of this process exhibits significant ecosystem dependency, and its quantitative characteristics have been systematically characterized on a global scale. A comprehensive meta-analysis of 398 global ^{15}N isotope dilution and tracer experiments revealed average immobilization rates of 1.93 ± 0.31 and 8.17 ± 0.94 mg N $kg^{-1} \cdot d^{-1}$ for NO_3^--N and NH_4^+-N, respectively [16]. These findings globally validate that most natural soils possess a higher affinity and immobilization capacity for NH_4^+. This preference dictates the primary pathway for the flux of inorganic nitrogen into organic pools. This mechanism serves as the cornerstone for nitrogen retention and buffering capacity within ecosystems.

In natural forest ecosystems, the continuous input of litter with high C/N ratios provides abundant carbon substrate and energy for microbes, thereby driving intensive microbial nitrogen immobilization. Global meta-analyses indicate that microbial nitrogen immobilization rates in forest soils reach 8.10 mg N $kg^{-1} \cdot d^{-1}$, significantly exceeding the average gross N mineralization rate of 6.65 mg N $kg^{-1} \cdot d^{-1}$. Furthermore, these rates exhibit significant positive correlations with soil organic carbon, microbial biomass carbon, and the C/N ratio [17].

In contrast, nitrogen immobilization in agricultural ecosystems is strongly influenced by anthropogenic management, exhibiting distinct intensities, mechanisms, and dynamics compared with forest soils dominated by natural litter inputs. Generally, intensive agricultural practices tend to suppress the soil's nitrogen immobilization capacity [16]. A comparative study demonstrated that NH_4^+ immobilization in forest soils can reach 6.67 mg N $kg^{-1} \cdot d^{-1}$, whereas it drops to only 0.34 mg N $kg^{-1} \cdot d^{-1}$ in adjacent croplands. This disparity underscores that land-use conversion from forest to cropland significantly weakens the soil's internal nitrogen retention capacity [18].

2.2. Nitrogen Immobilization in Growing Media

In substrate-based systems, the rhizosphere microenvironment—characterized by optimal aeration and water-holding capacity—promotes rapid microbial proliferation and metabolic activity, thereby intensifying the competitive uptake and temporary immobilization of exogenous inorganic nitrogen. Simultaneously, due to the confined rhizo-

https://doi.org/10.3390/plants15091298

sphere volume, limited physicochemical buffering capacity, and small nutrient reservoir of growing media, any short-term fluctuations in nitrogen availability are rapidly transmitted to the root system. Consequently, the impact of nitrogen immobilization on crop growth is more immediate and pronounced in soilless growing media than in traditional soil-based systems.

Nitrogen immobilization in growing media has been recognized since the 1990s and occurs mainly in organic substrates [19–21]. Its intensity depends largely on feedstock composition and maturity. Peat generally shows little immobilization because it is highly humified [22,23]. In contrast, coconut coir, typically characterized by a C/N ratio of 80–100 and high lignin content (approx. 35–45%), exhibits moderate immobilization. Based on its reported Nitrogen Drawdown Index (NDI), this may necessitate a 15–20% compensation in initial nitrogen loading compared to peat to ensure optimal seedling nitrogen uptake [24,25]. Research indicates that these woody substrates can necessitate an additional 50–100 mg·L^{-1} of supplemental inorganic nitrogen to compensate for rapid microbial assimilation. Furthermore, based on established diagnostic protocols, fresh woody materials often yield a Nitrogen Immobilization Index (NII) exceeding the critical threshold of 100 mg N·L^{-1} substrate, whereas stabilized or composted materials typically remain below 50 mg N·L^{-1}, highlighting the significant impact of material maturity on nitrogen dynamics [26,27].

Green waste typically possesses extremely high C/N ratios, which directly triggers intensive microbial nitrogen immobilization [19,20,23]. Quantitative assessments reveal that the nitrogen immobilization intensity in these growing media far exceeds that of mineral soils or conventional organic media: For instance, fresh pine wood growing media exhibit an nitrogen immobilization capacity of 10–20 mg N·L^{-1} within just four days, whereas specific wood fiber growing media can reach as high as 115 mg N·L^{-1} [22,28]. This intensive immobilization leads to a severe depletion of plant-available nitrogen. Studies indicate that tomato plants grown in pure wood fiber require an additional N supplementation of up to 100 mg N·L^{-1} to achieve biomass levels comparable to those grown in peat [14].

2.3. The Dual Role of Nitrogen Immobilization

Nitrogen immobilization is inherently dual-natured, and its effects are context-dependent, reflecting a trade-off between short-term limitation of plant-available N and long-term formation of a N reservoir. In natural or semi-natural ecosystems such as forests and croplands, nitrogen immobilization is generally regarded as an ecological buffering mechanism that helps reduce nitrogen losses and maintain nutrient cycling. However, in growing substrates dominated by garden waste with a high C/N ratio, this process can rapidly immobilize inorganic nitrogen in the short term, thereby restricting the immediate nutrient availability for crops.

In these ecosystems, nitrogen immobilization functions as a key ecological buffering mechanism. Under increasing atmospheric N deposition, microbial immobilization rapidly sequesters both exogenous and mineralized inorganic N, thereby delaying its transfer into leaching pathways [29]. Especially in carbon-rich systems, this process serves as a dominant mechanism for regulating the balance between N retention and loss in N-saturated ecosystems [30]. Agricultural studies have demonstrated that incorporating organic amendments with high C/N ratios (e.g., wheat or rice straw) promotes microbial immobilization and physicochemical adsorption, which sequesters excess mineral N in the short term while reducing nitrate leaching and N_2O emissions [31–33]. This mechanism remains effective under both flooded and arid conditions. Furthermore, it acts synergistically with chemical fertilizers to facilitate slow-release nitrogen supply, thereby improving nitrogen use efficiency [34,35].

https://doi.org/10.3390/plants15091298

In contrast, the negative impacts of nitrogen immobilization are significantly amplified in cultivation growing media dominated by high C/N ratio organic materials, such as green waste. Intense microbial N assimilation is typically initiated within days of transplanting and peaks around two weeks [22]. This process leads to a rapid decline in mineral N concentration in the substrate solution. This induces severe N deficiency during critical early growth stages, which manifests as growth retardation and foliar chlorosis and further impairs the plant's photosynthetic physiology [36]. During this phase, the organic substrate may transition from a potential nutrient source to a nutrient competitor for plants. However, the latent positive effects of nitrogen immobilization persist: the sequestered nitrogen is not permanently lost but is instead incorporated into microbial biomass and necromass. Structural organic components such as lignin may also sequester nitrogen through chemical bonding. Together, these processes form an organic N pool that gradually releases nutrients through mineralization over subsequent weeks or months [37]. Furthermore, intensive microbial nitrogen immobilization alters rhizosphere chemical signaling and nutrient gradients, potentially stimulating plant roots to excrete organic acids as a stress response. This may indirectly influence the bioavailability of elements such as phosphorus (P) and iron (Fe) and selectively enrich specific rhizosphere microbial communities [38]. Such processes may foster symbiotic relationships between plants and rhizosphere microbes, thereby enhancing abiotic stress tolerance and the nutrient acquisition capacity of plants.

3. Mechanisms of Nitrogen Immobilization

Nitrogen immobilization in organic media is governed by two fundamentally different but interacting mechanisms: microbial assimilation and abiotic chemical immobilization.

3.1. Microbial Nitrogen Immobilization

Microbial nitrogen immobilization is the most widely recognized mechanism driving N fixation in organic media, fundamentally stemming from microbial assimilation based on stoichiometric requirements [39]. Owing to their higher specific surface area and more rapid growth rates compared to plant roots, microbes typically possess a competitive advantage in the acquisition of available nitrogen [40]. In high-C/N media, the rapid proliferation of microorganisms triggers the rapid uptake of inorganic NH_4^+ and NO_3^- from the substrate solution to maintain a stable cellular C/N ratio (typically 5 to 10), thereby causing local nitrogen deficiency. Empirical studies demonstrate the high efficiency of this process. For instance, in acidic spruce forest soils, 50–60% of exogenous N is rapidly incorporated into microbial biomass within 48 h [30]. In agricultural systems, wheat straw facilitates the rapid immobilization of excess nitrogen, storing it as microbial biomass nitrogen (MBN), with rates reaching 42 kg N ha^{-1} within 7 days [31]. The sequestered nitrogen is primarily synthesized into intracellular components, including peptides, proteins, and nucleic acids [41]. Microbial nitrogen immobilization is mediated by a functional consortium of bacteria, fungi, and actinomycetes rather than by a single group. These microorganisms cooperate in lignocellulose degradation and nitrogen transformation, thereby sustaining nitrogen immobilization across decomposition stages [42–45]. This biotic process is characterized by its rapid onset and potential reversibility; as carbon sources are exhausted, the immobilized nitrogen is eventually returned to the plant-available pool through microbial turnover and remineralization.

3.2. Abiotic Nitrogen Immobilization

Beyond microbial nitrogen immobilization, abiotic chemical immobilization represents another crucial pathway for nitrogen sequestration in organic media, particularly in acidic environments with high organic matter content. It proceeds through direct reactions

https://doi.org/10.3390/plants15091298

between inorganic N and reactive organic functional groups and can occur rapidly, with NH_4^+, NO_3^-, and NO_2^- all participating under suitable chemical conditions [46,47].

Currently, our understanding of the mechanisms underlying abiotic nitrogen immobilization in organic materials remains quite limited. Both NH_4^+ and NO_2^- exhibit high chemical reactivity toward various soil organic compounds (e.g., humic and fulvic acid analogs, as well as lignin derivatives), thereby driving abiotic nitrogen immobilization within the soil matrix [48]. Specifically, NH_3 can undergo condensation reactions with activated phenols or quinone rings to form nitrogenous humates [49,50]. NH_4^+ may react chemically with lignin-derived phenols, facilitating the abiotic sequestration of nitrogen in the soil [51]. Additionally, NO_2^- reacts with lignin or its derivatives (e.g., phenolic compounds) to form recalcitrant organic N conjugates [52,53]. Using ^{15}N labeling, researchers found that NO_2^- primarily reacts abiotically with fulvic acids, sequestering ~6% of applied NO_2^--N into stable forms such as amides and pyrrolic "black nitrogen" [54]. Furthermore, NO_3^- can be converted into organic nitrogen species under specific conditions via the so-called "ferrous wheel mechanism" [55]. Lignin likely plays a pivotal role in abiotic nitrogen immobilization [56,57]. Collectively, abiotic nitrogen immobilization is characterized by rapid chemical reactions and the formation of stable, low-turnover organic nitrogen pools, particularly under conditions rich in reactive phenolics and acidic environments. Nevertheless, research on nitrogen immobilization has predominantly focused on organic-rich soils within forests, grasslands, oceans, and agricultural lands [58]. Consequently, for growing media such as green waste—which are uniquely characterized by high concentrations of lignin and polyphenols—the intensity, specific pathways, and relative contributions of abiotic nitrogen immobilization remain poorly understood. This constitutes a critical knowledge gap for the valorization and sustainable application of green waste as growing media.

3.3. Dominant Factors in Nitrogen Immobilization Pathways

The relative contributions of nitrogen immobilization pathways—both biotic and abiotic—are primarily determined by the C/N stoichiometric status of the system, with carbon (C) sources and N availability acting as co-regulators [59]. When organic matter is abundant in labile carbon (e.g., sugars, organic acids), microbial metabolism is strongly stimulated, thereby rendering biotic immobilization the dominant pathway. Conversely, if the organic matter is dominated by recalcitrant components (e.g., lignin-derived phenols), abiotic immobilization through chemical reaction pathways is favored. Nitrogen availability further modulates this balance: during the N-limitation phase, microbial assimilation governs the nitrogen immobilization process. As N sufficiency shifts microbial metabolism toward carbon limitation, the relative contribution of abiotic processes increases significantly, even becoming predominant under N-rich conditions [50].

4. Key Factors Influencing Nitrogen Immobilization in Growing Media

4.1. Material Composition

The substrate biochemical profile acts as the primary determinant; specifically, the balance between labile carbohydrates and recalcitrant lignin determines whether immobilization is dominated by rapid microbial uptake or stubborn chemical sequestration. Among various parameters, the carbon-to-nitrogen (C/N) ratio serves as the primary empirical indicator for assessing nitrogen immobilization potential [60,61]. It is generally accepted that when the C/N ratio exceeds 20–25, microbial growth becomes N-limited, prompting the assimilation of exogenous or mineralized inorganic N, which leads to net nitrogen immobilization [62]. However, as a macroscopic stoichiometric index, the C/N ratio possesses inherent limitations in its predictive capacity [63]. The bioavailability of

https://doi.org/10.3390/plants15091298

carbon sources often predicts the occurrence and intensity of nitrogen immobilization more accurately than total carbon content [58]. If a high C/N ratio originates from recalcitrant components such as lignin and cellulose, microbial metabolism remains sluggish, potentially delaying or attenuating nitrogen immobilization. Conversely, materials rich in labile carbon sources (e.g., soluble sugars, organic acids) can intensely stimulate microbial activity, even if the C/N ratio is not exceptionally high, triggering the rapid assimilation and immobilization of inorganic nitrogen [29,63,64].

For specific materials such as green waste, nitrogen immobilization is shaped by a unique suite of chemical attributes. First, its exceptionally high C/N ratio (often >100) provides the fundamental prerequisite for intensive nitrogen immobilization [65]. Second, the quality of its carbon (C) pool exhibits a distinct duality. On one hand, it is rich in lignin-derived phenolic compounds [3], which can selectively inhibit or modulate the activity of specific microbial communities through allelochemical effects, thereby regulating microbial nitrogen immobilization processes [66]. On the other hand, green waste is typically abundant in labile carbon fractions (e.g., organic acids) [67], which serve as preferred carbon sources for microbes and stimulate microbial proliferation [68,69]. This coexistence of inhibitory and stimulatory carbon patterns significantly complicates the microbial nitrogen immobilization process. Crucially, phenolics and their oxidation products (e.g., quinones) not only modulate biological processes but also directly participate in abiotic chemical immobilization. They chemically react with inorganic N (e.g., NH_4^+) to form stable quinone-ammonium complexes or aromatic organic N, sequestering nitrogen into a chemically stable pool with low turnover rates [65,70]. Consequently, nitrogen immobilization in green waste growing media is the synergistic result of microbial drivers and chemical reactions. Its intensity, dynamics, and the stability of the resulting N pool are jointly determined by the relative concentrations and interactions among labile carbon, recalcitrant lignocellulose, and reactive phenolics. This defines its unique nitrogen immobilization properties—distinct from those of other organic wastes—and serves as the fundamental basis for scientific management and resource valorization of green waste.

4.2. Medium Environment

Substrate environmental factors profoundly influence the intensity, pathways, and ultimate fate of nitrogen immobilization by modulating microbial activity and the microenvironmental conditions of chemical reactions. Rather than acting in isolation, these factors synergistically regulate microbial processes, inorganic N availability, mass transfer, and redox microenvironments to jointly shape the dynamic landscape of nitrogen immobilization.

4.2.1. The Role of pH

pH is a key environmental factor regulating both the pathways and intensity of nitrogen immobilization in growing media, affecting both biotic and abiotic processes. In the biotic pathway, pH influences nitrogen transformations by shaping microbial community structure and regulating enzyme activity and functional gene expression [71]. Acidic conditions ($pH < 5.5$) inhibit nitrification and promote NH_4^+ accumulation, thereby promoting microbial NH_4^+ assimilation and immobilization [72,73]. However, strongly acidic conditions also suppress organic matter decomposition, limiting energy supply and reducing microbial NO_3^- immobilization [59]. Conversely, neutral to slightly alkaline conditions generally stimulate microbial activity to enhance nitrogen immobilization.

pH is equally critical for abiotic pathways: In the abiotic pathway, pH controls the chemical reactivity of nitrogen species and organic matter. Lignin degradation products (e.g., phenolics) can react with amino acids or NH_3 via condensation reactions to form brown nitrogenous humic acids, a process facilitated under alkaline conditions [74]. Weakly

https://doi.org/10.3390/plants15091298

acidic conditions promote the abiotic adsorption of NH_3/NH_4^+ onto organic matter or clay minerals [75]. Furthermore, lower pH environments promote the generation of NO_2^- and its rapid reaction with reactive organic components like phenolics, forming stable "black nitrogen" compounds [54,76]. This abiotic process can reach completion within minutes without microbial involvement, serving as a vital N-retention mechanism in organic-rich acidic media [48,52]. Collectively, pH regulates the balance between biotic and abiotic N immobilization pathways, playing a pivotal role in regulating N cycling and availability in green-waste-based growing media.

4.2.2. Electrical Conductivity

High electrical conductivity (EC) acts as a physiological stressor that suppresses nitrogen immobilization primarily through osmotic imbalance and ion-specific toxicity [77,78]. These conditions inhibit microbial proliferation and metabolic vigor, effectively lowering the biological demand for inorganic nitrogen assimilation. Beyond direct immobilization, excessive salinity disrupts the kinetics of mineralization and nitrification, leading to an accumulation of unsequestered mineral nitrogen [79,80]. Consequently, the diminished microbial "buffering" capacity under high EC conditions significantly elevates the risk of nitrogen leaching and gaseous losses in green-waste-based media.

4.2.3. Temperature

Temperature regulates nitrogen immobilization by modulating microbial metabolic kinetics and enzymatic activity Within the biological range of 5–25 °C, warming typically accelerates organic matter decomposition; provided that labile carbon is sufficient, this leads to an intensified microbial demand for inorganic nitrogen. However, this stimulatory effect is often self-limiting. Sustained high temperatures can lead to the rapid exhaustion of available carbon sources or increase microbial maintenance energy requirements, eventually shifting the system toward the remineralization of previously immobilized nitrogen [81]. However, this stimulatory effect is subject to a threshold; sustained warming may lead to rapid substrate depletion or increased maintenance and metabolic costs for microorganisms. These factors can weaken long-term nitrogen immobilization capacity and potentially accelerate the remineralization of previously immobilized nitrogen [82]. Overall, temperature primarily regulates the rate and temporal dynamics of nitrogen immobilization, rather than its net accumulation.

4.2.4. Moisture and Oxygen Concentration

Moisture and oxygen concentrations jointly regulate the redox status of the substrate, profoundly influencing the pathways and efficiency of nitrogen immobilization. Insufficient moisture levels limit the diffusion of growing media and nutrients, thereby inhibiting microbial activity. Conversely, excessive moisture leads to pore–water saturation, creating anaerobic microenvironments that suppress aerobic microbial processes and potentially divert N flux toward loss pathways dominated by denitrification [83]. The pore structure and connectivity of the substrate directly determine the uniformity of oxygen transport; moreover, substrate compaction or structural degradation can exacerbate local hypoxia. Therefore, maintaining optimal moisture levels and a well-aerated structure is a critical prerequisite for ensuring adequate oxygen supply, promoting aerobic microbial assimilation-driven nitrogen immobilization, and minimizing nitrogen loss.

Overall, nitrogen immobilization in growing media is governed by the coupled effects of substrate properties and environmental conditions, rather than any single factor Table 1. Substrate composition—particularly C/N ratio and carbon quality—determines the baseline potential and dominant pathways of nitrogen immobilization. Labile carbon

https://doi.org/10.3390/plants15091298

generally promotes microbial assimilation, whereas lignin- and phenolic-rich components favor abiotic stabilization and the formation of more persistent nitrogen pools.

Environmental factors regulate the expression of this potential. Among them, pH acts as a key integrator by simultaneously influencing microbial activity, nitrogen speciation, and the chemical reactivity of organic matter. Moisture and oxygen conditions control mass transfer and redox status, thereby determining whether nitrogen is retained through immobilization or lost via processes such as denitrification. Temperature primarily regulates the rate and temporal dynamics of these processes, while elevated salinity (EC) generally suppresses microbial activity and weakens nitrogen retention capacity.

Taken together, nitrogen immobilization can be understood as a system-level process governed by the interaction between intrinsic substrate characteristics and external environmental controls. These interactions ultimately determine its intensity, dominant pathway (microbial vs. abiotic), and implications for nitrogen availability and loss risk in green-waste-based growing media.

Table 1. Key factors regulating nitrogen immobilization in growing media.

Influence Factor	Condition	Dominant Pathway	Mechanism
Substrate	Labile C	Microbial	Rapid microbial assimilation
	Recalcitrant C	Abiotic	Chemical stabilization
Nitrogen form	NH_4^+-dominated	Microbial/abiotic	Preferentially assimilated by microbes; can be directly adsorbed onto organic matter and clay minerals
	NO_3^--dominated	Microbial	Requires reduction before assimilation
	NO_2^- (intermediate)	Abiotic	Rapidly reacts with phenolic compounds to form stable "black nitrogen"
pH	Acidic	Mixed	Inhibits nitrification; promotes NH_4^+ and NO_2^- reactions
	Neutral–alkaline	Mixed	Stimulates microbes and phenolic reactions
Moisture & oxygen	Aerobic	Microbial	Supports microbial activity
	Anaerobic/excessive moisture	Loss pathways	Denitrification
Temperature	5–25 °C	Microbial	Enhances decomposition
	Sustained warming	Microbial	C depletion
EC	High EC	Suppressed microbial	Osmotic stress, ion toxicity

5. Strategies for Regulating Nitrogen Immobilization in Growing Media

Based on profound mechanistic insights into nitrogen immobilization and its key drivers, effectively regulating N sequestration during the transformation of green waste into growing media is pivotal for converting this environmental burden into a controllable resource.

5.1. Source Pretreatment to Modify Material Properties

Theoretically, a nitrogen pre-loading strategy can satisfy the immobilization demand of the substrate in advance. However, this approach substantially elevates total nitrogen inputs throughout the growing cycle and diminishes nitrogen-use efficiency [84]. Therefore, reducing the intrinsic nitrogen immobilization potential of the substrate at the source repre-

https://doi.org/10.3390/plants15091298

sents a more fundamental and efficient solution. The core of source-based pretreatment lies in directly altering the chemical properties of green waste to weaken the internal drivers of nitrogen immobilization. Aerobic composting remains the most conventional and widely used method. It relies on biochemical processes to promote the microbial consumption of labile carbon fractions, leading to the bio-stabilization of green waste. This process also significantly lowers the C/N ratio, thereby reducing the availability of carbon sources for microbial assimilation in the rhizosphere [85]. However, composting is typically time-consuming and may require substantial space, labor, and process control, which can limit its scalability and economic feasibility in large-scale applications. Moreover, incomplete or poorly managed composting may result in residual phytotoxic compounds or unstable organic matter, leading to variable performance in practice. Recently, ammonium salt incubation has emerged as a promising method for rapid detoxification and reduction of nitrogen immobilization potential [67]. Studies have confirmed that this method effectively lowers the nitrogen immobilization capacity of green waste, offering innovative insights for the development of high-efficiency pretreatment technologies. Nevertheless, the long-term stability of this treatment and its potential impacts on substrate salinity and microbial community structure remain insufficiently understood, which may constrain its broader application.

Thermal treatment is currently one of the most widely studied pre-conditioning strategies for wood fiber materials. It reduces the availability of readily degradable components in the fibers through thermal modification and increases the resistance of the material to microbial decomposition, thereby alleviating nitrogen immobilization [86]. Despite its effectiveness, thermal treatment is energy-intensive and may increase production costs, raising concerns regarding its sustainability and large-scale adoption. In addition, excessive thermal modification may negatively affect the physical properties of the substrate, such as porosity and water-holding capacity.

Lignin functionalization represents another, more targeted approach for mitigating nitrogen immobilization [87]. In this method, lignin is first dissolved in an alkaline solution and then mixed with wood fibers. Acid-induced deposition subsequently enables lignin to reattach to both inner and outer fiber surfaces, thereby reconstructing a degradation-resistant barrier similar to that of natural wood. This treatment significantly reduces microbial respiration and nitrogen immobilization in wood fibers. It suggests that limiting rapid microbial utilization of carbon through surface functionalization is an effective strategy for alleviating nitrogen immobilization in wood-fiber-based substrates. However, this approach involves multiple chemical processing steps and may increase operational complexity and cost. Furthermore, its feasibility at commercial scale and potential environmental impacts of chemical reagents require further evaluation.

5.2. Process Blending to Optimize Substrate Formulation

During the substrate formulation stage, the scientific blending of additives with diverse properties is intended to actively optimize the physical, chemical, and biological characteristics of the mixed substrate. This helps establish a rhizosphere environment that inhibits excessive nitrogen immobilization while favoring plant growth. For example, mixing pretreated green waste with low-C/N material (e.g., decomposed livestock manure or biogas residue) can rapidly adjust the overall C/N ratio of the blend to a safe range (typically <25). It also directly provides initial available N, thereby alleviating N stress during the crop seedling stage [88]. Simultaneously, incorporating materials with high cation exchange capacity—such as zeolite or biochar—can adsorb and slowly release nutrients like NH_4^+, which effectively buffers the sharp decline in available N concentrations triggered by rapid microbial immobilization. Such a multi-dimensional compounding

https://doi.org/10.3390/plants15091298

strategy systematically regulates the intensity and pathways of nitrogen immobilization, leading to optimized substrate performance. However, the effectiveness of blending strategies largely depends on the precise selection of material ratios and properties, and the substantial variability in composition among green waste from different sources further increases uncertainty in practical applications.

5.3. Intelligent Prediction for Managing Root-Zone Water and Fertilizer

Nitrogen immobilization in green-waste-based growing media is highly dynamic, and its intensity and dominant pathway vary with substrate properties, environmental conditions, and crop growth stage. In this context, intelligent water and fertilizer management, supported by sensor monitoring and model analysis, provides a promising approach for dynamically regulating nitrogen immobilization. To support such predictive management, several modeling approaches have been developed to describe nitrogen mineralization–immobilization dynamics. Classical models are generally based on carbon–nitrogen (C-N) balance principles, where nitrogen release or immobilization is governed by the stoichiometric relationship between substrate composition and microbial demand. For example, first-order decomposition models coupled with microbial C/N requirements simulate nitrogen fluxes as a function of organic matter decomposition and microbial assimilation [60]. More advanced frameworks incorporate microbial carbon use efficiency and enzyme-mediated processes. They emphasize that nitrogen immobilization arises from interactions among carbon availability, microbial metabolism, and environmental drivers [61].

However, despite these advances, existing models remain limited in their applicability to green-waste-based growing media. Most models have been developed for relatively homogeneous soil systems and often assume stable substrate properties. In contrast, green waste is highly heterogeneous in feedstock composition, lignocellulosic structure, and decomposition stage. Furthermore, current modeling frameworks primarily focus on microbial processes and rarely account for abiotic nitrogen immobilization pathways, such as reactions involving lignin-derived phenolics and reactive nitrogen species.

Building on these modeling foundations, the integration of IoT-based sensing technologies enables continuous monitoring of key root-zone variables, including temperature, moisture status, electrical conductivity, and nutrient concentrations. When coupled with nitrogen transformation kinetics, such data can support the development of adaptive water and fertilizer management strategies that are more closely aligned with crop nutrient demand. Nevertheless, nitrogen immobilization in green-waste-based media is jointly regulated by multiple interacting factors, including substrate composition, microbial activity and abiotic reactions. These interactions may vary substantially across feedstocks and cultivation systems. Therefore, predictive management should be based on a mechanistic understanding of substrate properties and nitrogen transformation processes, rather than through the direct application of generalized model parameters. In particular, in systems where microbial and abiotic immobilization coexist, identifying key process indicators and improving model adaptability across different materials and environmental conditions remain critical challenges.

6. Future Research Perspectives

The use of green waste as growing media should be considered within the broader context of urban nutrient cycling, the circular economy, and carbon neutrality, rather than being judged only by short-term horticultural performance. Among the many research directions, the top priority is to develop a mechanistic and quantitative understanding of nitrogen immobilization pathways and their relative contributions under different

https://doi.org/10.3390/plants15091298

substrate conditions. Future studies should therefore establish clear links between substrate properties, nitrogen transformation processes, and crop responses throughout the entire cultivation cycle.

A major knowledge gap is the lack of quantitative separation between microbial and abiotic nitrogen immobilization, especially in lignin-rich green waste systems. Addressing this issue should be a primary focus, as it underpins predictive regulation and substrate design. In addition, rapid and practical assessment methods are needed to evaluate nitrogen immobilization potential and its temporal dynamics under real production conditions.

Future work should also prioritize the identification of a small set of reliable indicators—such as pH, inorganic nitrogen form, dissolved organic matter composition, and oxygen status—that can effectively predict immobilization intensity. This approach is more useful than examining many variables with similar importance. It will support nitrogen risk assessment and help standardize different types of substrates.

Another key priority is to develop integrated and adaptable predictive models for green-waste-based media. Existing models, mostly developed for soil systems, do not account for substrate heterogeneity or abiotic processes. Future efforts should therefore focus on incorporating both microbial and abiotic pathways into unified modeling frameworks and validating them under commercial production conditions. These models can then be used to test whether process-based precision water and fertilizer management improves the match between nitrogen supply and crop demand, while reducing nitrogen loss.

Overall, future research should move away from descriptive and fragmented approaches and toward hypothesis-driven, quantitatively integrated studies, with a clear focus on the most impactful mechanisms and applications. Only by clarifying the mechanisms behind the "double-edged sword" effect of nitrogen immobilization can green waste be transformed into a stable, controllable, and agronomically valuable growing-media resource.

Author Contributions: Conceptualization, W.Z.; methodology, W.C. and M.Z.; writing—original draft preparation, R.L.; writing—review and editing, W.Z.; supervision, Z.Q.; funding acquisition, W.Z. and Z.Q. All authors have read and agreed to the published version of the manuscript.

Funding: This work was supported by National Natural Science Foundation of China [Grant No. 32541086], Sichuan Provincial Natural Science Foundation [Grant No. 2025ZNSESC0439]; the Project of the Institute of Urban Agriculture, Chinese Academy of Agricultural Sciences [Grant No. SZ202402, 2025KY05]; the Science and Technology Innovation Program of the Chinese Academy of Agricultural Sciences [Grant No. ASTIP-34-IUA-04]; and the Local Financial Special Fund Project of Chengdu Agricultural Science and Technology Center [Grant No. NASC2023ST09, NASC2024KY21].

Data Availability Statement: No new data were created or analyzed in this study. Data sharing is not applicable to this article.

Conflicts of Interest: The authors declare no conflicts of interest.

References

1. Blok, C.; Eveleens, B.; Winkel, A.V. Growing media for food and quality of life in the period 2020-2050. *Acta Hortic.* **2021**, *1305*, 341–356. [CrossRef]
2. Bart, V.; Hilde, M.; Imke, D.W.; Joris, V.A.; Nele, A.; Kasper, M.; Peter, C.; Jane, D. Plant fibers for renewable growing media: Potential of defibration, acidification or inoculation with biocontrol fungi to reduce the N drawdown and plant pathogens. *J. Clean. Prod.* **2018**, *203*, 1143–1154. [CrossRef]
3. Chemetova, C.; Mota, D.; Fabião, A.; Gominho, J.; Ribeiro, H. Low-temperature hydrothermally treated *Eucalyptus globulus* bark: From by-product to horticultural fiber-based growing media viability. *J. Clean. Prod.* **2021**, *319*, 128–805. [CrossRef]
4. Nocentini, M.; Panettieri, M.; García de Castro Barragán, J.M.; Mastrolonardo, G.; Knicker, H. Recycling pyrolyzed organic waste from plant nurseries, rice production and shrimp industry as peat substitute in potting substrates. *J. Environ. Manag.* **2021**, *277*, 111436. [CrossRef] [PubMed]

https://doi.org/10.3390/plants15091298

5. Farru, G.; Dang, C.H.; Schultze, M.; Kern, J.; Cappai, G.; Libra, J.A. Benefits and Limitations of Using Hydrochars from Organic Residues as Replacement for Peat on Growing Media. *Horticulturae* **2022**, *8*, 325. [CrossRef]
6. United Nations Environment Programme. *Global Waste Management Outlook 2024*; United Nations Environment Programme: Nairobi, Kenya, 2024.
7. Song, B.; Han, X.; Li, Y.; Huang, W.; Fang, Q.; Ren, X.; Li, H. Dynamic spatio-temporal patterns and driving factors of green waste generation based on provincial panel data in China. *Front. Environ. Sci.* **2025**, *13*, 1530769. [CrossRef]
8. Eurostat. Municipal Waste Statistics. Available online: https://ec.europa.eu/eurostat/statistics-explained/index.php?title=Municipal_waste_statistics (accessed on 28 March 2026).
9. United States Environmental Protection Agency. Yard Trimmings: Material-Specific Data. Available online: https://www.epa.gov/facts-and-figures-about-materials-waste-and-recycling/yard-trimmings-material-specific-data (accessed on 19 March 2026).
10. Chemetova, C.; Quilhó, T.; Braga, S.; Fabio, A.M.D.; Ribeiro, H. Aged *Acacia melanoxylon* bark as an organic peat replacement in container media. *J. Clean. Prod.* **2019**, *232*, 1103–1111. [CrossRef]
11. Witcher, A.L.; Blythe, E.K.; Fain, G.B.; Curry, K.J. Suitability of Whole Pine Tree Substrates for Seed Propagation. *J. Environ. Hortic.* **2015**, *33*, 20–28. [CrossRef]
12. Handreck, K.A. Use of the nitrogen drawdown index to predict fertilizer nitrogen requirements in soilless potting media. *Commun. Soil Sci. Plant Anal.* **1993**, *24*, 2137–2151. [CrossRef]
13. Jackson, B.E.; Wright, R.D.; Barnes, M.C. Pine Tree Substrate, Nitrogen Rate, Particle Size, and Peat Amendment Affect Poinsettia Growth and Substrate Physical Properties. *HortScience* **2008**, *43*, 2155–2161. [CrossRef]
14. Wright, R.D.; Jackson, B.E.; Browder, J.F.; Latimer, J.G. Growth of Chrysanthemum in a Pine Tree Substrate Requires Additional Fertilizer. *HortTechnology* **2008**, *18*, 111–115. [CrossRef]
15. Anas, M.; Liao, F.; Verma, K.K.; Sarwar, M.A.; Mahmood, A.; Chen, Z.-L.; Li, Q.; Zeng, X.-P.; Liu, Y.; Li, Y.-R. Fate of nitrogen in agriculture and environment: Agronomic, eco-physiological and molecular approaches to improve nitrogen use efficiency. *Biol. Res.* **2020**, *53*, 47. [CrossRef]
16. Elrys, A.S.; Uwiragiye, Y.; Zhang, Y.; Abdel-Fattah, M.K.; Chen, Z.-X.; Zhang, H.-M.; Meng, L.; Wang, J.; Zhu, T.-B.; Cheng, Y.; et al. Expanding agroforestry can increase nitrate retention and mitigate the global impact of a leaky nitrogen cycle in croplands. *Nat. Food* **2023**, *4*, 109–121. [CrossRef]
17. Ran, J.; He, S.; Luo, S.; Wang, Y.; Mao, C. Global patterns and drivers of soil gross nitrogen transformations of forest ecosystems. *Chin. J. Plant Ecol.* **2024**, *48*, 0292. [CrossRef]
18. Ren, X.; Zhang, J.; Bah, H.; Müller, C.; Cai, Z.; Zhu, B. Soil gross nitrogen transformations in forestland and cropland of Regosols. *Sci. Rep.* **2021**, *11*, 223. [CrossRef]
19. Buamscha, M.G.; Altland, J.E.; Sullivan, D.M.; Horneck, D.A.; Mcqueen, J.P.G. Nitrogen Availability in Fresh and Aged Douglas Fir Bark. *Horttechnology* **2008**, *18*, 619–623. [CrossRef]
20. Pineda, J.; Castillo, F.S.d.; Pérez, E.d.C.M.; Aguilar, L.A.V.; González, A.M.C.; Árias, A.R.; Canales, J.M.V. Immobilization and nutrient retention by pine sawdust for soilless culture. *Rev. Terra Latinoam.* **2019**, *37*, 261–271. [CrossRef]
21. Harris, C.N.; Dickson, R.W.; Fisher, P.R.; Jackson, B.E.; Poleatewich, A.M. Evaluating Peat Substrates Amended with Pine Wood Fiber for Nitrogen Immobilization and Effects on Plant Performance with Container-grown Petunia. *Hort Technol.* **2020**, *30*, 107. [CrossRef]
22. Jackson, B.E.; Wright, R.D.; Alley, M.M. Comparison of Fertilizer Nitrogen Availability, Nitrogen Immobilization, Substrate Carbon Dioxide Efflux, and Nutrient Leaching in Peat-lite, Pine Bark, and Pine Tree Substrates. *Hortscience* **2009**, *44*, 781–790. [CrossRef]
23. Boyer, C.R.; Torbert, H.A.; Gilliam, C.H.; Fain, G.B.; Gallagher, T.V.; Sibley, J.L. Nitrogen Immobilization in Plant Growth Substrates: Clean Chip Residual, Pine Bark, and Peatmoss. *Int. J. Agron.* **2012**, *2012*, 978528. [CrossRef]
24. Mariotti, B.; Martini, S.; Raddi, S.; Tani, A.; Jacobs, D.F.; Oliet, J.A.; Maltoni, A. Coconut Coir as a Sustainable Nursery Growing Media for Seedling Production of the Ecologically Diverse *Quercus* Species. *Forests* **2020**, *11*, 522. [CrossRef]
25. Prasad, M. Physical, chemical and biological properties of coir dust. *Acta Hortic.* **1997**, *450*, 21–30. [CrossRef]
26. Owen, W.; Jackson, B.; Whipker, B.; Fonteno, W. Pine Wood Chips as an Alternative to Perlite in Greenhouse Substrates: Nitrogen Requirements. *HortTechnology* **2016**, *26*, 199–205. [CrossRef]
27. Vandecasteele, B.; Viaene, J.; Debode, J.; Muylle, H.; Perneel, M. Selecting raw materials for sustainable growing media: A protocol to assess the risk for N immobilization. *Acta Hortic.* **2017**, *1168*, 149–160. [CrossRef]
28. Wu, L.; Li, R.; Liu, J.; Cui, W.; Qi, Z.; Zhou, W. Nitrogen Immobilization by Wood Fiber Substrates Strongly Affects the Photosynthetic Performance of Lettuce. *Plants* **2025**, *14*, 1518. [CrossRef]
29. Pei, G.; Liu, J.; Peng, B.; Gao, D.; Wang, C.; Dai, W.; Jiang, P.; Bai, E. Nitrogen, lignin, C/N as important regulators of gross nitrogen release and immobilization during litter decomposition in a temperate forest ecosystem. *For. Ecol. Manag.* **2019**, *440*, 61–69. [CrossRef]

https://doi.org/10.3390/plants15091298

30. Tahovská, K.; Kaňa, J.; Bárta, J.; Oulehle, F.; Richter, A.; Šantrůčková, H. Microbial N immobilization is of great importance in acidified mountain spruce forest soils. *Soil Biol. Biochem.* **2013**, *59*, 58–71. [CrossRef]
31. Reichel, R.; Wei, J.; Islam, M.S.; Schmid, C.; Wissel, H.; Schröder, P.; Schloter, M.; Brüggemann, N. Potential of Wheat Straw, Spruce Sawdust, and Lignin as High Organic Carbon Soil Amendments to Improve Agricultural Nitrogen Retention Capacity: An Incubation Study. *Front. Plant Sci.* **2018**, *9*, 900. [CrossRef]
32. Burger, M.; Jackson, L.E. Microbial immobilization of ammonium and nitrate in relation to ammonification and nitrification rates in organic and conventional cropping systems. *Soil Biol. Biochem.* **2003**, *35*, 29–36. [CrossRef]
33. Wei, J.; Reichel, R.; Islam, M.S.; Wissel, H.; Amelung, W.; Brüggemann, N. Chemical Composition of High Organic Carbon Soil Amendments Affects Fertilizer-Derived N_2O Emission and Nitrogen Immobilization in an Oxic Sandy Loam. *Front. Environ. Sci.* **2020**, *8*, 15. [CrossRef]
34. Said-Pullicino, D.; Cucu, M.A.; Sodano, M.; Birk, J.J.; Glaser, B.; Celi, L. Nitrogen immobilization in paddy soils as affected by redox conditions and rice straw incorporation. *Geoderma* **2014**, *228–229*, 44–53. [CrossRef]
35. Burke, I.C.; Bontti, E.E.; Barrett, J.E.; Lowe, P.N.; Lauenroth, W.K.; Riggle, R. Impact of labile and recalcitrant carbon treatments on available nitrogen and plant communities in a semiarid ecosystem. *Ecol. Appl.* **2013**, *23*, 537–545. [CrossRef]
36. Hu, Y.; Zhu, B. Effects of Organic Amendments with Different C/N Ratios and Application Rates on N Mineralization–Immobilization in Calcareous Cropland Soil. *Agronomy* **2025**, *15*, 2795. [CrossRef]
37. Chen, B.; Liu, E.K.; Tian, Q.; Yan, C.; Zhang, Y. Soil nitrogen dynamics and crop residues. A review. *Agron. Sustain. Dev.* **2014**, *34*, 429–442. [CrossRef]
38. Lu, Y.; Shi, W. Rhizospheric Chemical Signals and Soil Nutrient Transformation. *Biotechnol. Bull.* **2020**, *36*, 14–24.
39. Li, Z.; Zeng, Z.; Song, Z.; Wang, F.; Tian, D.; Mi, W.; Huang, X.; Wang, J.; Song, L.; Yang, Z.; et al. Vital roles of soil microbes in driving terrestrial nitrogen immobilization. *Glob. Change Biol.* **2021**, *27*, 1848–1858. [CrossRef] [PubMed]
40. Hodge, A.; Robinson, D.; Fitter, A. Are microorganisms more effective than plants at competing for nitrogen? *Trends Plant Sci.* **2000**, *5*, 304–308. [CrossRef]
41. Bengtsson, G.; Bengtson, P.; Månsson, K.F. Gross nitrogen mineralization-, immobilization-, and nitrification rates as a function of soil C/N ratio and microbial activity. *Soil Biol. Biochem.* **2003**, *35*, 143–154. [CrossRef]
42. Pot, S.; Tender, C.D.; Ommeslag, S.; Delcour, I.; Ceusters, J.; Vandecasteele, B.; Debode, J.; Vancampenhout, K. Elucidating the microbiome of the sustainable peat replacers composts and nature management residues. *Front. Microbiol.* **2022**, *13*, 983855. [CrossRef]
43. Aguilar-Paredes, A.; Valdés, G.; Araneda, N.; Valdebenito, E.; Hansen, F.; Nuti, M. Microbial Community in the Composting Process and Its Positive Impact on the Soil Biota in Sustainable Agriculture. *Agronomy* **2023**, *13*, 542. [CrossRef]
44. Liu, Q.; He, X.; Luo, G.; Wang, K.; Li, D. Deciphering the dominant components and functions of bacterial communities for lignocellulose degradation at the composting thermophilic phase. *Bioresour. Technol.* **2022**, *348*, 126808. [CrossRef] [PubMed]
45. Biyada, S.; Merzouki, M.; Dėmčėnko, T.; Vasiliauskienė, D.; Ivanec-Goranina, R.; Urbonavičius, J.; Marčiulaitienė, E.; Vasarevičius, S.; Benlemlih, M. Microbial community dynamics in the mesophilic and thermophilic phases of textile waste composting identified through next-generation sequencing. *Sci. Rep.* **2021**, *11*, 23624. [CrossRef] [PubMed]
46. Islam, A.; Chen, D.; White, R.E.; Weatherley, A.J. Chemical decomposition and fixation of nitrite in acidic pasture soils and implications for measurement of nitrification. *Soil Biol. Biochem.* **2008**, *40*, 262–265. [CrossRef]
47. Isobe, K.; Koba, K.; Suwa, Y.; Ikutani, J.; Kuroiwa, M.; Fang, Y.; Yoh, M.; Mo, J.; Otsuka, S.; Senoo, K. Nitrite transformations in an N-saturated forest soil. *Soil Biol. Biochem.* **2012**, *52*, 61–63. [CrossRef]
48. Fitzhugh, R.D.; Lovett, G.M.; Venterea, R.T. Biotic and abiotic immobilization of ammonium, nitrite, and nitrate in soils developed under different tree species in the Catskill Mountains, New York, USA. *Glob. Change Biol.* **2003**, *9*, 1591–1601. [CrossRef]
49. Schimel, J.P.; Firestone, M.K. Inorganic N incorporation by coniferous forest floor material. *Soil Biol. Biochem.* **1989**, *21*, 41–46. [CrossRef]
50. Johnson, D.W.; Cheng, W.; Burke, I.C. Biotic and Abiotic Nitrogen Retention in a Variety of Forest Soils. *Soil Sci. Soc. Am. J.* **2000**, *64*, 1503–1514. [CrossRef]
51. Schmidt-Rohr, K.; Mao, J.-D.; Olk, D.C. Nitrogen-bonded aromatics in soil organic matter and their implications for a yield decline in intensive rice cropping. *Proc. Natl. Acad. Sci. USA* **2004**, *101*, 6351–6354. [CrossRef]
52. Dail, D.B.; Davidson, E.A.; Chorover, J. Rapid abiotic transformation of nitrate in an acid forest soil. *Biogeochemistry* **2001**, *54*, 131–146. [CrossRef]
53. Miyajima, T. Abiotic versus biotic immobilization of inorganic nitrogen in sediment as a potential pathway of nitrogen sequestration from coastal marine ecosystems. *Geochem. J.* **2015**, *49*, 453–468. [CrossRef]
54. Wei, J.; Knicker, H.; Zhou, Z.; Eckhardt, K.U.; Leinweber, P.; Wissel, H.; Yuan, W.; Brüggemann, N. Nitrogen immobilization caused by chemical formation of black- and amide-N in soil. *Geoderma* **2023**, *429*, 116274. [CrossRef]
55. Davidson, E.A.; Chorover, J.; Dail, D.B. A mechanism of abiotic immobilization of nitrate in forest ecosystems: The ferrous wheel hypothesis. *Glob. Change Biol.* **2003**, *9*, 228–236. [CrossRef]

https://doi.org/10.3390/plants15091298

56. Wei, J.; Zhou, M.; Vereecken, H.; Brüggemann, N. Large variability in CO_2 and N_2O emissions and in 15N site preference of N_2O from reactions of nitrite with lignin and its derivatives at different pH. *Rapid Commun. Mass Spectrom.* **2017**, *31*, 1333–1343. [CrossRef] [PubMed]
57. Wei, J.; Amelung, W.; Lehndorff, E.; Schloter, M.; Vereecken, H.; Brüggemann, N. N_2O and NOx emissions by reactions of nitrite with soil organic matter of a Norway spruce forest. *Biogeochemistry* **2017**, *132*, 325–342. [CrossRef]
58. Cao, Y.; He, Z.; Zhu, T.; Zhao, F. Organic-C quality as a key driver of microbial nitrogen immobilization in soil: A meta-analysis. *Geoderma* **2021**, *383*, 114784. [CrossRef]
59. Cao, Y.; Zhao, F.; Zhang, Z.; Zhu, T.; Xiao, H. Biotic and abiotic nitrogen immobilization in soil incorporated with crop residue. *Soil Tillage Res.* **2020**, *202*, 104664. [CrossRef]
60. Nicolardot, B.; Recous, S.; Mary, B. Simulation of C and N mineralisation during crop residue decomposition: A simple dynamic model based on the C:N ratio of the residues. *Plant Soil* **2001**, *228*, 83–103. [CrossRef]
61. Manzoni, S.; Jackson, R.B.; Trofymow, J.A.; Porporato, A. The Global Stoichiometry of Litter Nitrogen Mineralization. *Science* **2008**, *321*, 684–686. [CrossRef]
62. Mooshammer, M.; Wanek, W.; Hämmerle, I.; Fuchslueger, L.; Hofhansl, F.; Knoltsch, A.; Schnecker, J.; Takriti, M.; Watzka, M.; Wild, B.; et al. Adjustment of microbial nitrogen use efficiency to carbon:nitrogen imbalances regulates soil nitrogen cycling. *Nat. Commun.* **2014**, *5*, 3694. [CrossRef]
63. Bonanomi, G.; Sarker, T.C.; Zotti, M.; Cesarano, G.; Allevato, E.; Mazzoleni, S. Predicting nitrogen mineralization from organic amendments: Beyond C/N ratio by 13C-CPMAS NMR approach. *Plant Soil* **2019**, *441*, 129–146. [CrossRef]
64. Ma, Q.; Wu, Z.; Pan, F.; Wang, J.; Zhou, H.; Jiang, C.; Xu, Y.; Yu, W. Effect of glucose addition on the fate of urea-15N in fixed ammonium and soil microbial biomass N pools. *Eur. J. Soil Biol.* **2016**, *75*, 168–173. [CrossRef]
65. Doane, T.A. The Abiotic Nitrogen Cycle. *ACS Earth Space Chem.* **2017**, *1*, 411–421. [CrossRef]
66. Li, Z.H.; Wang, Q.; Ruan, X.; Pan, C.D.; Jiang, D.A. Phenolics and Plant Allelopathy. *Molecules* **2010**, *15*, 8933–8952. [CrossRef]
67. Zhou, W.L.; Liao, J.X.; Zhou, B.; Yang, R.; Lin, W.; Zhang, D.D.; Wang, H.; Qi, Z.Y. Rapidly reducing phytotoxicity of green waste for growing media by incubation with ammonium. *Environ. Technol. Innov.* **2023**, *31*, 103136. [CrossRef]
68. Shi, S.; Richardson, A.E.; O'Callaghan, M.; DeAngelis, K.M.; Jones, E.E.; Stewart, A.; Firestone, M.K.; Condron, L.M. Effects of selected root exudate components on soil bacterial communities. *FEMS Microbiol. Ecol.* **2011**, *77*, 600–610. [CrossRef] [PubMed]
69. Wiesenbauer, J.; Gorka, S.; Jenab, K.; Schuster, R.; Kumar, N.; Rottensteiner, C.; König, A.; Kraemer, S.; Inselsbacher, E.; Kaiser, C. Preferential use of organic acids over sugars by soil microbes in simulated root exudation. *Soil Biol. Biochem.* **2025**, *203*, 109738. [CrossRef]
70. Chen, Z.; Zhang, X.; Huang, Y.; Shi, Z.; Yao, H. Mechanisms of plant phenolic compounds affecting soil nitrogen transformation. *Alex. Eng. J.* **2025**, *120*, 173–184. [CrossRef]
71. Nicol, G.W.; Leininger, S.; Schleper, C.; Prosser, J.I. The influence of soil pH on the diversity, abundance and transcriptional activity of ammonia oxidizing archaea and bacteria. *Environ. Microbiol.* **2008**, *10*, 2966–2978. [CrossRef]
72. Faisal, Z.; Ziqi, L.; Xiuzheng, F.; Weidong, M.; Xingjie, W.; Jing, T.; Jingjing, P.; Werner, L.; Zhenling, C. Soil pH modulates microbial nitrogen cycling under integrated soil-crop system management across agroecosystems. *Soil Ecol. Lett.* **2025**, *7*, 250360. [CrossRef]
73. Liu, S.-Y.; Wu, D.; Ju, X.-T.; Shen, J.-P.; Cheng, Y.; Deng, N.; Song, X.-T.; Di, H.-J.; Li, P.-P.; Han, L.-L.; et al. Nitrification inhibitor induced microbial NH_4^+-N immobilization improves maize nitrogen use efficiency in strong ammonia oxidation soil. *Soil Biol. Biochem.* **2025**, *202*, 109687. [CrossRef]
74. Nommik, H.; Vahtras, K. Retention and fixation of ammonium and ammonia in soils. *Nitrogen Agric. Soil* **1982**, *22*, 123–172. [CrossRef]
75. Yu, A J.; Lin, X.; Zhu, J.; He, H.; Li, L. Environmental effects on ammonium adsorption onto clay minerals: Experimental constraints and applications. *Appl. Clay Sci.* **2023**, *246*, 107165. [CrossRef]
76. Van Cleemput, O.; Samater, A.H. Nitrite in soils: Accumulation and role in the formation of gaseous N compounds. *Fertil. Res.* **1995**, *45*, 81–89. [CrossRef]
77. Malal, H.; Garcia, J.A.; Marrs, A.; Ait Hamza, M.; Emerson, C.; Nocco, M.; Lakhtar, H.; Lazcano, C. Organic and inorganic fertilizers modulate the response of the soil microbiome to salinity stress. *Front. Microbiol.* **2025**, *16*, 1551586. [CrossRef]
78. Yan, N.; Marschner, P.; Cao, W.; Zuo, C.; Qin, W. Influence of salinity and water content on soil microorganisms. *Int. Soil Water Conserv. Res.* **2015**, *3*, 316–323. [CrossRef]
79. Tao, Y.; Xie, W.; Xu, L.; Zhang, L.; Wang, G.; Wang, X.; Shi, C. The characteristics of soil salinization effects on nitrogen mineralization and nitrification in upland fields. *Front. Environ. Sci.* **2024**, *12*, 1369554. [CrossRef]
80. Xie, R., Rao, P.; Pang, Y.; Shi, C.; Li, J.; Shen, D. Salt intrusion alters nitrogen cycling in tidal reaches as determined in field and laboratory investigations. *Sci. Total Environ.* **2020**, *729*, 138803. [CrossRef] [PubMed]
81. Wang, J.; Zhang, J.; Müller, C.; Cai, Z. Temperature sensitivity of gross N transformation rates in an alpine meadow on the Qinghai-Tibetan Plateau. *J. Soils Sediments* **2017**, *17*, 423–431. [CrossRef]

https://doi.org/10.3390/plants15091298

82. Davidson, E.A.; Janssens, I.A. Temperature sensitivity of soil carbon decomposition and feedbacks to climate change. *Nature* **2006**, *440*, 165–173. [CrossRef] [PubMed]
83. Schlüter, S.; Zawallich, J.; Vogel, H.J.; Dörsch, P. Physical constraints for respiration in microbial hotspots in soil and their importance for denitrification. *Biogeosciences* **2019**, *16*, 3665–3678. [CrossRef]
84. Gruda, N.S. Soilless culture systems and growing media in horticulture: An overview. In *Advances in Horticultural Soilless Culture*; Gruda, N.S., Ed.; Burleigh Dodds Science Publishing Limited: Cambridge, UK, 2021; pp. 1–20.
85. Shen, B.; Zheng, L.; Zheng, X.; Yang, Y.; Xiao, D.; Wang, Y.; Sheng, Z.; Ai, B. Insights from meta-analysis on carbon to nitrogen ratios in aerobic composting of agricultural residues. *Bioresour. Technol.* **2024**, *413*, 131416. [CrossRef] [PubMed]
86. Germer, S.; Reineke, T.; Muhammed, H.H.; Frerichs, C.; Anlauf, R.; Mayer, A.K.; Mai, C.; Daum, D. Approaches to Improve the Properties of Wood Fibers in Peat-Reduced Substrates for Tree Nursery Production. In Proceedings of the II International Symposium on Growing Media, Compost Utilization and Substrate Analysis for Soilless Cultivation, Freising, Germany, 7–12 September 2025.
87. Frerichs, C.; Feldner, A.; Daum, D. Lignin Addition to Wood Fibers—A Way to Reduce Nitrogen Immobilization in Growing Media? In Proceedings of the II International Symposium on Growing Media, Compost Utilization and Substrate Analysis for Soilless Cultivation, Freising, Germany, 7–12 September 2025.
88. Kumar, M.; Ou, Y.; Lin, J. Co-composting of green waste and food waste at low C/N ratio. *Waste Manag.* **2010**, *30*, 602–609. [CrossRef] [PubMed]

https://doi.org/10.3390/plants15091298

plants

MDPI

Article

Fulvic Acid, Amino Acids, and Vermicompost Enhanced Yield and Improved Nutrient Profile of Soilless Iceberg Lettuce

Beyza Keskin [1], Yelderem Akhoundnejad [2], Hayriye Yildiz Dasgan [1,*] and Nazim S. Gruda [3]

1 Department of Horticulture, Faculty of Agriculture, University of Cukurova, 01330 Adana, Türkiye; beyzaucel@gmail.com

2 Department of Horticulture, Faculty of Agriculture, University of Sirnak, 73000 Sirnak, Türkiye; yakhoundnejad@sirnak.edu.tr

3 Institute of Plant Sciences and Resource Conservation, Department of Horticultural Sciences, University of Bonn, D-53113 Bonn, Germany; ngruda@uni-bonn.de

* Correspondence: dasgan@cu.edu.tr

Abstract: Soilless cultivation systems are sustainable innovations in modern agriculture, promoting high efficiency per unit area, supporting food sustainability, and addressing the growing demand for high-quality produce with minimal environmental impact. This study evaluates the effects of fulvic acid, amino acid, and vermicompost biostimulants on the growth, yield, and nutrient profile of soilless-grown iceberg lettuce (*Lactuca sativa var. capitata)* in floating culture under controlled glasshouse conditions. Two experiments were conducted to determine the most effective concentrations and combinations of biostimulants. In the first experiment, varying doses of fulvic acid (40 and 80 ppm), amino acid (75 and 100 ppm), and vermicompost (1 and 2 mL L^{-1}) were tested alongside a control. Optimal doses were identified based on their positive effects on lettuce growth and yield. The second experiment examined combinations of fulvic acid, amino acid, and vermicompost extract compared to a control. Biostimulants improved lettuce growth, nutrient uptake, and antioxidants. Vermicompost boosted root biomass and leaf area, while fulvic acid and amino acid reduced nitrates and increased dry matter. Fulvic acid and vermicompost resulted in the highest yield (17.15 kg/m^2, 18.2% increase), and the combined treatment maximized antioxidants, increasing vitamin C by 17.16%, total phenols by 52.54%, and flavonoids by 52.38%. These findings highlight the potential of biostimulants as eco-friendly solutions for optimizing lettuce production in soilless systems.

Keywords: biostimulants; floating hydroponic culture; crop quality improvement; eco-friendly practices; *Lactuca sativa* var. *capitata*; nutritional quality; produce quality

1. Introduction

Soilless agriculture offers numerous advantages, including improved water-use efficiency, precise nutrient management, higher yields and quality, the ability to cultivate in regions with challenging conditions, reduced labor needs, enhanced irrigation practices, and streamlined production [1,2]. Soilless is a soil-free cultivation method extensively utilized for modern crop production on a global scale. This field encompasses various cultivation methods, such as deep-water culture, flowing water culture, nutrient film technique (NFT), and aeroponic culture [3,4]. In deep-water culture, plant roots are grown on a floating material like styrofoam, which is submerged in a nutrient-rich solution; this method is also known as floating hydroponic [5,6].

Lettuce (*Lactuca sativa* L.) belongs to the *Lactuca* genus within the Asteraceae family and is an annual vegetable. Its production occurs primarily in autumn, winter, and spring,

https://doi.org/10.3390/plants14040609

as it is classified as a cool-season crop. Iceberg lettuce (*Lactuca sativa var. capitata*), commonly known as crisp iceberg lettuce, is a widely cultivated variety characterized by its compact, spherical heads and pale green, crisp leaves. It is unlike leaf lettuces (*Lactuca sativa var. crispa*), which have loose, ruffled leaves, or romaine lettuces (*Lactuca sativa var. longifolia*), known for their elongated, firm leaves. Iceberg lettuce is prized for its dense, crunchy texture and mild flavor [7]. These attributes make it particularly popular for adding texture to salads, sandwiches, and wraps. Iceberg lettuce is low in calories and rich in water. It is valued for its crisp texture and dietary fiber, vitamin A, vitamin K, folate (B9), vitamin C, potassium, and calcium [8]. It also contains antioxidants such as beta-carotene, lutein, and zeaxanthin [9]. Among lettuce types grown in soilless greenhouses and vertical indoor farming systems, iceberg lettuce stands out as particularly noteworthy.

Biostimulants are natural or synthetic substances that enhance plant growth, yield, and stress tolerance by stimulating physiological processes without acting as conventional nutrients or pesticides [10]. In agriculture, their primary purpose is to improve nutrient-use efficiency, enhance crop quality, and mitigate the effects of abiotic stresses, offering an innovative and sustainable approach to increasing productivity while reducing environmental impact [11–13].

Humic substances, among the most prominent biostimulants, are organic compounds naturally formed through the decomposition of plant and animal matter. They are categorized into humin, humic acid, and fulvic acid (FA), standing out due to their lower molecular weight and abundance of oxygen-rich functional groups [10]. These characteristics enable fulvic acids to pass through biological membrane micropores, unlike humic acids, which have larger molecular weights [14]. Fulvic acids also exhibit higher total acidity, a significant number of carboxyl groups, and superior adsorption and cation-exchange capacities, allowing them to act as natural chelators that facilitate the mobilization and transport of micronutrients and pass through cell membranes [15–17]. They support plant growth by enhancing photosynthesis, secretion of growth hormones, nutrient absorption, availability of pH, and boosting resistance to biotic and abiotic stresses [18]. Fulvic acids are a mixture of weak aliphatic and aromatic organic acids that are water-soluble across a broad pH range, including acidic, neutral, and alkaline conditions. The composition of fulvic acid is highly variable and is recognized as one of the most active carbon-based chelating agents.

Amino acids (AAs), often considered the fundamental building blocks of proteins, are nitrogen-containing compounds essential for protein synthesis [19]. They play a crucial role in metabolic processes by providing key enzymes that stimulate cell growth [20]. They are widely recognized as effective biostimulants that significantly enhance plant growth and development. Extensive research has demonstrated their multifaceted benefits across various plant species. AAs have been shown to improve fertilizer assimilation, optimize nutrient and water uptake, and enhance photosynthetic efficiency in numerous vegetable crops. These effects collectively contribute to increased flower production, improved fruit sets, and higher fruit yield, highlighting their vital role in boosting agricultural productivity [21].

Vermicompost is an organic fertilizer produced by earthworms by decomposing organic waste. It consists of the nutrient-rich excreta of earthworms, which enhances soil health and fertility [22]. This natural process transforms organic matter enriched with essential macro and micronutrients, beneficial microorganisms such as bacteria, fungi, and actinomycetes, and growth hormones like auxins, cytokinins, and gibberellins [23]. Vermicompost has been proven to act as both a growth promoter and a natural protector for crop plants [24]. These properties make it a growth promoter and a natural protector for crop plants. Vermicompost also improves soil structure, water retention, and aeration while gradually releasing nutrients to plants, making it an eco-friendly and sustainable biostimulant [25]. Vermicompost has emerged as a highly effective and promising biostimulant, demonstrating remarkable success in traditional soil-based cultivation and soilless techniques for vegetable production in greenhouses [26].

The extended cultivation period of iceberg lettuce, which is 2–3 times longer than that of Batavia-type lettuce due to the requirement for head formation, highlights the importance of improving cultivation practices. Faster growth is particularly crucial in soilless hydroponic water culture systems as it allows for increased production cycles per year. To address this, our study investigates using biostimulants to enhance plant nutrition, growth, quality, and yield. Although several studies have investigated the mentioned biostimulants, to our knowledge, no studies have systematically explored their application in soilless floating water culture systems for cultivating iceberg lettuce. By addressing this gap, the study contributes to advancing sustainable agricultural practices and explores the potential of biostimulants as innovative solutions for optimizing crop production in soilless systems.

This study evaluates the effects of fulvic acid, vermicompost, and amino acid biostimulants on the growth, yield, and quality of iceberg lettuce (*Lactuca sativa* var. *capitata*) grown in a hydroponic floating culture system. The research is based on the hypothesis that these eco-friendly biostimulants, applied individually or in combination, can synergistically enhance plant performance.

2. Results

2.1. Iceberg Lettuce Yield

The application of biostimulants significantly influenced iceberg lettuce yield in the first trial (Figure 1). The highest yield was observed with the application of vermicompost at 2 mL (VC 2 mL), reaching 12.58 kg m^{-2}, corresponding to a 13.96% increase compared to the control treatment (11.04 kg m^{-2}). Fulvic acid at 40 ppm (FA 40) also resulted in a notable yield increase, achieving 12.12 kg m^{-2}, corresponding to a 9.78% increase compared to the control. However, no significant difference was observed between FA 40 and VC 2 mL. Other treatments, including FA 80, amino acid applications (AA 75 and AA 100), and vermicompost at 1 mL (VC 1 mL), did not significantly increase compared to the control. VC 1 mL exhibited a 4.35% increase, which, although higher than the control, was not statistically significant. In the first experiment, VC 2 mL and FA 40 demonstrated the most promising results for enhancing iceberg lettuce yield under experimental conditions.

The application of biostimulants significantly influenced iceberg lettuce yield during the second trial (Figure 2). Among the treatments, FA 40 + VC 2 mL exhibited the highest total yield, representing a 17.94% increase compared to the control. Similarly, FA 40 + AA 100 achieved a yield corresponding to a 17.11% increase relative to the control. The yields observed in FA 40 + AA 100 (17.16 kg m^{-2}) and FA 40 + AA 100 + VC 2 mL (17.04 kg m^{-2}) were statistically similar. The combination treatment of AA 100 + VC 2 mL produced a moderate yield, reflecting a 9.42% increase over the control. In contrast, the treatment FA 40 + AA 100 + VC 2 mL resulted in the lowest yield, indicating a 20.14% reduction compared to the control (14.55 kg m^{-2}). Other treatments, such as AA 100 + VC 2 mL, yielded 15.92 kg m^{-2} and significantly differed from the control and the higher-yielding treatments.

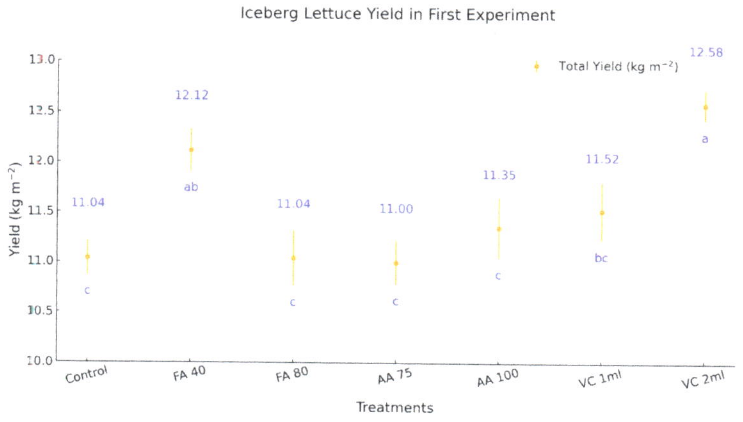

Figure 1. The effect of biostimulants on iceberg lettuce yield in the first winter trial. FA = Fulvic acid, AA = Aminoacid, VC = Vermicompost. No significant difference was observed between means represented by the same letter on the point ($p < 0.05$).

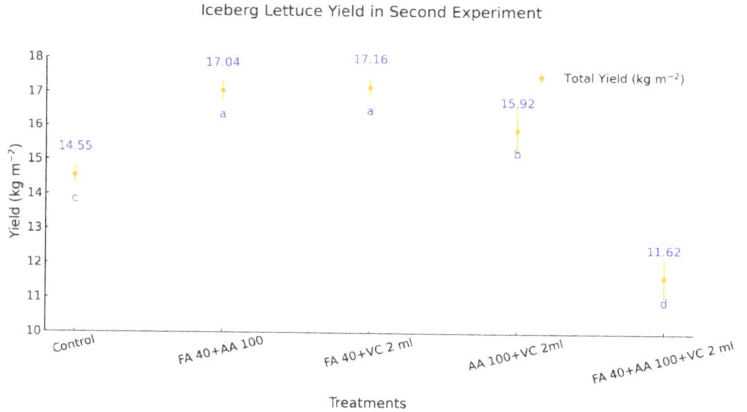

Figure 2. The effect of biostimulants on iceberg lettuce yield in the second spring trial. FA = Fulvic acid, AA = Aminoacid, VC = Vermicompost. No significant difference was observed between means represented by the same letter on the point ($p < 0.05$).

2.2. Iceberg Lettuce Growth Parameters in First Experiment

Plant Height: The FA 80 led to a 6.3% increase in plant height compared to the control, with FA 40 showing a 5.6% improvement. VC 2 mL also contributed positively, achieving a 2.8% increase, indicating the beneficial effects of fulvic acid and vermicompost on plant height (Table 1).

Perimeter Dimension: FA 40 exhibited a slight advantage with a 0.6% increase over the control, and FA 80 displayed a 0.4% improvement.

Head Firmness: AA 75 was the only one to increase head firmness, achieving a 4.5% improvement compared to the control. All other treatments were lower than the control, highlighting AA 75 as the most effective parameter enhancement.

Root Weight: VC 2 mL achieved a notable 26% increase in root weight over the control, while AA 100 provided a 22% improvement and VC 1 mL yielded a 20.2% increase. These results underscore the efficacy of vermicompost and amino acid in promoting root biomass accumulation.

Table 1. The effects of biostimulants on growth parameters of iceberg lettuce in the first trial-I.

Treatments	Plant Height (cm)	Perimeter (cm)	Head Firmness (kg cm^3)	Root Weight (g)
Control	38.85 abc	44.85 a	8.24 ab	45.40
FA 40	41.05 a	45.10 a	6.88 abc	50.00
FA 80	41.30 a	45.05 a	7.73 ab	48.50
AA 75	37.30 bc	40.85 b	8.61 a	44.70
AA 100	33.60 d	43.55 ab	6.75 bc	55.40
VC 1 mL	35.55 cd	41.75 ab	5.82 c	54.60
VC 2 mL	39.95 ab	44.00 ab	6.67 bc	57.20
p	<0.0001	0.0011	<0.0001	0.1211
$Tukey_{0.05}$	3.516	3.575	1.733	NS

FA = Fulvic acid, AA = Aminoacid, VC = Vermicompost. No significant difference was observed between means sharing the same letter within the same column ($p < 0.05$). NS: Not Significant.

Leaf Fresh Weight: The VC 2 mL resulted in the highest leaf fresh weight, with a 14% increase compared to the control. FA 40 also showed a notable improvement, with a 10% increase, while AA 100 demonstrated a 3% rise (Table 2). These results suggest that vermicompost and fulvic acid applications positively impacted leaf fresh weight.

Table 2. The effects of biostimulants on growth parameters of iceberg lettuce in the first trial-II.

Treatments	Leaf Fresh Weight (g Plant^{-1})	Leaf Area (cm^2 Plant^{-1})	Number of Leaves per Plant	Dry Matter (%)	Chlorophyll (SPAD)
Control	372 d	4793 b	33.15 c	7.31	16.44 ab
FA 40	409 b	5952 ab	34.25 bc	7.55	17.86 ab
FA 80	372 d	5791 ab	34.80 bc	7.63	17.74 ab
AA 75	371 d	7585 a	33.05 c	7.41	15.04 b
AA 100	383 c	6748 ab	39.85 a	7.68	21.61 a
VC 1 mL	388 c	6071 ab	36.95 ab	7.72	17.53 ab
VC 2 mL	424 a	692 ab	37.15 ab	7.83	18.46 ab
p	<0.0001	0.0398	<0.0001	0.1951	0.0215
$Tukey_{0.05}$	5.719	2515	3.654	NS	5.329

FA = Fulvic acid, AA = Aminoacid, VC = Vermicompost. NS: Non-significant. No significant difference was observed between means sharing the same letter within the same column ($p < 0.05$). NS: Not Significant.

Leaf Area: AA 75 achieved the largest leaf area, showing a substantial 58% increase over the control. AA 100 followed with a 41% increase, and VC 2 mL exhibited a 45% improvement.

Number of Leaves: AA 100 application led to the highest leaf count, with a 20% increase relative to the control. VC 2 mL and VC 1 mL also showed positive impacts, with increases of 12% and 11.5%, respectively. This suggests that amino acids and vermicompost applications effectively enhance leaf production.

Dry Matter: In comparison to the control, dry matter content increased by 7.1%, 5.6%, 5.1%, 4.4%, 3.3%, and 1.4%, respectively, for the VC 2 mL, VC 1 mL, AA 100, FA 80, FA 40, and AA 75 treatments, although these increases were not statistically significant.

Chlorophyll Content (SPAD): The AA 100 demonstrated the highest chlorophyll content, increasing SPAD values by 31% relative to the control. VC 2 mL and FA 40 followed with 12% and 9% improvements, respectively, indicating that amino acid and vermicompost contributed positively to chlorophyll concentration.

2.3. Iceberg Lettuce Growth Parameters in Second Experiment

Plant height: The combined FA 40 + AA 100 yielded the highest plant height, with an increase of 6.4% compared to the control. FA 40 + VC 2 mL was closely followed, achieving a 6.2% increase, and AA 100 + VC 2 mL showed a 1.3% improvement over the control. (Table 3).

Table 3. The effects of biostimulants on growth parameters of iceberg lettuce in the second trial-I.

Treatments	Plant Height (cm)	Perimeter (cm)	Head Firmness (kg·cm^{-3})	Root Weight (g)
Control	29.42 b	38.85 bc	9.47 b	70.90 c
FA 40 + AA 100	31.30 a	45.45 a	8.96 b	80.70 b
FA 40 + VC 2 mL	31.25 a	42.15 ab	8.75 b	95.00 a
AA 100 + VC 2 mL	29.80 ab	42.80 ab	8.85 b	87.00 ab
FA 40 + AA 100 + VC 2 mL	29.20 b	37.55 c	22.17 a	72.10 c
p	0.0008	<0.0001	<0.0001	<0.0001
Tukey$_{0.05}$	1.743	4.1311	1.744	8.294

FA = Fulvic acid, AA = Aminoacid, VC = Vermicompost. No significant difference was observed between means sharing the same letter within the same column ($p < 0.05$).

Perimeter: FA 40 + AA 100 demonstrated a 17% increase in perimeter dimension over the control, while AA 100 + VC 2 mL achieved a 10.2% increase. The FA 40 + VC 2 mL combination showed an 8.5% enhancement compared to the control.

Head firmness: In terms of head firmness, the FA 40 + AA 100 + VC 2 mL substantially increased, providing a 134.2% improvement over the control, making it the most effective treatment in this parameter. Other treatments, including FA 40 + AA 100, FA 40 + VC 2 mL, and AA 100 + VC 2 mL, demonstrated values below the control.

Root weight: FA 40 + VC 2 mL yielded the highest root weight, with a 33.9% increase compared to the control. The AA 100 + VC 2 mL combination also showed a substantial increase of 22.7%, and FA 40 + AA 100 demonstrated a 13.8% enhancement.

Leaf fresh weight: In the second trial, the combinations led to a higher increase in leaf fresh weight than in the first trial (Table 4). Specifically, the FA 40 + VC 2 mL and FA 40 + AA 100 produced 17–18% higher leaf fresh weight than the control.

Table 4. The effects of biostimulants on growth parameters of iceberg lettuce in the second trial.

Treatments	Leaf Fresh Weight (g Plant^{-1})	Leaf Area (cm^2 Plant^{-1})	Number of Leaves per Plant	Dry Matter (%)	Chlorophyll (SPAD)
Control	490 b	10,943 bc	33.90 ab	9.92 c	20.45
FA 40 + AA 100	574 a	12,005 b	35.70 ab	11.16 bc	19.67
FA 40 + VC 2 mL	579 a	13,580 a	35.60 ab	11.04 bc	17.80
AA 100 + VC 2 mL	537 ab	10,492 c	37.15 a	13.48 ab	20.02
FA40 + AA100 + VC 2 mL	392 c	10,308 c	32.20 b	14.26 a	19.97
p	<0.0001	<0.0001	0.0025	<0.0001	0.3022
Tukey$_{0.05}$	82.545	1389	3.549	1.916	NS

FA = Fulvic acid, AA = Aminoacid, VC = Vermicompost. NS: Non-significant. No significant difference was observed between means sharing the same letter within the same column ($p < 0.05$). NS: Not Significant.

Leaf area: The FA 40 + VC 2 ml treatment increased leaf area by 24.1%, while the FA 40 + AA 100 treatment resulted in a 9.7% increase compared to the control.

Number of leaves: The AA 100 + VC 2 mL combination yielded the highest leaf count, with a 9.6% increase compared to the control. This increase is more significant than observed in the first trial, suggesting that combinations were more supportive of leaf production than single biostimulant applications.

Dry matter: For dry matter content, the most substantial increases were observed in the FA 40 + AA 100 + VC 2 mL and AA 100 + VC 2 mL treatments, with enhancements of 43.8% and 35.9%, respectively, compared to the control.

Chlorophyll content (SPAD): Chlorophyll content did not show statistically significant differences among treatments.

2.4. Macro-Nutrient Profiles of Iceberg Lettuce

In the first trial of iceberg lettuce cultivation under soilless culture, significant variations in macro-nutrient concentrations were observed among treatments with different biostimulants ($p < 0.05$) (Table 5). Nitrogen concentration was highest in VC 2 mL, showing a substantial increase of 37.4% compared to the control. VC 1 mL and FA 40 followed, with increases of 23.0% and 15.7%, respectively. Phosphorus levels reached their maximum with VC 2 mL, surpassing the control by 19.2%. FA 80 showed a minor increase of 11.5%, while FA 40 minimally contributed to phosphorus enhancement. Potassium concentration peaked with FA 80, achieving a rise of 57.6% over the control. Strong performances were also observed with VC 2 mL and AA 75, which recorded increases of 45.4% and 42.4%, respectively. The highest calcium content was recorded with FA 40, representing an improvement of 18.0% relative to the control. FA 80 followed closely with a 15.0% increase, and AA 75 contributed moderately to calcium enhancement. Magnesium concentrations exhibited the most significant increase with AA 75, rising by 144.3% compared to the control. FA 80 and FA 40 also demonstrated notable improvements, with increases of 102.5% and 69.6%, respectively.

Table 5. Macronutrients of iceberg lettuce leaves grown with different biostimulants in the first trial (%).

Treatment	N	P	K	Ca	Mg
Control	2.30 b	0.26 b	2.71 c	1.94 ab	0.79 c
FA 40	2.66 ab	0.20 c	3.34 abc	2.29 a	1.34 b
FA 80	2.57 ab	0.23 bc	4.27 a	2.23 a	1.60 b
AA 75	2.36 b	0.23 bc	3.86 ab	1.68 b	1.93 a
AA 100	2.37 b	0.24 bc	3.00 bc	1.67 b	0.40 d
VC 1 mL	2.83 ab	0.25 bc	3.53 abc	1.50 bc	0.34 d
VC 2 mL	3.16 a	0.31 a	3.94 ab	1.10 c	0.53 cd
p	0.0039	<0.0001	<0.0001	<0.0001	<0.0001
$Tukey_{0.05}$	0.653	0.0524	0.976	0.514	0.304

FA = Fulvic acid, AA = Aminoacid, VC = Vermicompost. No significant difference was observed between means sharing the same letter within the same column ($p < 0.05$).

In the second trial of lettuce cultivation under soilless conditions, the macronutrient concentrations showed significant differences among treatments, with some combinations outperforming others (Table 6). Nitrogen concentration reached its highest levels in FA 40 + VC 2 mL, demonstrating a considerable improvement of 37.4% over the control. FA 40 + AA 100 and AA 100 + VC 2 mL followed closely, showing similar enhancements in nitrogen levels. Phosphorus concentration was most elevated in FA 40 + VC 2 mL, achieving an increase of 69.2% compared to the control. FA 40 + AA 100 ranked second with a 65.4% increase, while the AA 100 + VC 2 mL combination also performed well, with a moderate enhancement. Although statistically not significant, potassium concentration peaked in FA 40 + VC 2 mL, which improved by 43.5% compared to the control. AA 100 + VC 2 mL followed with an increase of 43.2%, and FA 40 + AA 100 ranked third with a 33.2% improvement. FA 40 + VC 2 mL recorded the highest concentration for calcium, reflecting an improvement of 68.6% over the control. FA 40 + AA 100 followed with a 50.0% increase, while AA 100 + VC 2 mL showed an enhancement of 48.5%. Magnesium levels were highest in AA 100 + VC 2 mL, marking a 92.4% increase over the control. FA 40 + AA 100 contributed a 69.6% improvement, while FA 40 + VC 2 mL achieved a rise of 65.8%.

Table 6. Macronutrients of iceberg lettuce leaves grown with different biostimulants in the second trial (%).

Treatment	N	P	K	Ca	Mg
Control	2.67 bc	0.42 ab	3.61	1.03 ab	0.79 a
FA 40 + AA 100	3.03 ab	0.43 a	3.70	0.97 ab	0.70 ab
FA 40 + VC 2 mL	3.16 a	0.44 a	3.89	1.14 a	0.67 ab
AA 100 + VC 2 mL	3.00 a ab	0.40 ab	3.88	0.96 ab	0.73 a
FA40 + AA 100 + VC 2 mL	2.50 b c	0.36 b	3.44	0.64 b	0.50 b
p	0.0011	0.0087	0.5855	0.0321	0.0063
$Tukey_{0.05}$	0.404	0.0654	NS	0.438	0.218

FA = Fulvic acid, AA = Aminoacid, VC = Vermicompost, NS: Non-Significant. No significant difference was observed between means sharing the same letter within the same column ($p < 0.05$). NS: Not Significant.

2.5. Antioxidant Contents of Iceberg Lettuce

In the first trial of iceberg lettuce cultivation under soilless culture, significant differences were observed in antioxidant parameters (Table 7). Vitamin C content was highest in VC 1 mL, showing a 33.9% increase compared to the control. VC 2 mL ranked second with a 28.1% improvement, followed by AA 100, which increased Vitamin C levels by 25.6%. The application of biostimulants resulted in an increase in total phenol content compared to the control for the following treatments: FA 40 (25.34%), AA 100 (8.96%), and VC 2 mL (7.54%). Other treatments did not show an increase in total phenol content. In terms of total flavonoid content, the treatments that exhibited an increase compared to the control were FA 40 (60.64%), AA 100 (31.38%), VC 2 mL (29.26%), and AA 75 (15.96%). Other treatments did not increase total flavonoid content.

Table 7. Effects of biostimulants on antioxidant of iceberg lettuce in first trial aquaculture.

Treatments	Vitamin C (mg 100 g FW^{-1})	Total Phenols (mg GA 100 g FW^{-1})	Total Flavonoids (mg RU 100 g FW^{-1})
Control	5.40 d	41.39 ab	188 b
FA 40	6.37 c	51.88 a	302 a
FA 80	6.57 bc	39.08 b	182 b
AA 75	6.76 abc	41.19 ab	218 b
AA 100	6.78 abc	45.10 ab	247 ab
VC 1 mL	7.23 a	32.94 b	178 b
VC 2 mL	6.92 ab	44.51 ab	243 ab
p	<0.0001	0.0055	0.0008
$Tukey_{0.05}$	0.493	12.188	74.574

FA = Fulvic acid, AA = Aminoacid, VC = Vermicompost. No significant difference was observed between means sharing the same letter within the same column ($p < 0.05$). FW: Fresh weight, GA: Gallic acid, RU: Rutin.

In the second trial in the spring, the combination treatments demonstrated notable increases in antioxidant content compared to the control (Table 8). For vitamin C, the highest increase was observed with the FA 40 + AA 100 + VC 2 mL treatment, which showed a 17.16% improvement, followed by FA 40 + AA 100 (15.06%), FA 40 + VC 2 mL (9.11%), and AA 100 + VC 2 mL (4.20%). Regarding total phenol content, FA 40 + AA 100 + VC 2 mL exhibited the most significant increase at 52.54%, while FA 40 + AA 100 and AA 100 + VC 2 mL also showed substantial increases of 40.50% and 43.06%, respectively. The FA 40 + VC 2 mL treatment resulted in a 22.28% increase. For total flavonoid content, FA 40 + AA 100 + VC 2 mL again led with a 52.38% increase, followed by FA 40 + AA 100 at 46.32%, AA 100 + VC 2 mL at 42.42%, and FA 40 + VC 2 mL at 35.50%.

Table 8. Effects of biostimulants on antioxidant of iceberg lettuce in second trial of aquaculture.

Treatments	Vitamin C (mg 100 g FW^{-1})	Total Phenols (mg GA 100 g FW^{-1})	Total Flavonoids (mg RU 100 g FW^{-1})
Control	5.71 d	57.71 c	231 c
FA 40 + AA 100	6.57 ab	81.08 ab	338 ab
FA 40 + VC 2 mL	6.23 bc	70.57 bc	313 b
AA 100 + VC 2 mL	5.95 cd	82.56 ab	329 ab
FA 40 + AA 100 + VC 2 mL	6.69 a a	88.03 a	352 a a
p	0.0003	0.0013	<0.0001
$Tukey_{0.05}$	0.447	16.142	35.437

FA = Fulvic acid, AA = Aminoacid, VC = Vermicompost. No significant difference was observed between means sharing the same letter within the same column ($p < 0.05$). FW: Fresh weight, GA: Gallic acid, RU: Rutin.

2.6. *Leaf Quality Parameters of BRIX, Acidity, pH, and EC Values of Iceberg Lettuce*

In the first trial of iceberg lettuce cultivation under soilless conditions, the effects of biostimulants on brix, acidity, pH, and EC values were evaluated. In both trials, the Brix values in lettuce leaves did not show any statistically significant differences among the treatments. Among these parameters, pH showed statistically significant differences ($p < 0.05$), while other parameters exhibited no variation (Table 9). pH levels showed significant variation, with the highest values recorded in FA 80, FA 40, and the control, all maintaining statistically similar levels. The lowest pH levels were observed in the VC 1 mL and AA 100, significantly reduced compared to the control.

Table 9. Effects of biostimulants on some quality parameters of iceberg lettuce in the first trial.

Treatments	Brix (%)	pH	EC (µS cm^{-1})
Control	1.77	6.64 abc	780 b
FA 40	1.55	6.66 ab	1150 a
FA 80	1.60	6.69 a	1182 a
AA 75	1.30	6.73 a	1081 a
AA 100	1.40	6.39 bc	1123 a
VC 1 mL	1.42	6.37 c	966 ab
VC 2 mL	1.90	6.68 a	1123 a
p	0.4389	0.0013	0.0008
$Tukey_{0.05}$	NS	0.286	260.679

FA = Fulvic acid, AA = Aminoacid, VC = Vermicompost. No significant difference was observed between means sharing the same letter within the same column ($p < 0.05$). NS: Not Significant.

In the second trial of iceberg lettuce under soilless culture, statistically significant differences were observed for pH ($p < 0.05$) and EC ($p < 0.05$). At the same time, brix and acidity values did not show significant variations (Table 10). pH levels demonstrated significant differences among treatments. The highest pH values were recorded in FA 40 + VC 2 mL and FA 40 + AA 100 + VC 2 mL, maintaining slightly higher levels than the control. Conversely, FA 40 + AA 100 and AA 100 + VC 2 mL recorded lower pH values, indicating a notable reduction. Electrical conductivity values were significantly affected by the treatments. In comparison to the control, the highest increases in EC values were observed with the FA 80, FA 40, and AA 100 treatments, showing increases of 51.5%, 47.4%, and 44.0%, respectively. These results indicate that fulvic acid and amino acid applications significantly enhanced the EC levels of the growing medium. The control recorded one of the lowest EC values among all treatments.

Table 10. Effects of biostimulants on some quality parameters of iceberg lettuce in the second trial.

Treatments	Brix (%)	pH	EC (μS cm^{-1})
Control	2.25	6.56 ab	1344 b
FA 40 + AA 100	2.50	6.43 b	1697 ab
FA 40 + VC 2 mL	2.30	6.62 a	1569 ab
AA 100 + VC 2 mL	2.50	6.46 ab	1952 a
FA 40 + AA 100 + VC 2 mL	2.54	6.60 ab	1618 ab
p	0.4382	0.0263	0.0440
$Tukey_{0.05}$	NS	0.188	537.117

FA = Fulvic Acid, AA = Aminoacid, VC = Vermicompost. No significant difference was observed between means sharing the same letter within the same column ($p < 0.05$). NS: Not Significant.

2.7. Nitrate Content in Iceberg Lettuce

In the first trial of iceberg lettuce cultivation, the nitrate reduction capacity of biostimulants was evident, with specific treatments significantly decreasing nitrate accumulation in leaf tissues (Figure 3). The lowest nitrate content was recorded in AA 100 (690 mg/kg), demonstrating a remarkable reduction of 39.1% compared to the control (1133 mg/kg). This highlights AA 100 as the most effective biostimulant in mitigating nitrate accumulation, making it highly desirable for enhancing lettuce quality. Among the vermicompost, VC 2 mL (1078 mg/kg) and VC 1 mL (1085 mg/kg) achieved moderate reductions of 4.9% and 4.2%, respectively, compared to the control. While these reductions are less pronounced, they still reflect the potential of vermicompost to limit nitrate levels. Conversely, FA-based, including FA 80 (1238 mg/kg), FA 40 (1229 mg/kg), and AA 75 (1189 mg/kg), showed increased nitrate content, with levels exceeding the control. These results suggest that FA-based are less effective or may even promote nitrate accumulation under the conditions tested. Overall, the findings demonstrate that the nitrate-reducing capacity of biostimulants varies significantly, with AA 100 emerging as the most effective treatment for reducing nitrate content in lettuce leaves, a highly desirable outcome for improved quality and safety.

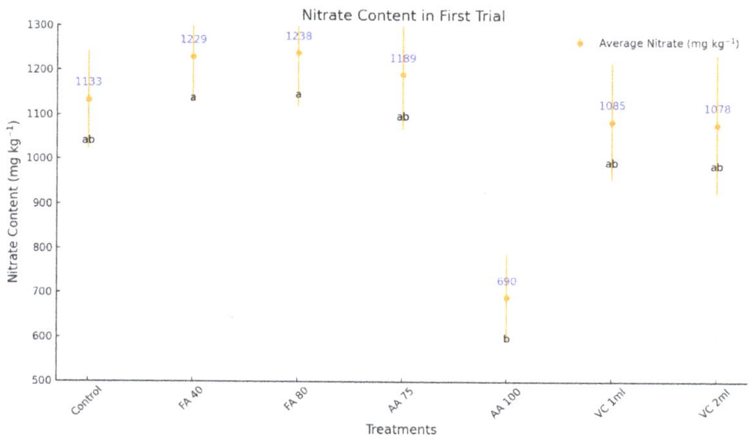

Figure 3. Effects of biostimulants on nitrate content in soilless-grown iceberg lettuce in first winter experiment. FA = Fulvic acid, AA = Aminoacid, VC = Vermicompost. No significant difference was observed between means represented by the same letter on the point ($p < 0.05$).

In the second trial of iceberg lettuce cultivation, biostimulants demonstrated varying capacities to reduce nitrate content in leaf tissues, highlighting their potential to enhance lettuce quality (Figure 4). The lowest nitrate content was recorded in FA 40 + AA 100 + VC 2 mL (275 mg/kg), representing a significant reduction of 56.2% compared to the control (628 mg/kg). This treatment emerged as the most effective in mitigating nitrate accumulation, underscoring the synergistic effects of combining multiple biostimulants. AA 100 + VC 2 mL also showed a substantial reduction in nitrate content, with a 44.9% decrease compared to the control. This highlights its effectiveness in reducing nitrate levels while maintaining leaf quality. In contrast, FA 40 + AA 100 (646 mg/kg) and FA 40 + VC 2 mL (596 mg/kg) recorded nitrate contents similar to or slightly higher than the control, indicating limited or no nitrate-reducing capacity under the tested conditions. The results demonstrate that combining biostimulants reduces nitrate levels in lettuce leaves. FA 40 + AA 100 + VC 2 mL is particularly effective, making it a desirable option for achieving high-quality lettuce with reduced nitrate content.

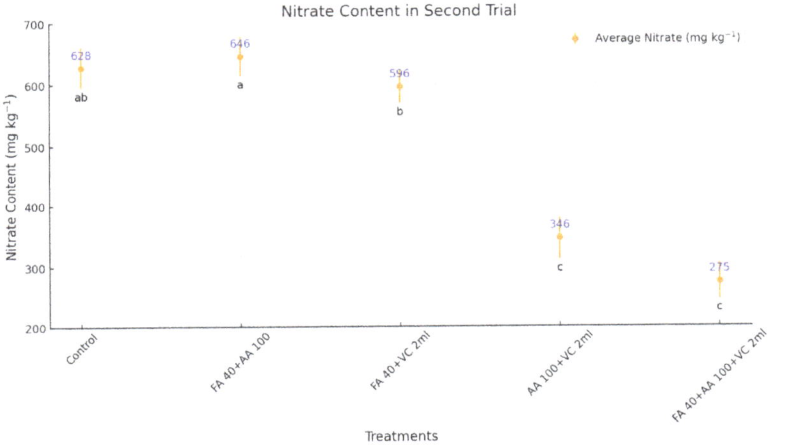

Figure 4. Effects of biostimulants on nitrate content in soilless-grown iceberg lettuce in second spring experiment. FA = Fulvic acid, AA = Aminoacid, VC = Vermicompost. No significant difference was observed between means represented by the same letter on the point ($p < 0.05$).

3. Discussion

The exogenous application of biostimulants, including amino acids, vermicompost, and fulvic acid, represents an innovative and eco-friendly agricultural approach that complements mineral fertilization, enhances plant productivity, promotes growth, increases yield, and improves product quality.

3.1. *Effects of Biostimulants on Plant Growth Parameters and Iceberg Lettuce Yield*

The findings in this study reveal that biostimulants enhanced lettuce performance, promoting growth parameters, chlorophyll levels, and nutrient absorption. Vermicompost achieved the most substantial increases in root biomass and leaf area. In contrast, combinations of amino acids and fulvic acid effectively reduced nitrate content while increasing antioxidants and dry matter accumulation. The combined application of FA 40 + VC 2 mL resulted in an 18.12% increase in iceberg lettuce yield compared to the control. At the same time, the combination of FA 40 + AA 100 led to a 17.4% increase.

Biostimulants are rich in growth regulators, including humic substances, cytokinins, and auxins, contributing to increases in root and leaf biomass, plant height, leaf number, and shoot and root dry weight. Canellas et al. [27] and Taha et al. [16] reported that the application of fulvic acid significantly enhanced lettuce growth parameters compared to the control. The foliar application at 3.1 g m^{-2} resulted in a 126% increase in fresh weight, 127% in dry weight, and 30% in plant height. Similarly, irrigation application at 3.1 g m^{-2} increased by 105% in fresh weight, 103% in dry weight, and 14% in plant height. These results demonstrate the effectiveness of fulvic acid, mainly through foliar application, in promoting lettuce growth. Lüdtke et al. [28] showed that the application of FA + NPK significantly improved lettuce growth compared to the control. By day 48, the number of leaves increased by 51%, plant diameter by 40%, plant height by 33%, and root length by 45%. Additionally, the aerial fresh mass of lettuce reached approximately 280 g/plant under FA + NPK treatment, representing a 180% increase compared to the control. These findings highlight the potential of FA + NPK to enhance overall lettuce growth, contributing to both shoot and root development and biomass production.

Fulvic acid application plants increased the net photosynthetic rate and chlorophyll fluorescence characteristics, including initial fluorescence (Fo), steady-state fluorescence (Fs), maximum fluorescence (Fm), variable fluorescence (Fv), photochemical quenching coefficient (qP), and the effective photochemical quantum yield of PSII (Fv′/Fm′) and promoted photosynthesis [29]. In addition, fulvic acids enhance photosynthesis by modulating signaling pathways associated with plant hormones. They improve essential physiological parameters, including transpiration rate, stomatal conductance, water-use efficiency, intercellular CO_2 concentration, chlorophyll content, and relative water content. These findings underscore the significant potential of fulvic acids to enhance plant growth and productivity by optimizing critical physiological processes [30].

Al-Karaki and Othman [31] demonstrated that in soilless cultivation using a tuff-zeolite medium, foliar application of an amino acid biostimulant at a concentration of 4 mL/L significantly increased yields, with a 26% improvement in iceberg lettuce and a 50% increase in romaine lettuce compared to the control. Amino acids are building blocks for protein biosynthesis and precursors for various biosynthetic pathways and key regulators in signaling processes and stress responses [32]. Their role in promoting a more robust root system enhances water and nutrient uptake efficiency, ultimately driving better plant growth and yield [33]. Haghighi et al. [21] further reported that amino acid application improved the fresh and dry weights of cabbage shoots and the fresh weight of roots by enhancing key physiological processes such as photosynthesis, transpiration, and stomatal conductance. These improvements were linked to enhanced nitrogen uptake and assimilation, contributing to more efficient CO_2 assimilation and the effective transport of soluble sugars via the phloem to sink tissues. The biostimulatory effects of amino acids are also associated with activating critical enzymes, such as nitrate reductase and glutamine synthase, which are pivotal in nitrogen metabolism [34]. These findings underscore the multifaceted benefits of amino acid biostimulants in improving plant growth and productivity, particularly under soilless culture systems.

According to Yassen et al. [34], the application of 0.95 kg m^{-2} o fvermicompost combined with 150 mL L^{-1} of vermiwash, a liquid extract derived from vermicomposting processes, significantly improved lettuce yield, increasing it from 2.48 kg m^{-2} in the control to 4.72 kg m^{-2}, a remarkable 91% increase. Similarly, Kilic and Saracoglu [35] reported substantial yield enhancements with vermicompost application, achieving a 121% increase in iceberg lettuce and an 184% increase in Lollo Rosso lettuce. These findings emphasize the efficacy of vermicompost and its combination with vermiwash in enhancing lettuce productivity. The beneficial effects of vermicompost on plant growth are attributed to its role in improving photosynthesis and chlorophyll fluorescence. Zou et al. [36] demonstrated that vermicompost significantly increased key physiological parameters, including net photosynthetic rate, stomatal conductance, transpiration rate, and intercellular CO_2 concentration. Furthermore, chlorophyll fluorescence traits, such as maximum fluorescence

(Fm), maximum photochemical efficiency (Fv/Fm), and potential photochemical activity (Fv/F_0), were also enhanced, contributing to improved plant growth and yield [37].

The positive impacts of vermicompost are linked to its high nutrient content and bioactive composition. Vermicompost contains essential macronutrients (N, P, K, Ca, Mg, and S) and micronutrients (Fe, Mn, Zn, Cu, and B), which are critical for supporting root, shoot, and fruit development [38]. Additionally, it is enriched with beneficial microorganisms, vitamins, growth hormones, and enzymes such as proteases, amylases, cellulases, and chitinases, which promote plant development [23]. The microbial diversity in vermicompost, particularly bacterial and fungal populations, further enhances its growth-promoting properties [39]. Another key advantage of vermicompost is its abundance of plant growth regulators, including humic substances, cytokinins, and auxins. These compounds are likely responsible for the observed increases in leaf and root weights, leaf number, plant height, leaf area, and dry weight of shoots and roots [10]. Toor et al. [22] highlighted the effectiveness of vermicompost in enhancing lettuce growth, reporting a 100% increase in total chlorophyll content, a 118% increase in fresh weight, and a 20% improvement in plant height with the optimal application dose of 2%.

3.2. *Effects of Biostimulants on Mineral Nutrient Contents*

Biostimulants, particularly fulvic acid, amino acids, and vermicompost, play a significant role in enhancing the mineral content of plants. Biostimulants primarily stimulate root development, increasing the plant's nutrient uptake capacity, as highlighted in [40]. According to Yassen et al. [34], the application of 0.95 kg m^{-2} vermicompost + 150 mL L^{-1} vermiwash significantly improved the N, P, K, Ca, and Mg contents in the head of lettuce compared to the control. The increases were 46%, 43%, 45%, 16%, and 62%, respectively. These results demonstrate the effectiveness of this optimal vermicompost dose in enhancing both the macronutrient and micronutrient composition of lettuce heads, contributing to their overall nutritional quality. A more robust root system enables the efficient absorption and transport of nutrients to plant tissues. Moreover, biostimulants like vermicompost enhance nutrient availability by providing organic compounds and beneficial microorganisms that facilitate nutrient uptake [41]. Additionally, biostimulants influence hormone balance and signaling pathways to support nutrient mobilization. Elevated cytokinin levels promote nutrient transport to source tissues, such as leaves, enhancing plant growth [42].

Al-Karaki and Othman [31] demonstrated that in the soilless culture of iceberg lettuce, the foliar application of an amino acid-based biostimulant led to notable increases in leaf mineral concentrations compared to the control. Specifically, for iceberg lettuce, nitrogen (N) content increased by approximately 29%, phosphorus (P) by 25%, potassium (K) by 22%, and magnesium (Mg) by 53% [31]. Lüdtke et al. [28] demonstrated that the application of FA + NPK significantly improved the macronutrient and micronutrient contents of lettuce compared to the control. N, P, K, Ca, and Mg contents increased by 11%, 14%, 67%, 15%, and 2%, respectively. These findings underscore the effectiveness of FA + NPK in enhancing plant nutrient levels, contributing to the nutritional quality of lettuce. Biostimulants also improve photosynthesis, supporting energy metabolism and facilitating active nutrient transport and ion uptake in roots [41]. Amino acids contribute to increased carbohydrate production, meeting the energy demands of roots and improving nutrient acquisition [43]. Vermicompost supports beneficial microbial populations, further enhancing the bioavailability of essential minerals [36].

3.3. *Effects of Biostimulants on Iceberg Lettuce Quality and Antioxidant Contents*

In this study, biostimulants, particularly fulvic acid, amino acids, and vermicompost, have proven effective in enhancing vitamin content, antioxidant capacity, phenolic compounds, and flavonoid levels in lettuce. Plants treated with amino acids exhibited higher phenolic content and antioxidant activity, which can be partially attributed to the localized beneficial effects of amino acids on plant processes. Amino acids are precursors to a wide range of secondary metabolites and are closely linked to synthesizing proteins, hormones, and carbohydrates [31]. Haghighi et al. [21] demonstrated that the exogenous application of AAs enhances the nutritional profile of cabbage, including increased levels of phenols, total protein, proline, and essential amino acids such as glutamic acid, glutamine, and asparagine, along with improved antioxidant capacity.

Vermicompost has enhanced key quality compounds in strawberries, including soluble sugar and vitamin C content [36]. Al Jaouni et al. [44] reported that vermicompost exhibited the highest levels of antioxidants and bioactive compounds, including phenolics, flavonoids, steroids, and ascorbate. Abdel–Baky et al. [30] demonstrated that applying fulvic acid significantly increased total carbohydrate content, crude protein, and essential minerals. Additionally, it elevated the levels of key amino acids such as arginine, lysine, phenylalanine, and tryptophan. The foliar application of fulvic acid significantly enhanced secondary metabolites, including total phenolic content, anthocyanins, and total carotenoids, underscoring its potential to improve quality attributes [14].

3.4. *Effects of Biostimulants on Iceberg Lettuce Nitrate Content*

During the second experimental period (February, March, and April), the light intensity inside the greenhouse ranged between 350 and 550 W·m^{-2}, and the day length extended from 10.4 to 13.7 h. These conditions may have contributed to the observed reduction in nitrate levels, as increasing light intensity and extended day length play a critical role in nitrate metabolism in leafy vegetables [45]. In this study, applying the biostimulant combination FA 40 + AA 100 + VC 2 mL resulted in the highest reduction in nitrate content, decreasing levels by 56.2% (275 mg/kg) compared to the control (628 mg/kg). Similarly, AA 100 + VC 2 mL significantly reduced nitrate content by 44.9%. These findings align with the role of nitrate reductase (NR) in plant nitrogen metabolism, where NR catalyzes the reduction of nitrate to ammonium, a precursor for amino acid synthesis [46].

In the second spring trial, biostimulant applications significantly decreased nitrate content in lettuce leaf tissues. This is consistent with the findings of Coronel et al. [47], who reported that soilless lettuce cultivation enhances NR activity, making it a promising method for producing lettuce with reduced nitrate levels. Biostimulants promote plant growth by increasing amino acid concentrations, which are directly utilized in protein biosynthesis. Elevated carbohydrate levels in the leaves further support this process by enhancing nitrogen assimilation through the nitrate pathway. Carbohydrates provide the carbon skeletons required for converting reduced nitrate (ammonia) into amino acids, facilitating protein synthesis and overall growth [10]. Additionally, applying exogenous amino acids has reduced nitrate concentrations in cabbage leaves, suggesting that externally supplied amino acids may regulate nitrate uptake [21]. In our study, nitrate levels of both trials were significantly lower than the safety limits for human health. According to the European Commission regulation (EC Reg. No. 1258/2011), the maximum allowable concentration is 5000 mg kg^{-1} FW for lettuce grown in protected conditions under cover. Our research complies with this standard, as it was conducted between November and April [45].

4. Materials and Methods

The research was conducted in a glasshouse at the Department of Horticulture, Faculty of Agriculture, Çukurova University, during the winter and early spring seasons. This study aimed to assess the impact of biostimulants on iceberg lettuce cultivation within a floating culture soilless system (Figure 5). The plant material used was the "Bombola" iceberg lettuce variety, sourced from the seed company "Enza Zaden". Biostimulants were applied to enhance plant growth, yield, and nutrient content in iceberg lettuce. To establish effective biostimulant dosages, we organized two distinct experimental trials. The first trial was conducted during the winter months (November, December, and January) for 74 days, while the second trial was conducted in the early spring months (February, March, and April) for 70 days in a glasshouse environment.

Figure 5. Iceberg lettuce grown in a floating hydroponic system, showing plant growth, system setup, and root development.

4.1. Biostimulants

Fulvic acid, amino acid, and vermicompost were the biostimulants [26]. The fulvic acid and amino acid products were obtained from "Köklü Group" and marketed under the brand names "Sacaka WS" and "Aminoset," respectively. "Sacaka WS" has the following specifications: 80% organic matter content, 70% total (humic + fulvic) acid content, 70% fulvic acid content, and a pH range of 2–4. "Aminoset" contains 50% organic matter, 20% organic carbon, 4% organic nitrogen, 30% free amino acids, and a pH range of 3.1–5.1. The vermicompost, branded as "EkosolFarm", 100% organic liquid vermicompost, was sourced from the "Ekosol Tarim" company. Vermicompost products are typically rich in organic matter (up to 70%), humic acid (up to 30%), and fulvic acid (up to 20%). They also contain essential macronutrients and micronutrients, such as nitrogen, phosphorus, potassium, calcium, magnesium, iron, zinc, enzymes, beneficial microorganisms, and natural growth hormones, which promote plant growth, resilience, and yield. The vermicompost used in this study was produced from agricultural waste and manure.

4.2. Experimental Conditions

The dimensions of the containers used in the hydroponic system were 73.6 × 46 × 20 cm (L × W × H), with lid-shaped plates measuring 105 cm × 55 cm (L × W). The spacing between rows of iceberg lettuce plants was maintained at 18.4 cm × 18.4 cm, resulting in a plant density of 29.54 plants m^{-2}. Yield (kg/m^2) was calculated based on the harvest of 10 iceberg lettuce plants grown in an area of 0.33856 m^2.

The first trial included seven treatments: a control, two doses of fulvic acid (40 and 80 ppm), amino acid (75 and 100 ppm), and vermicompost (liquid; 1 and 2 mL L^{-1}). The second trial was established using combinations of biostimulant doses that demonstrated the best performance in the first trial. In the second experiment, five treatments were applied, including a control, fulvic acid at 40 ppm + amino acid at 100 ppm, fulvic acid at 40 ppm + vermicompost at 2 mL L^{-1}, amino acid at 100 ppm + vermicompost at 2 mL L^{-1}, and a combination of fulvic acid at 40 ppm + amino acid at 100 ppm + vermicompost at 2 mL L^{-1}.

The nutrient solution was prepared to supply essential nutrients for plant growth [48]. The elemental composition of the nutrient solution for lettuce plants included (in mg L^{-1}): nitrogen (230), phosphorus (50), potassium (320), calcium (220), magnesium (60), iron (4.0), zinc (0.50), boron (0.51), copper (0.23), molybdenum (0.18), and manganese (0.78).

The nutrient solution's electrical conductivity (EC) level was gradually increased from 1.5 to 2.4 dS·m^{-1} as the iceberg lettuce plants progressed through their phenological growth stages. This adjustment was made to meet the increasing mineral nutrient requirements of the plants, particularly during head formation. The nutrient solution was replaced with a fresh nutrient solution every 10 days. The pH values of the nutrient solution were maintained between 6.0 and 6.5. The nutrient solution was continuously aerated, with oxygen levels fluctuating between 6 and 8 ppm throughout the experiments. During the first trial, the greenhouse temperatures ranged between 15 °C and 21 °C, while in the second trial, they ranged between 17 °C and 24 °C. During the first experimental period (November, December, and January), the light intensity inside the greenhouse ranged between 250 and 350 W·m^{-2}, and the day length varied from 9.5 to 10.6 h. In the second experimental period (February, March, and April), the light intensity inside the greenhouse ranged between 350 and 550 W·m^{-2}, while the day length extended from 10.4 to 13.7 h.

The "Bombola" iceberg lettuce variety reached the harvest stage 74 days after transplanting in the first experiment, while the second harvest occurred 70 days post-transplanting. Parameters measured across both experiments included total yield (kg m^{-2}), plant height (cm), perimeter length (cm), head firmness (kg cm^{-3}), root weight (g), leaf fresh weight (g $plant^{-1}$), leaf area (cm^2 $plant^{-1}$), leaf count (leaves $plant^{-1}$), dry matter content (%), chlorophyll content (SPAD), and dry weight (g $plant^{-1}$). Leaf samples were analyzed for nitrogen (N), phosphorus (P), potassium (K), magnesium (Mg), calcium (Ca), iron (Fe), manganese (Mn), copper (Cu), and zinc (Zn) concentrations. Nitrate concentration (ppm), vitamin C (L-ascorbic acid), total phenolic, and flavonoid contents were also measured. For lettuce leaves, total soluble solids (TSS, %), titratable acidity (%), pH, and electrical conductivity (EC, mS cm^{-1}) were also determined.

4.3. Measurements of Plant Growth Parameters

At harvest time, the yield (kg m^{-2}) was computed by summing the total weight of 10 plants within each replicate, with leaf counts also documented. Lettuce head firmness was evaluated using a digital penetrometer (Bareiss HPE-III-Fff, ABQ Industrial, Woodlands, TX, USA), with results recorded in kilograms [49]. Leaf area per plant (cm^{-2}) was measured using a leaf area meter (Li-3100, LICOR, Lincoln, NE, USA). After harvest, each plant was weighed on a digital scale to determine fresh leaf weight per plant (g). Chlorophyll content in the leaves was measured by a SPAD chlorophyll meter (Minolta 502, Tokyo, Japan). The harvested leaves' luminosity (L) and chromaticity coordinates [a (red–green) and b (blue–yellow)] were recorded digitally via a handheld color spectrophotometer (HunterLab, Reston, VA, USA) [50]. Following this, fresh leaves were dried at 65 °C for 48 h to obtain dry weight per plant.

4.4. Determination of Total Soluble Solids (TSS), pH, and Electrical Conductivity (EC) in Lettuce Leaves

The juice extracted from lettuce leaves determined total soluble solids (TSS), pH, and electrical conductivity (EC). TSS was measured with a digital refractometer (Atago PR-101, Tokyo, Japan). At the same time, pH and EC values were assessed using a pH and EC meter (WTW pH/Cond 3320, Weilheim, Germany) [11].

4.5. Determination of Nitrate Concentration

Nitrate concentration in lettuce leaves was assessed using the salicylic acid method outlined by Cataldo et al. (1975), with measurements taken colorimetrically at 410 nm. Results were reported in milligrams per kilogram of fresh weight [50].

4.6. Determination of Ascorbic Acid Content (Vitamin C)

The ascorbic acid content in lettuce leaves was analyzed following a modified method by [51]. Lettuce leaves were processed using a juicer for extraction. Five milliliters of the extracted juice were mixed with 45 milliliters of 0.4% oxalic acid and then filtered. From the filtered solution, one milliliter was combined with nine milliliters of 2,6-dichlorophenolindophenol and thoroughly mixed. The samples were then measured at 520 nm using a UV spectrophotometer (UV-1700 PharmoSpec Shimadzu, Kyoto, Japan).

4.7. Determination of Total Phenolic and Flavonoid Compounds

The total phenolic content in lettuce leaves was measured with a modified spectrophotometric method based on Spanos and Wrolstad [52]. Absorbance readings were taken at 765 nm using a spectrophotometer (UV-1700 PharmoSpec Shimadzu, Kyoto, Japan) with a gallic acid calibration curve for quantification. Total flavonoid content was assessed at 415 nm, following the method by Quettier–Deleu et al. [53] and quantified using a calibration curve prepared with rutin.

4.8. Leaf Mineral Nutrient Analysis

Leaf mineral nutrient analysis was performed by sampling one-quarter of three plants from each replicate at harvest, using the method outlined by Dasgan et al. [49]. The analysis included macronutrients (N, P, K, Mg, Ca) and micronutrients (Fe, Mn, Cu, Zn). Lettuce leaf samples were thoroughly cleaned, rinsed three times with distilled water to avoid contamination, and dried at 65 °C for 48 h using oven-drying. The dried leaves were ground to a 40-mesh particle size using a leaf grinder. For K, Ca, Mg, Na, Fe, Mn, Zn, and Cu analysis, 0.2 g of the ground sample were incinerated at 550 °C for 5 h, and the resulting ash was dissolved in 3.3% HCl (*v*/*v*) and filtered [49]. Potassium, calcium, magnesium, and sodium were analyzed in emission mode, while iron, manganese, zinc, and copper were analyzed in absorbance mode with an Atomic Absorption Spectrophotometer (Varian FS220, Palo Alto, CA, USA). Leaf nitrogen and phosphorus were determined using the Kjeldahl and Barton methods, respectively [54].

4.9. Statistical Analysis

This study conducted both experiments using a randomized block design with four replicates, each containing 10 plants. The data obtained were analyzed using the JMP statistical software 13 version, and the means were compared using the Tukey test at a 5% significance level.

5. Conclusions

This study demonstrates the significant potential of biostimulants such as fulvic acid, amino acids, and vermicompost in enhancing the growth, yield, and nutritional quality of soilless-grown iceberg lettuce. Applying these eco-friendly biostimulants, individually and in combination, improved key plant parameters, including biomass production, nutrient uptake, and antioxidant content, while reducing nitrate levels in lettuce leaves. Notably, FA 40 + VC 2 mL achieved the highest yield (17.15 kg/m^2), an 18.2% increase over the control, followed closely by FA 40 + AA 100 (17.03 kg/m^2, 17.4% increase). Both combinations significantly enhanced lettuce yield. These findings underscore the value of biostimulants as sustainable solutions for modern agriculture, reducing dependency on conventional fertilizers and mitigating environmental impacts. Biostimulants are a promising approach for advancing sustainable soilless culture systems and addressing the growing global demand for high-quality, nutrient-rich vegetables.

Biostimulants provide sustainable solutions for soilless systems by enhancing nutrient-use efficiency and amplifying the effectiveness of conventional fertilizers. Future research should aim to optimize biostimulant application strategies by exploring novel biostimulants, innovative combinations, and optimal dosing regimens to maximize their efficacy and investigate their long-term benefits for sustainable soilless culture production.

Author Contributions: All authors contributed significantly to this research. H.Y.D. and B.K. designed the experiment and conducted the investigation. B.K. and Y.A. were responsible for the conceptualization, data curation, formal analysis, and drafting of the manuscript. Supervision, funding acquisition, and manuscript review and editing were conducted by H.Y.D. and N.S.G. All authors have read and approved the final version of the manuscript for publication. All authors have read and agreed to the published version of the manuscript.

Funding: This work was supported by the Cukurova University Research Foundation (BAP) under project number FYL-2021-13435.

Data Availability Statement: The data supporting the findings of this study are included within the article.

Acknowledgments: We thank the Research Foundation Office of Cukurova University (BAP).

Conflicts of Interest: The authors declare no conflict of interest.

References

1. Gruda, N.S. Increasing sustainability of growing media constituents and stand-alone substrates in soilless culture systems. *Agronomy* **2019**, *9*, 298. [CrossRef]
2. Albadwawi, M.A.O.K.; Ahmed, Z.F.R.; Kurup, S.S.; Alyafei, M.A.; Jaleel, A. A comparative evaluation of aquaponic and soil systems on yield and antioxidant levels in basil, an important food plant in Lamiaceae. *Agronomy* **2022**, *12*, 3007. [CrossRef]
3. Savvas, D.; Gruda, N. Application of soilless culture technologies in the modern greenhouse industry—A review. *Eur. J. Hortic. Sci.* **2018**, *83*, 280–293. [CrossRef]
4. Scaltrito, E.; Cristiano, G.; Sdao, A.E.; Gruda, N.S.; Loconsole, D.; De Lucia, B. Influence of Water Spraying Intervals and Indole-3-Butyric Acid Concentrations on *Salvia* Rooted Cuttings Quality in a Closed Aeroponics System. *Sci. Hortic.* **2024**, *337*, 1134524. [CrossRef]
5. Gruda, N.S. Advances in Soilless Culture and Growing Media in Today's Horticulture—An Editorial. *Agronomy* **2022**, *12*, 2773. [CrossRef]
6. Niu, G.; Masabni, J. Chapter 9—Hydroponics. In *Plant Factory Basics, Applications and Advances*; Elsevier: Amsterdam, The Netherlands, 2022; pp. 153–166.
7. Pink, D.A.C.; Keane, E.M. Lettuce: *Lactuca sativa* L. In *Genetic Improvement of Vegetable Crops*; Pergamon Press: Oxford, UK, 1993; pp. 543–571.
8. Murray, J.J.; Basset, G.; Sandoya, G. Nutritional Benefits of Lettuce Consumed at Recommended Portion Sizes. *Edis* **2021**, *HS1416*, 1–8. [CrossRef]
9. Medina-Lozano, I.; Bertolín, J.R.; Díaz, A. Nutritional value of commercial and traditional lettuce (*Lactuca sativa* L.) and wild relatives: Vitamin C and anthocyanin content. *Food Chem.* **2021**, *359*, 129864. [CrossRef]
10. Bulgari, R.; Cocetta, G.; Trivellini, A.; Vernieri, P.; Ferrante, A. Biostimulants and crop responses: A review. *Biol. Agric. Hortic.* **2015**, *31*, 1–17. [CrossRef]

11. Dasgan, H.Y.; Aldiyab, A.; Elgudayem, F.; Ikiz, B.; Gruda, N.S. Effect of biofertilizers on leaf yield, nitrate amount, mineral content, and antioxidants of basil (*Ocimum basilicum* L.) in a floating culture. *Sci. Rep.* **2022**, *12*, 20917. [CrossRef] [PubMed]
12. Boubaker, H.; Saadaoui, W.; Dasgan, H.Y.; Tarchoun, N.; Gruda, N.S. Enhancing Seed Potato Production from In Vitro Plantlets and Microtubers through Biofertilizer Application: Investigating Effects on Plant Growth, Tuber Yield, Size, and Quality. *Agronomy* **2023**, *13*, 2541. [CrossRef]
13. İkiz, B.; Dasgan, H.Y.; Balik, S.; Kusvuran, S.; Gruda, N.S. The Use of Biostimulants as a Key to Sustainable Hydroponic Lettuce Farming under Saline Water Stress. *BMC Plant Biol.* **2024**, *24*, 808. [CrossRef] [PubMed]
14. Ali, E.F.; Al-Yasi, H.M.; Issa, A.A.; Hessini, K.; Hassan, F.A.S. Ginger Extract and Fulvic Acid Foliar Applications as Novel Practical Approaches to Improve the Growth and Productivity of Damask Rose. *Plants* **2022**, *11*, 412. [CrossRef] [PubMed]
15. Calvo, P.; Nelson, L.; Kloepper, J.W. Agricultural Uses of Plant Biostimulants. *Plant Soil.* **2014**, *383*, 3–41. [CrossRef]
16. Taha, A.A.; Omar, M.M.; Ghazy, M.A. Effect of Humic and Fulvic Acids on Growth and Yield of Lettuce Plant. *J. Soil. Sci. Agric. Eng. Mansoura Univ.* **2016**, *7*, 517–522. [CrossRef]
17. Zhang, P.; Zhang, H.; Wu, G.; Chen, X.; Gruda, N.; Li, X.; Dong, J.; Duan, Z. Dose-Dependent Application of Straw-Derived Fulvic Acid on Yield and Quality of Tomato Plants Grown in Greenhouse. *Front. Plant Sci.* **2021**, *12*, 736613. [CrossRef]
18. Wang, Y.; Yang, R.; Zheng, J.; Shen, Z.; Xu, X. Exogenous foliar application of fulvic acid alleviate cadmium toxicity in lettuce (*Lactuca sativa* L.). *Ecotoxicol. Environ. Saf.* **2019**, *167*, 10–19. [CrossRef] [PubMed]
19. Okumoto, S.; Funck, D.; Trovato, M.; Forlani, G. Editorial: Amino Acids of the Glutamate Family: Functions Beyond Primary Metabolism. *Front. Plant Sci.* **2016**, *7*, 318. [CrossRef] [PubMed]
20. Tyskiewicz, R.; Nowak, A.; Ozimek, E.; Jaroszuk-Ściśel, J. *Trichoderma*: The Current Status of Its Application in Agriculture for the Biocontrol of Fungal Phytopathogens and Stimulation of Plant Growth. *Int. J. Mol. Sci.* **2022**, *23*, 2329. [CrossRef]
21. Haghighi, M.; Saadat, S.; Abbey, L. Effect of Exogenous Amino Acids Application on Growth and Nutritional Value of Cabbage under Drought Stress. *Sci. Hortic.* **2020**, *272*, 109561. [CrossRef]
22. Toor, M.D.; Eldesoky, G.E. Effects of Vermicompost on Soil Microbiological Properties in Lettuce Rhizosphere: An Environmentally Friendly Approach for Sustainable Green Future. *Environ. Res.* **2024**, *243*, 117737. [CrossRef] [PubMed]
23. Kamar Zaman, A.M.; Yaacob, J.S. Exploring the Potential of Vermicompost as a Sustainable Strategy in Circular Economy: Improving Plants' Bioactive Properties and Boosting Agricultural Yield and Quality. *Environ. Sci. Pollut. Res.* **2022**, *29*, 12948–12964. [CrossRef] [PubMed]
24. Singh, A.; Karmegam, N.; Singh, G.S.; Bhadauria, T.; Chang, S.W.; Awasthi, M.K.; Sudhakar, S.; Arunachalam, K.D.; Biruntha, M.; Ravindran, B. Earthworms and Vermicompost: An Eco-Friendly Approach for Repaying Nature's Debt. *Environ. Geochem. Health* **2020**, *42*, 1617–1642. [CrossRef] [PubMed]
25. Blouin, M.; Barrere, J.; Meyer, N.; Lartigue, S.; Barot, S.; Mathieu, J. Vermicompost Significantly Affects Plant Growth: A Meta-Analysis. *Agron. Sustain. Dev.* **2019**, *39*, 34. [CrossRef]
26. Dasgan, H.Y.; Aksu, K.S.; Zikaria, K.; Gruda, N.S. Biostimulants Enhance the Nutritional Quality of Soilless Greenhouse Tomatoes. *Plants* **2024**, *13*, 2587. [CrossRef] [PubMed]
27. Canellas, L.P.; Olivares, F.L.; Aguiar, N.O.; Jones, D.L.; Nebbioso, A.; Mazzei, P.; Piccolo, A. Humic and fulvic acids as biostimulants in horticulture. *Sci. Hortic.* **2015**, *196*, 15–27. [CrossRef]
28. Lüdtke, A.C.; Dick, D.P.; Morosino, L.; Kraemer, V. Productivity of Lettuce in Greenhouse as Affected by Humic and Fulvic Acids Application in Association to Mineral Fertilizer. *Hortic. Bras.* **2021**, *39*, 444–450. [CrossRef]
29. Huang, S.; Xiong, B.; Sun, G.; He, S.; Liao, L.; Wang, J.; Wang, Z. Effects of Fulvic Acid on Photosynthetic Characteristics of Citrus Seedlings under Drought Stress. *IOP Conf. Ser. Earth Environ. Sci.* **2020**, *474*, 032007. [CrossRef]
30. Abdel-Baky, Y.R.; Abouziena, H.F.; Amin, A.A.; Rashad El-Sh, M.; Abd El-Sttar, A.M. Improve Quality and Productivity of Some Faba Bean Cultivars with Foliar Application of Fulvic Acid. *Bull. Natl. Res. Cent.* **2019**, *43*, 2. [CrossRef]
31. Al-Karaki, G.N.; Othman, Y. Effect of Foliar Application of Amino Acid Biostimulants on Growth, Macronutrient, Total Phenol Contents and Antioxidant Activity of Soilless Grown Lettuce Cultivars. *S. Afr. J. Bot.* **2023**, *154*, 225–231. [CrossRef]
32. Hildebrandt, T.M.; Nesi, A.N.; Araújo, W.L.; Braun, H.P. Amino Acid Catabolism in Plants. *Mol. Plant* **2015**, *8*, 1563–1579. [CrossRef]
33. Colla, G.; Rouphael, Y.; Canaguier, R.; Svecova, E.; Cardarelli, M. Biostimulant Action of a Plant-Derived Protein Hydrolysate Produced Through Enzymatic Hydrolysis. *Front. Plant Sci.* **2014**, *5*, 448. [CrossRef] [PubMed]
34. Yassen, A.A.; Essa, E.M.; Marzouk, N.M.; Zaghloul, S.M. Impact of Vermicompost and Foliar Spray of Vermiwash on Growth, Yield and Nutritional Status of Lettuce Plants. *Plant Arch.* **2020**, *20*, 449–455.
35. Kilic, C.C.; Saraçoglu, Ö.A. Effect of Liquid Vermicompost on Quality Properties of Different Lettuce Varieties. *Turk. J. Agric. Food Sci. Technol.* **2022**, *10*, 1716–1723.
36. Zuo, Y.; Zhang, J.; Zhao, R.; Dai, H.; Zhang, Z. Application of Vermicompost Improves Strawberry Growth and Quality through Increased Photosynthesis Rate, Free Radical Scavenging, and Soil Enzymatic Activity. *Sci. Hortic.* **2018**, *233*, 132–140. [CrossRef]

37. Liu, X.; Zhang, J.; Wang, Q.; Chang, T.; Shaghaleh, H.; Hamoud, Y.A. Improvement of Photosynthesis by Biochar and Vermicompost to Enhance Tomato (*Solanum lycopersicum* L.) Yield under Greenhouse Conditions. *Plants* **2022**, *11*, 3214. [CrossRef]
38. Mahmud, M.; Abdullah, R.; Yaacob, J.S. Effect of Vermicompost Amendment on Nutritional Status of Sandy Loam Soil, Growth Performance, and Yield of Pineapple (*Ananas comosus* var. MD2) under Field Conditions. *Agronomy* **2018**, *8*, 183. [CrossRef]
39. Grantina-Ievina, L.; Andersone, U.; Berkolde-Pīre, D.; Nikolajeva, V.; Ievinsh, G. Critical Tests for Determination of Microbiological Quality and Biological Activity in Commercial Vermicompost Samples of Different Origins. *Appl. Microbiol. Biotechnol.* **2013**, *97*, 10541–10554. [CrossRef]
40. Rouphael, Y.; Colla, G. Biostimulants in agriculture. *Front. Plant Sci.* **2020**, *11*, 40. [CrossRef] [PubMed]
41. Halpern, M.; Bar-Tal, A.; Ofek, M.; Minz, D.; Muller, T.; Yermiyahu, U. The Use of Biostimulants for Enhancing Nutrient Uptake. *Adv. Agron.* **2015**, *130*, 141–174.
42. Bulgari, R.; Franzoni, G.; Ferrante, A. Biostimulants Application in Horticultural Crops under Abiotic Stress Conditions. *Agronomy* **2019**, *9*, 306. [CrossRef]
43. Nardi, S.; Pizzeghello, D.; Schiavon, M.; Ertani, A. Plant Biostimulants: Physiological Responses Induced by Protein Hydrolysate-Based Products and Humic Substances in Plant Metabolism. *Sci. Agric.* **2016**, *73*, 18–23. [CrossRef]
44. Al Jaouni, S.; Selim, S.; Hassan, S.H.; Mohamad, H.S.; Wadaan, M.A.; Hozzein, W.N.; Asard, H.; AbdElgawad, H. Vermicompost Supply Modifies Chemical Composition and Improves Nutritive and Medicinal Properties of Date Palm Fruits from Saudi Arabia. *Front. Plant Sci.* **2019**, *10*, 424. [CrossRef] [PubMed]
45. Alexander, J.; Benford, D.; Cockburn, A.; Cravedi, J.-P.; Dogliotti, E.; Di Domenico, A.; Fernández-Cruz, M.L.; Fink-Gremmels, J.; Fürst, P.; Galli, C.; et al. Nitrate in vegetables—Scientific opinion of the panel on contaminants in the food chain. *EFSA J.* **2008**, *6*, 689.
46. Zhang, X.; Taylor, Z.; Goatley, M.; Booth, J.; Brown, I.; Kosiarski, K. Seaweed Extract-Based Biostimulant Impacts on Nitrate Reductase Activity and Root Viability of Ultradwarf Bermudagrass Subjected to Heat and Drought Stress. *HortScience* **2022**, *57*, 1328–1333. [CrossRef]
47. Coronel, G.; Chang, M.; Rodríguez-Delfín, A. Nitrate Reductase Activity and Chlorophyll Content in Lettuce Plants Grown Hydroponically and Organically. *Acta Hortic.* **2009**, *843*, 137–144. [CrossRef]
48. Dasgan, H.Y.; Yilmaz, D.; Zikaria, K.; Ikiz, B.; Gruda, N.S. Enhancing the yield, quality, and antioxidant content of lettuce through innovative and eco-friendly biofertilizer practices in hydroponics. *Horticulturae* **2023**, *9*, 1274. [CrossRef]
49. Dasgan, H.Y.; Yilmaz, M.; Dere, S.; Ikiz, B.; Gruda, N.S. Bio-Fertilizers Reduced the Need for Mineral Fertilizers in Soilless-Grown Capia Pepper. *Horticulturae* **2023**, *9*, 188. [CrossRef]
50. Dasgan, H.Y.; Kacmaz, S.; Arpaci, B.B.; Ikiz, B.; Gruda, N.S. Biofertilizers Improve the Leaf Quality of Hydroponically Grown Baby Spinach (*Spinacia oleracea* L.). *Agronomy* **2023**, *13*, 575. [CrossRef]
51. Elgailani, I.E.H.; Elkareem, M.A.M.G.; Noh, E.A.A.; Adam, O.E.A.; Alghamdi, A.M.A. Comparison of Two Methods for the Determination of Vitamin C (Ascorbic Acid) in Some Fruits. *Am. J. Chem.* **2017**, *2*, 1–7. [CrossRef]
52. Spanos, G.A.; Wrolstad, R.E. Influence of Processing and Storage on the Phenolic Composition of Thompson Seedless Grape Juice. *J. Agric. Food Chem.* **1990**, *38*, 1565–1571. [CrossRef]
53. Quettier-Deleu, C.; Gressier, B.; Vasseur, J.; Dine, T.; Brunet, C.; Luyckx, M.; Cazin, M.; Cazin, J.-C.; Bailleul, F.; Trotin, F. Phenolic Compounds and Antioxidant Activities of Buckwheat (*Fagopyrum esculentum* Moench) Hulls and Flour. *J. Ethnopharmacol.* **2000**, *72*, 35–42. [CrossRef]
54. Ikiz, B.; Dasgan, H.Y.; Gruda, N.S. Utilizing the power of plant growth promoting rhizobacteria on reducing mineral fertilizer, improved yield, and nutritional quality of Batavia lettuce in a floating culture. *Sci. Rep.* **2024**, *14*, 1616. [CrossRef] [PubMed]

MDPI AG
Grosspeteranlage 5
4052 Basel
Switzerland
Tel.: +41 61 683 77 34

Plants Editorial Office
E-mail: plants@mdpi.com
www.mdpi.com/journal/plants

www.ingramcontent.com/pod-product-compliance
Lightning Source LLC
Chambersburg PA
CBHW040821250926
51850CB00007BA/1231
9783725880072